NMR of Macromolecules

The Practical Approach Series

SERIES EDITORS

D. RICKWOOD
Department of Biology, University of Essex
Wivenhoe Park, Colchester, Essex CO4 3SQ, UK

B. D. HAMES
Department of Biochemistry and Molecular Biology
University of Leeds, Leeds LS2 9JT, UK

Affinity Chromatography
Anaerobic Microbiology
Animal Cell Culture
(2nd Edition)
Animal Virus Pathogenesis
Antibodies I and II
Behavioural Neuroscience
Biochemical Toxicology
Biological Data Analysis
Biological Membranes
Biomechanics—Materials
Biomechanics—Structures and
Systems
Biosensors
Carbohydrate Analysis
Cell–Cell Interactions
The Cell Cycle
Cell Growth and Division
Cellular Calcium
Cellular Interactions in
Development
Cellular Neurobiology
Centrifugation (2nd Edition)
Clinical Immunology

Computers in Microbiology
Crystallization of Nucleic Acids
and Proteins
Cytokines
The Cytoskeleton
Diagnostic Molecular Pathology
I and II
Directed Mutagenesis
DNA Cloning I, II, and III
Drosophila
Electron Microscopy in Biology
Electron Microscopy in
Molecular Biology
Electrophysiology
Enzyme Assays
Essential Developmental
Biology
Essential Molecular Biology I
and II
Experimental Neuroanatomy
Fermentation
Flow Cytometry
Gas Chromatography
Gel Electrophoresis of Nucleic
Acids (2nd Edition)

Gel Electrophoresis of Proteins (2nd Edition)
Gene Targeting
Gene Transcription
Genome Analysis
Glycobiology
Growth Factors
Haemopoiesis
Histocompatibility Testing
HPLC of Macromolecules
HPLC of Small Molecules
Human Cytogenetics I and II (2nd Edition)
Human Genetic Disease Analysis
Immobilised Cells and Enzymes
Immunocytochemistry
In Situ Hybridization
Iodinated Density Gradient Media
Light Microscopy in Biology
Lipid Analysis
Lipid Modification of Proteins
Lipoprotein Analysis
Liposomes
Lymphocytes
Mammalian Cell Biotechnology
Mammalian Development
Medical Bacteriology
Medical Mycology
Microcomputers in Biochemistry
Microcomputers in Biology
Microcomputers in Physiology
Mitochondria
Molecular Genetic Analysis of Populations

Molecular Imaging in Neuroscience
Molecular Neurobiology
Molecular Plant Pathology I and II
Molecular Virology
Monitoring Neuronal Activity
Mutagenicity Testing
Neural Transplantation
Neurochemistry
Neuronal Cell Lines
NMR of Biological Macromolecules
Nucleic Acid and Protein Sequence Analysis
Nucleic Acid Hybridisation
Nucleic Acids Sequencing
Oligonucleotides and Analogues
Oligonucleotide Synthesis
PCR
Peptide Hormone Action
Peptide Hormone Secretion
Photosynthesis: Energy Transduction
Plant Cell Culture
Plant Molecular Biology
Plasmids
Pollination Ecology
Postimplantation Mammalian Embryos
Preparative Centrifugation
Prostaglandins and Related Substances
Protein Architecture
Protein Engineering
Protein Function

Protein Phosphorylation

Protein Purification
  Applications

Protein Purification Methods

Protein Sequencing

Protein Structure

Protein Targeting

Proteolytic Enzymes

Radioisotopes in Biology

Receptor Biochemistry

Receptor–Effector Coupling

Receptor–Ligand Interactions

Ribosomes and Protein
  Synthesis

RNA Processing

Signal Transduction

Solid Phase Peptide
  Synthesis

Spectrophotometry and
  Spectrofluorimetry

Steroid Hormones

Teratocarcinomas and
  Embryonic Stem Cells

Transcription Factors

Transcription and Translation

Tumour Immunobiology

Virology

Yeast

NMR of Macromolecules

A Practical Approach

Edited by
GORDON C. K. ROBERTS

Biological NMR Centre,
University of Leicester

OXFORD UNIVERSITY PRESS
Oxford New York Tokyo

Oxford University Press, Walton Street, Oxford OX2 6DP
Oxford   New York   Toronto
Delhi   Bombay   Calcutta   Madras   Karachi
Kuala Lumpur   Singapore   Hong Kong   Tokyo
Nairobi   Dar es Salaam   Cape Town
Melbourne   Auckland   Madrid
and associated companies in
Berlin   Ibadan

Oxford is a trade mark of Oxford University Press

A Practical Approach 🛆 is a registered trade mark
of the Chancellor, Masters, and Scholars of the University of Oxford
trading as Oxford University Press

Published in the United States
by Oxford University Press Inc., New York

A catalogue record for this book is available from the British Library

Library of Congress Cataloging in Publication Data
NMR of macromolecules : a practical approach / edited by Gordon C.K. Roberts.
(The Practical approach series ; 134)
Includes bibliographical references.
1. Nuclear magnetic resonance spectroscopy.   2. Macromolecules–
–Analysis.   I. Roberts, G. C. K. (Gordon Carl Kenmure)   II. Series.
[DNLM: 1. Nuclear Magnetic Resonance—methods.   2. Macromolecular
Systems.   3. Protein Confirmation.   4. Proteins—physiology. QU 55
N738 1993]
QP519.9.N83N69   1993   547.7'046—dc20   93–17466
ISBN 0–19–963225–1 (h/bk)
0–19–963224–3 (p/bk)

Typeset by Footnote Graphics, Warminster, Wilts
Printed in Great Britain by Information Press Ltd, Eynsham, Oxon

Contents

List of contributors   xv

Nomenclature and abbreviations   xvii

1. Introduction   1
Gordon C. K. Roberts

References   5

2. Sample preparation   7
William U. Primrose

   1. Introduction   7

   2. Requirements for an NMR sample   7
   Sample quantity   7
   Solubility   8
   Stability   9
   Molecular weight   9

   3. The NMR tube   10
   Choice of NMR tube   10
   Cleaning the NMR tube   11
   Drying the NMR tube   11
   Sample volume   13

   4. Sample handling   14
   Methods for concentration   14
   Desalting and buffer exchange   16
   Methods of filtration   20
   Degassing the sample—removal of dissolved oxygen   21

   5. Sample parameters   22
   Definition of pH in ^2H_2O—the isotope effect   22
   pH adjustment   23
   Choice of buffer   25
   Ionic strength effects   26
   Organic solvents   29
   Detergents   30
   Temperature   30
   Chemical shift references   31

6. Preventing sample contamination   32
Paramagnetic contamination   32
Microbial contamination   33

General reading   33

References   33

3. Instrumentation and pulse sequences   35
T. A. Frenkiel

1. Introduction   35

2. NMR instrumentation   36
The NMR spectrometer   36
Sample temperature control   40
Environmental factors affecting spectral quality   43

3. Obtaining one-dimensional spectra   44
Proton spectra   45
Observing nuclei other than protons   49

4. Working with samples dissolved in H_2O 51
Setting-up the spectrometer   51
Solvent irradiation methods   53
Solvent non-excitation methods   55

5. Two-dimensional NMR   57
Introduction to 2D NMR   57
General considerations in setting-up 2D experiments   58
Simple ^{1}H-^{1}H experiments   60
Rotating-frame experiments   63
Heteronuclear correlation experiments   66

References   68

4. Resonance assignment strategies for small proteins   71
Christina Redfield

1. Introduction   71

2. Assessing the quality of 2D spectra   72

3. Stage 1: spin system identification   73
Glycine spin system identification   76
Alanine and threonine spin system identification   77
Valine, isoleucine, and leucine spin system identification   78
Type-J spin system identification   80
Type-U spin system identification   85
Type-X cross-peak category   87

Contents

4. Stage 2: sequence-specific assignment   88

5. Alternatives to the sequential assignment approach   97
Mainchain directed approach (MCD)   97
Use of X-ray diffraction data   98

References   99

5. Stable isotope labelling and resonance
assignments in larger proteins   101
John L. Markley and Masatsune Kainosho

1. Introduction   101

2. Labelling strategies   102
Isotopes of interest   102
Applications of protein labelling   103
Labelling patterns   104
Labelling methods   105

3. Data acquisition and analysis   111
Introduction   111
Data collection   114
Data processing and display   114

4. Isotope-assisted assignment strategies   115
Goals and expectations (full or partial assignment)   115
Assignment procedures   115

5. Structural information from labelled proteins   121
NOEs   121
Coupling constants   123
Chemical shifts   125
Solvent accessibility   125

6. Strategies for very large proteins   126
Properties of carbonyl ^{13}C resonances   126
The CN labelling strategy   127
Carbonyl carbon signals as probes for chemical modification   129

7. Thermodynamic information from labelled proteins   136
Titration curves   136
Ligand binding   139
Conformational equilibria   139

8. Protein dynamics and hydrogen exchange rates from
labelled proteins   140
Relaxation measurements   140
Chemical exchange   141

Acknowledgements   147

References   148

6. Effects of chemical exchange on NMR spectra
153

Lu-Yun Lian and Gordon C. K. Roberts

1. Introduction—general considerations 153

2. Definitions 155

3. Initial qualitative analysis 157

4. Lineshape analysis 159
  General 159
  Slow exchange 161
  Fast exchange 164
  Problems of fast exchange 167

5. Magnetization transfer experiments 168
  General theory 169
  One-dimensional magnetization transfer 169
  Two-dimensional exchange spectroscopy 175
  Comparison of magnetization transfer experiments 178

Acknowledgement 181

References 181

7. NMR studies of protein–ligand interactions
183

J. Feeney and B. Birdsall

1. Introduction 183

2. Preparation of the sample and preliminary binding studies 185

3. Analysing the NMR spectra 186
  Fast exchange 187
  Slow exchange 189

4. Assignment of resonances 190
  Protein resonances 190
  Ligand resonances 190

5. NMR parameters of the bound species 192

6. Determination of ionization states 194
  Protein ionization states 194
  Ligand ionization states 198

7. Determination of conformations of protein–ligand complexes 201
  Protein–DNA complexes 201
  Isotope-editing using labelled ligands 202
  Transferred NOEs and bound conformations 202

8. Detection of multiple conformations 206

9. Dynamic processes in protein–ligand complexes   208
    Dissociation rate constants from transfer of saturation studies   208
    Rapid segmental motions in protein–ligand complexes   208
    Determination of rates of ring flipping   209
    Measurements of exchange rates with solvent   211

Acknowledgement   212

References   212

8. NMR of nucleic acids; from spectrum to structure   217

Sybren S. Wijmenga, Margret M. W. Mooren, and Cornelis W. Hilbers

1. Introduction   217
    Specific features of NMR structural information   218

2. Structural parameters   218
    Nomenclature and definition of structural parameters   218
    Description of the furanose ring   223
    Description of the mononucleotide conformation   227
    Distances in secondary elements of nucleic acid structures   228

3. Recording of spectra   229
    Water suppression   231
    NOESY and ROESY spectra   231
    J-correlated spectra   235

4. Spectrum interpretation   237
    Distribution of resonances   237
    Assignment strategies   243
    Assignment via 3D NMR   251

5. Local conformational analysis   253
    NOEs   254
    J-couplings   258
    Conformation of the furanose ring   258
    Rotation about the χ angle   267
    The backbone angles   270

6. Determination of the overall structure of nucleic acids   276
    Restrained molecular dynamics   278
    The embed algorithm   278
    Conclusions   280

References   281

Appendix   283

Contents

9. ^{1}H NMR studies of oligosaccharides   289

S. W. Homans

1. Introduction   289

2. Sample preparation   289
Aqueous solvents   289
Other solvents   290
Chemical shift references   291

3. Assignment methods   291
Overall morphology of oligosaccharide NMR spectra   291
Resonance assignment   291
An assignment case study   301

4. Determination of primary sequence   304
Fingerprinting   304
'*Ab-initio*' method   305

5. Determination of three-dimensional structure   307
Choice of method—experimental or theoretical?   307
Restrained energy minimization   308
Adiabatic mapping   309
Restrained and free molecular dynamics simulations   310
Simulated annealing   311

Acknowledgements   311

References   311

10. Structure determination from NMR data
I. Analysis of NMR data   315

Igor L. Barsukov and Lu-Yun Lian

1. Introduction   315

2. Scalar coupling constants and amide hydrogen exchange
rates   317
Determination of $^3J_{H^N\text{-}H^\alpha}$ coupling constants   317
Determination of $^3J_{H^\alpha\text{-}H^\beta}$ coupling constants and stereospecific
assignments   323
Heteronuclear scalar coupling constants   328
Amide hydrogen exchange rates   329

3. Secondary structure identification   330

4. Nuclear Overhauser enhancement and distance restraints   333
Measurement of NOE cross-peak intensities   335
Assignment of NOEs   339
From NOEs to distances   342
Summary: the use of NOEs for the structure calculation   353

References   355

Contents

11. Structure determination from NMR data II. Computational approaches

359

Michael J. Sutcliffe

1. Introduction   359

2. Determination of constraints   359
   Proton–proton distance constraints   360
   Dihedral angle constraints   361
   Secondary structural constraints   361

3. Protein secondary structure determination   365
   Short-range NOEs   365
   H^N-H^α coupling constants   367
   Amide proton exchange rates   367

4. Tertiary structure determination   369
   Model building   369
   A probabilistic approach   371
   Distance geometry   373
   Restrained molecular dynamics   379
   Distance geometry followed by restrained MD   381
   Hybrid distance geometry–dynamical simulated annealing   382
   Allowing for internal motion   384
   Analysis of structures   385

5. Future developments   387

Acknowledgements   387

References   387

Appendix   391

Index   393

Contributors

IGOR L. BARSUKOV
Biological NMR Centre, University of Leicester, PO Box 138, Medical Sciences Building, University Road, Leicester LE1 9HN, UK.

B. BIRDSALL
Laboratory of Molecular Structure, National Institute for Medical Research, Mill Hill, London NW7 1AA, UK.

J. FEENEY
Laboratory of Molecular Structure, National Institute for Medical Research, Mill Hill, London NW7 1AA, UK.

T. A. FRENKIEL
MRC Biomedical NMR Centre, National Institute for Medical Research, Mill Hill, London NW7 1AA, UK.

CORNELIS W. HILBERS
Nijmegen SON Research Centre for Molecular Design, Structure, and Synthesis, Laboratory of Biophysical Chemistry, University of Nijmegen, Nijmegen, The Netherlands.

S. W. HOMANS
Department of Biochemistry, Medical Sciences Institute, University of Dundee, Dundee DD1 4HN, Scotland, UK.

MASATSUNE KAINOSHO
Department of Chemistry, Faculty of Science, Tokyo Metropolitan University, Minami Ohsawa 1-1, Hachioji-shi, Tokyo 192–03, Japan.

LU-YUN LIAN
Biological NMR Centre and Department of Biochemistry, University of Leicester, PO Box 138, Medical Sciences Building, University Road, Leicester LE1 9HN, UK.

JOHN L. MARKLEY
Department of Biochemistry, University of Wisconsin, 420 Henry Mall, Madison, Wisconsin 53706, USA.

MARGRET M. W. MOOREN
Nijmegen SON Research Centre for Molecular Design, Structure, and Synthesis, Laboratory of Biophysical Chemistry, University of Nijmegen, Nijmegen, The Netherlands.

WILLIAM U. PRIMROSE
Department of Biochemistry, University of Leicester, University Road, Leicester LE1 7RH, UK.

Contributors

CHRISTINA REDFIELD
Oxford Centre for Molecular Sciences, University of Oxford, South Parks Road, Oxford OX1 3QR, UK.

GORDON C. K. ROBERTS
Biological NMR Centre and Department of Biochemistry, University of Leicester, PO Box 138, Medical Sciences Building, University Road, Leicester LE1 9HN, UK.

MICHAEL J. SUTCLIFFE
Biological NMR Centre and Department of Biochemistry, University of Leicester, PO Box 138, Medical Sciences Building, University Road, Leicester LE1 9HN, UK.

SYBREN S. WIJMENGA
Nijmegen SON Research Centre for Molecular Design, Structure, and Synthesis, Laboratory of Biophysical Chemistry, University of Nijmegen, Nijmegen, The Netherlands.

Nomenclature and abbreviations

1. Atoms, nuclei, and conformations

A number of different schemes for identification of individual atoms in biological macromolecules are in use in the literature. For NMR, it is especially important to identify individual hydrogen atoms (and their nuclei). In this book, the following system of nomenclature, based on the established standards, is used.

Proteins. Hydrogen atoms (and protons) are identified by following superscripts, using the standard Greek letters for the carbon or nitrogen atom to which the hydrogen is attached, with the exception of N for the proton attached to the backbone amide nitrogen. Thus: H^N, H^α, H^β, etc. An analogous nomenclature is used for carbon and nitrogen, with the addition of a preceding superscript for the isotope; e.g., $^{15}N^\alpha$ (the backbone nitrogen), $^{13}C^\beta$. The standard notation is used for conformational angles: ω (peptide bond), ϕ (N^α—C^α), ψ (C^α—C'), χ_1 (C^α—C^β), etc.

Oligonucleotides. Protons in the (deoxy)ribose rings are denoted H1', H2', H2'', etc. Those in the bases are denoted Xn-Hm, where X is the base (A, G, C, T), n its number in the sequence, and m the number of the proton within the ring. Further details, and definitions of conformational angles, are given in Chapter 8.

Oligosaccharides. Protons are identified with numerical superscripts: H^1, H^2, H^3, etc., with the residue involved being explicitly identified where necessary.

2. NMR parameters

Chemical shift: δ

Scalar coupling constants. ^1H-^1H coupling constants are denoted $^nJ_{H^x\text{-}H^y}$, where n is the number of bonds separating the coupled nuclei, and the subscript defines the protons involved. An analogous nomenclature is used for ^1H-^{15}N, ^1H-^{13}C, and ^{13}C-^{15}N coupling constants. Two shorter forms are also used: first, if n is omitted, the coupling constant is over three bonds. Second, the designation of the protons involved is abbreviated where there is no ambiguity; thus, in proteins, $J_{N\alpha}$ for $^3J_{H^N\text{-}H^\alpha}$, or in oligonucleotides $J_{1'2'}$ for $^3J_{H1'\text{-}H2'}$.

Nuclear Overhauser effects (NOEs). In proteins, these are either referred to explicitly, as 'the H^α-H^β-NOE', or by using the 'd' nomenclature introduced by Wüthrich: $d_{H^N\text{-}H^\alpha}$ for the NOE between H^N and H^α. When the NOE is between nuclei in different residues, this nomenclature is extended thus: $d_{H^N(i)\text{-}H^\alpha}(i + 1)$ for the NOE between H^N of one residue and H^α of the

following residue in the sequence. For oligonucleotides, the notation is $d_x(i;j)$, where x distinguishes, for example, interstrand and intrastrand NOEs, and i and j are the protons involved; for detailed definitions, see Chapter 8.
Relaxation parameters. T_1, the spin-lattice or longitudinal relaxation time and T_2, the spin-spin or transverse relaxation time, with R_1 and R_2 the corresponding relaxation rates ($R_1 = 1/T_1$, $R_2 = 1/T_2$). The resonance linewidth, denoted Δ or LW, is given by $\Delta = 1/\pi T_2$. The cross-relaxation rate is denoted σ_{ij}, where i and j are the protons involved.

3. General NMR abbreviations

Pulse sequences

The proliferation of new pulse sequences for 2D and 3D NMR has been accompanied by a proliferation of (occasionally bizarre) acronyms. These are defined in the chapter where they are used, but a small number are sufficiently widely used to be worth defining here. For details, see Chapter 3.

COSY: correlated spectroscopy; through-bond connectivities between protons separated by two or three bonds. Also DQF-COSY, double quantum filtered COSY.

TOCSY: total correlation spectroscopy *or* HOHAHA: homonuclear Hartmann–Hahn spectroscopy; through-bond connectivities between protons, not restricted to two or three bonds, but extending under favourable conditions throughout a spin system such as an amino acid side chain or a deoxyribose ring.

NOESY: nuclear Overhauser effect spectroscopy; through-space connectivities between protons close in space. Also ROESY, rotating-frame Overhauser spectroscopy.

HMQC: heteronuclear multiple quantum correlation *and* HSQC: heteronuclear single quantum correlation; connectivities between proton resonances and those of directly bonded heteronuclei (^{13}C or ^{15}N).

Pulses in these sequences are designated using the angle by which they cause the magnetization to flip, together with the axis (in the rotating-frame) about which this flip occurs: thus, e.g., $90_x°$. In multi-dimensional spectra the following parameters are used:

t_1, t_2, t_3: the time delays which are incremented so as to form, after Fourier transformation, the frequency axes of 2D and 3D spectra; one of these will be the acquisition time.

ω_1, ω_2, ω_3 or F_1, F_2, F_3: the corresponding frequency axes.

t_m: the 'mixing time', an additional delay in, for example, the NOESY pulse sequence.

Other abbreviations and symbols are defined in the chapter(s) in which they are used.

1

Introduction

GORDON C. K. ROBERTS

Nuclear magnetic resonance (NMR) spectroscopy is based on the fact that atomic nuclei oriented by a strong magnetic field (2–14 Tesla) absorb radiation at characteristic frequencies (typically a few hundred megahertz, in the radiofrequency part of the spectrum). The usefulness of NMR to the chemist and biologist results largely from the fact that nuclei of the same element in different environments give rise to distinct spectral lines. This makes it possible to observe signals from individual atoms even in complex biological macromolecules in solution. The parameters that can be measured from the resulting spectra can be interpreted in terms of molecular structure, conformation, and dynamics.

A very wide range of different elements have nuclei which are amenable to study by NMR spectroscopy; those which are most relevant to the study of biological macromolecules are listed in *Table 1*. The nucleus which is most sensitive to detection by NMR is that of hydrogen (^{1}H; the proton), and this is by far the most important nucleus for the study of biological macromolecules; other nuclei such as ^{13}C and ^{15}N are nowadays often detected through their attached protons to take advantage of this sensitivity. It is also important for many applications of NMR to complex molecules that isotopes of these elements (^{1}H/^{2}H; ^{12}C/^{13}C; ^{14}N/^{15}N) have very different NMR properties, so that isotope labelling can be used to change the NMR properties of a molecule.

The first published NMR spectrum of a biological macromolecule was the 40 MHz ^{1}H spectrum of pancreatic ribonuclease reported in 1957; the most that could be deduced from this spectrum was that it was consistent with the amino acid composition of the protein. The subsequent years—perhaps particularly the last ten years—have seen astonishing developments in instrumentation and methodology which have enormously increased the power of NMR, notably in its application to studies of the conformations and interactions of biological macromolecules. The most important of these developments include the following:

(a) The construction of higher field spectrometers, with a consequent increase in sensitivity and spectral dispersion.

Table 1. Properties of nuclei of interest in NMR studies of proteins

Isotope	Spin	Frequency (MHz) at 11.74 T	Natural abundance (%)	Relative sensitivity[a]
^1H	1/2	500.0	99.98	1.00
^2H	1	76.7	1.5×10^{-2}	9.65×10^{-3}
^3H	1/2	533.3	0	1.21
^{12}C	0	—	98.89	—
^{13}C	1/2	125.7	1.108	1.59×10^{-2}
^{14}N	1	36.1	99.63	1.01×10^{-3}
^{15}N	1/2	50.7	0.37	1.04×10^{-3}
^{16}O	0	—	~100	—
^{17}O	5/2	67.8	3.7×10^{-2}	2.91×10^{-2}
^{19}F	1/2	470.4	100	0.83
^{31}P	1/2	202.4	100	6.63×10^{-2}

[a] For equivalent numbers of nuclei (i.e. 100% isotope).

(b) The development of pulse Fourier transform methods, in which the radiofrequency radiation is applied to the sample in the form of a more or less complex sequence of pulses, and the spectrum obtained by Fourier transformation of the response of the nuclear spins to these pulse trains.

(c) The development of multi-dimensional NMR, in which resonance intensity is recorded as a function of two, three, or four frequency variables.

The majority of modern applications of NMR to the study of biological macromolecules involve the analysis of two- (or higher) dimensional NMR spectra. The majority of these have a common structure: the diagonal corresponds to the conventional 'one-dimensional' spectrum, while the off-diagonal peaks, or cross-peaks, contain information about *connections* between resonances on the diagonal. The nature of these connections depend on the kind of two-dimensional experiment being carried out; one can observe *through-bond connections* between the resonances of nuclei separated by two or three (or sometimes four) bonds, *through-space connections*, between the resonances of nuclei which are close together in space, and *exchange connections*, between the resonances of the same nucleus in two different environments (e.g. of a ligand bound to a protein or free in solution).

These methodological developments have been accompanied by a substantial increase in the use of NMR for the study of macromolecular structure and interactions in solution. The assignment of resonances to individual nuclei is an essential first step in any NMR study, and this has been made very much easier by the advent of multi-dimensional NMR experiments. Nonetheless, for all but the smallest macromolecules this is still the rate-limiting step. The

2

structural information is then obtained principally from measurements of the nuclear Overhauser effect (NOE) which provide constraints on internuclear distances. Two-dimensional (2D) ^{1}H NMR experiments have provided three-dimensional structures for a substantial number of small proteins (and isolated domains of larger proteins), oligonucleotides, and oligosaccharides. The problems of overlap in the spectrum limit this approach to proteins of less than about 100–120 residues and oligonucleotides of up to about 20 base pairs. Selective substitution with ^{2}H can simplify the ^{1}H spectrum and overcome overlap problems. The most powerful approach is, however, isotope labelling with ^{13}C and ^{15}N, in conjunction with three- and four-dimensional (3D and 4D) NMR; this is crucial in extending the usefulness of NMR to larger molecules. This isotope labelling, together with the provision of the fairly substantial quantities of protein (tens of milligrams) required for NMR spectroscopy, has been greatly facilitated by the developments in molecular genetics and the ability to construct over-expression systems. In 3D and 4D NMR the two-dimensional ^{1}H spectra are, in effect, 'spread out' in a third or fourth dimension by the ^{13}C or/and ^{15}N chemical shift. Individual 'slices' at a particular ^{13}C or ^{15}N chemical shift are thus much less crowded and easier to analyse. The combination of isotope labelling and multi-dimensional NMR has been used for the determination of three-dimensional structures of proteins of around 170 residues, and extension up to 300 residues should be possible.

In addition to the determination of complete three-dimensional structures for macromolecules in solution, NMR can provide very valuable information on local structure, on conformational dynamics, and on both structural and kinetic aspects of interactions with small molecules. NMR can be used to characterize the charge state, conformation, and dissociation rates of bound ligands, and to identify contacts between individual atoms of the ligand, and individual atoms of the protein. This kind of information can be obtained, again with the help of isotope labelling, for much larger systems—certainly up to M_r 100 000. The ability to combine structural and dynamic information is perhaps the single most important attribute of NMR in the context of structural molecular biology.

The aim of this book is to provide the newcomer to these applications of NMR, whether a graduate student or a more senior biologist who wishes to apply NMR to his or her favourite protein, with practical guidance on how to choose the right experiment to obtain the desired information, how to carry out the experiment, and how to analyse the resulting spectra. We hope that those who are familiar with chemical applications of NMR but not those to biological macromolecules will also find this book helpful in describing the special requirements of NMR studies of these large molecules. One of the consequences of the considerable proliferation of papers describing the application of NMR to biological macromolecules has been the increasing reluctance of journals to provide space for a detailed description of the methods involved in, for example, resonance assignment or the analysis of NOEs for structure determination. It is our hope this book will have filled this gap, so as to allow the reader both to understand what is involved in the experiments described in the literature and to embark on the application of these powerful methods to his or her own system.

Chapters 2 and 3 provide the first essential information—advice on the preparation of the sample, and on obtaining the spectrum. The details of spectrometer operation differ from instruments of one manufacturer to those of another, and thus exact step-by-step protocols which will be generally applicable can not be given. Rather, Chapter 3 concentrates on describing general procedures to follow and precautions to be taken in order to obtain the best quality spectrum—including, for example, methods for overcoming the dynamic range problem involved in obtaining ^1H spectra in H_2O solution—and on an introduction to two-dimensional NMR spectroscopy. Chapter 4 describes in detail procedures for resonance assignment in the ^1H NMR spectra of small proteins. Chapter 5 continues this general theme by describing methods for labelling proteins with stable isotopes (^2H, ^{13}C, and ^{15}N), and for using these labels to make resonance assignments in larger molecules; it also provides illustrations of the use of ^{13}C and ^{15}N to obtain structural and dynamic information. Many (perhaps most) of the situations one encounters in the biological applications of NMR involve the exchange of atoms from one state to another—for example, pH titration (exchange between protonated and unprotonated states), conformational equilibria, and ligand binding. Chapter 6 describes the effects of these exchange processes on NMR spectra and ways to analyse them qualitatively and quantitatively. This leads on to a detailed description, in Chapter 7, of the kinds of information NMR can provide about ligand binding. Chapters 8 and 9 describe the use of NMR in studying oligonucleotides and oligosaccharides, from resonance assignment right through to the interpretation of the data in structural terms. Finally, Chapters 10 and 11 describe, respectively, the experimental and computational aspects of the use of NMR to determine the three-dimensional structures of proteins in solution.

Since it is quite impractical to take an encyclopaedic approach to this extensive and very active field, it is important to emphasize those aspects which this book does not cover, and to indicate where information on these aspects can be found. First, we provide neither an introduction to the basic physical principles of NMR, nor details of the theory behind the modern multi-dimensional NMR experiments. Useful introductions to the principles of NMR are provided in references 1–5; the 'bible' for the theory of modern multi-dimensional NMR is the treatise by Ernst *et al.* (6). A detailed discussion of the nuclear Overhauser effect (NOE), which is crucial to the structural uses of NMR, is given in reference 7. Secondly, no attempt is made to provide practical information on all the possible kinds of applications of NMR to biological macromolecules. These are now so extensive that this would be an impossible task; rather the authors of the chapters in this book have provided detailed information on what might be termed 'core' experiments, which should readily be extrapolated to the many variants which can be found in the literature. Some useful surveys of applications can be found in references 8–13.

Finally, there are other valuable sources of practical information. References 14–16 provide an introduction to modern NMR methods with a practical slant, although focusing primarily on chemical problems, and two recent volumes in the *Methods in Enzymology* series (17, 18) provide valuable methodological information on a range of applications of NMR to biological macromolecules.

References

1. Gunther, H. (1980). *NMR spectroscopy: an introduction*. Wiley, Chichester.
2. Harris, R. K. (1983). *Nuclear magnetic resonance spectroscopy: a physicochemical view*. Pitman, London.
3. Freeman, R. (1987). *A handbook of nuclear magnetic resonance*. Longman, Harlow.
4. Homans, S. (1992). *Dictionary of concepts in NMR*. Oxford University Press, Oxford.
5. Campbell, I. D. and Dwek, R. A. (1984). *Biological spectroscopy* (Chapter 6). Benjamin-Cummings, Menlo Park, CA.
6. Ernst, R. R., Bodenhausen, G., and Wokaun, A. (1987). *Principles of nuclear magnetic resonance in one and two dimensions*. Oxford University Press (Clarendon Press), Oxford.
7. Neuhaus, D. and Williamson, M. (1989). *The nuclear Overhauser effect in structural and conformational analysis*. VCH, Weinheim.
8. Jardetzky, O. and Roberts, G. C. K. (1981). *NMR in molecular biology*. Academic Press, New York.
9. Wemmer, D. E. and Reid, B. R. (1985). *Annu. Rev. Phys. Chem.*, **36**, 105.
10. Wüthrich, K. (1986). *NMR of proteins and nucleic acids*. Wiley, New York.
11. Bax, A. (1989). *Annu. Rev. Biochem.*, **58**, 223.
12. Wüthrich, K. (1989). *Science*, **243**, 45.
13. Clore, G. M. and Gronenborn, A. M. (1991). *Prog. NMR Spectrosc.*, **23**, 46.
14. Derome, A. (1987). *Modern NMR techniques for chemistry research*. Pergamon Press, Oxford.
15. Saunders, J. K. M. and Hunter, B. K. (1987). *Modern NMR spectroscopy: a guide for chemists*. Oxford University Press, Oxford.
16. Croasmun, W. R. and Carlson, R. M. K. (ed.) (1987). *2D NMR spectroscopy*. VCH, Weinheim.
17. Oppenheimer, N. J. and James, T. L. (ed.) (1989). *Methods in enzymology*, Vol. 176. Academic Press, New York.
18. Oppenheimer, N. J. and James, T. L. (ed.) (1989). *Methods in enzymology*, Vol. 177. Academic Press, New York.

2

Sample preparation

WILLIAM U. PRIMROSE

1. Introduction

Over the last few years, there have been rapid and sustained advances in the hardware and software associated with the NMR analysis of biological macro-molecules. Larger magnets, faster computers, and new and improved pulse sequences have been combined with isotopic labelling to make NMR the unrivalled method for answering biological questions at the atomic level in the solution state. These dramatic advances have perhaps tended to obscure the importance of correct preparation of the sample, which tends to be a rather poorly documented subject. The purpose of this chapter is to aid you in the correct choice of conditions for preparation of your vital and hard-won sample. This will then allow you to obtain the maximum benefit from the sophisticated spectroscopic methods described in other chapters in this book. The importance of correct handling of the sample can not be over-emphasized, as no amount of expensive spectrometer and computer equipment can compensate for an intrinsically poor sample. A Formula One racing car will not break any lap records running on two-stroke fuel!

2. Requirements for an NMR sample

2.1 Sample quantity

The fundamental problem in the biological applications of NMR is its intrinsically poor sensitivity. The intensity of (area under) the NMR signal is proportional to the amount of material within the 'sensitive volume' of the spectrometer. For high resolution NMR, the homogeneity of the magnetic field must be extremely good. In practice, this means that the sample is confined to a small volume within the central bore of the magnet, commonly in a tube measuring 5 mm o.d. The different probe designs used by the various instrument manufacturers mean that the optimum height of the liquid sample within the tube, and hence the optimum sample volume, varies somewhat, but should be around 400 µl (see *Figure 1*). In order to improve the signal-to-noise ratio, the signal from the sample is commonly averaged many

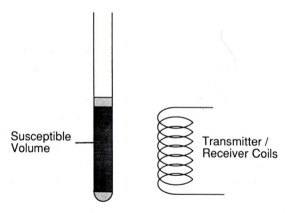

Figure 1. The amount of sample required for optimal sensitivity in a high resolution NMR tube.

times in the course of an experiment. The signal-to-noise ratio increases as the square root of the number of times the signal is averaged. Thus for a sample concentration of 0.5 mM, it will take four times as long to obtain a spectrum with a given signal-to-noise ratio as for a 1 mM concentration. Very weak samples may require an impractical length of time in the spectrometer to obtain a satisfactory spectrum. For structural studies, a solution concentration of the macromolecule of interest of at least 1 mM is required, and for multi-dimensional experiments 3–5 mM will be required.

Taken together, these concentration and volume requirements mean that amounts of the order of tens of milligrams of a small protein will be required. Whilst the technique is by its very nature non-destructive, the sample must spend tens of hours in the spectrometer, often at elevated temperatures, and this can lead to some breakdown of even the most perfectly prepared sample. Some material will also be lost during preliminary experiments designed to optimize the sample conditions. Thus, the amount of material available may well be the limiting factor in the application of NMR to a specific system. It is not within the scope of this chapter to discuss ways in which these material limitations may be overcome, but the application of recombinant DNA technology (1) has enormously increased the ease with which large amounts of protein may be isolated, while oligonucleotides may be produced synthetically in large quantities (2).

2.2 Solubility

If ample quantities of material are available, solubility may become the limiting factor. Techniques are described later in this chapter which can be used to identify the conditions for maximum solubility of the biological macromolecule of interest.

2.3 Stability

The macromolecule must be stable during the course of the NMR experiment. Air oxidation, microbial contamination, and hydrolytic breakdown must all be eliminated to maximize the lifetime of the sample in the probe. Even after all necessary precautions have been taken, there remain molecules which are intrinsically unstable, for which it may be impossible to obtain high resolution spectra. To determine the stability of a particular macromolecular solution, the 'nativeness' of the sample should be tested before and after a long term data acquisition—for example by enzyme activity assay. In practice, the best way of determining if your sample is in the same state at the end of the experiment as at the beginning is to run a quick 1D ^{1}H NMR spectrum. An identical spectrum means an identical sample.

2.4 Molecular weight

Increasing the size of a molecule has two main effects on its NMR spectrum:

(a) The complexity of the spectrum increases. A larger molecule has more chemically and magnetically distinct nuclei than a smaller one. Each unique nucleus will contribute at least one line to the NMR spectrum, so the total number of lines will be proportional to the molecular weight (see *Table 1*). As the number of lines in a given frequency interval increases, the greater the overlap between them, and the harder it is to distinguish the resonances of individual residues.

(b) The linewidths of the individual resonances increase. Linewidth is proportional to the correlation time[a] of the molecule. Larger molecules have longer correlation times and therefore broader lines. Broad lines are both intrinsically less easily observed than narrow ones and more subject to overlap.

The complexity of the spectrum can be reduced by isotope labelling and employing editing techniques (see Chapter 5). The problem of broad lines and spectral overlap can be alleviated by multi-dimensional NMR techniques (see Chapter 3). At the time of writing, the largest monomeric protein for which near total assignment of the ^{1}H NMR spectrum has been achieved has M_r 19 000 (3), obtained at a concentration of 3–4 mM.

[a] The rotational correlation time, τ_C, is the time taken for the particle to rotate through an angle of one radian (57°). For a spherical particle, an upper limit to this value can be approximated by the Stokes–Einstein equation;

$$\tau_C = \frac{4\Pi\eta a^3}{3kT}$$

where a is the radius of the particle, η is the solvent viscosity, and k is Boltzmann's constant.

Table 1. Average number of unique resonances expected for biopolymers of various molecular weights

	No. of residues	M$_r$ ($\times 10^{-3}$)	Amount required[a]	Number of magnetically distinct nuclei				
				^{1}H[b] (^{2}H$_2$O)	^{1}H[c] (H$_2$O)	^{13}C	^{15}N	^{31}P
Protein	50	5.5	2.2	210	270	240	52	0
	100	11.0	4.4	420	540	470	105	0
	200	22.0	8.8	840	1080	940	210	0
DNA[d]	10	3.4	1.4	90	100	100	40	10
	20	6.7	2.7	175	200	195	75	20
	50	16.8	6.8	440	500	490	190	50
RNA[d]	10	3.5	1.4	80	90	95	40	10
	20	7.0	2.8	155	180	190	75	20
	50	17.5	7.0	390	450	475	190	50
Carbohydrate	5	1.0	0.4	30	35	35	2	0
	10	2.0	0.8	60	70	70	4	0

Note. All the data in this table is dependent on the exact sequence of the biopolymer involved. For proteins, the statistical distribution of the amino acids in known proteins was used. For RNA and DNA, an equivalent number of each of the four bases was assumed. The data for carbohydrates is extremely dependent on sequence and the numbers quoted should be used as guidelines only.

[a] Amount required (mg) for 400 μl of a 1 mM solution.
[b] Assuming that for proteins, all β-protons are non-equivalent, as are γ-protons for Glu and Gln residues; γ, δ, etc., protons for Lys, Arg, and Pro are equivalent, and that the aromatic rings of Tyr and Phe residues are flipping freely.
[c] Only exchangeable protons bound to nitrogen are additionally included.
[d] Calculated for single stranded oligonucleotides.

3. The NMR tube

3.1 Choice of NMR tube

Get a good one! High resolution NMR requires high resolution tubes. Although these may appear to be quite expensive, they can be reused many times if properly maintained. The cost of the tube reflects the dimensional tolerances to which the side walls have been ground and polished. Cheap tubes have low tolerances and do not possess perfect cylindrical symmetry. This is most obviously manifested as unacceptably large side bands in the spectrum when the tube is spun in the spectrometer; it is common practice to obtain two-dimensional spectra of macromolecules without sample spinning, but the low tolerance tubes should still be avoided as they limit the attainable field homogeneity.

The traditional tube has a round bottom, but this can be wasteful of sample, as the part of the tube where the side walls are not parallel should not be within the receiver coil (see *Figure 1*). The hemispherical dead volume can be filled with a PTFE spacer, but flat-bottomed tubes are much more practical.

For tubes greater than 5 mm diameter, the bottom can be made flat and perpendicular to the side walls, resulting in the absolute minimum dead volume. For 5 mm tubes, this process is technically more difficult, and there tends to be slight variation in the actual shape of the bottom of each tube, even within the same batch. This necessitates adjustments in the exact placing of each tube within the probe of the spectrometer to achieve optimal resolution. The standard length of an NMR tube is seven inches. Other sizes are available, but may not then fit specialized pieces of maintenance equipment.

3.2 Cleaning the NMR tube

NMR tubes should not be used straight from the pack without cleaning, as they may contain chemicals associated with their manufacture. Most tubes (especially those >5 mm in diameter) can be cleaned by simple soaking in cleaning solutions and rinsing. Particular care should be taken, however, not to use cleaners containing metal ions, which, if not rigorously removed will give rise to paramagnetic line-broadening. This is especially true of chromic acid, which should be avoided at all costs, in cleaning glassware used in sample preparation as well as with the NMR tube itself. If soaking in a strong acid is really necessary, metal-free sulfuric acid preparations are commercially available. Strong acid should not, however, be used as the first resort, as it can lead to formation of stubborn precipitates of biological macromolecules. These may be removed by long pipe cleaners (though special care should be taken to avoid scratching the glass, which will degrade spectral resolution), or by soaking in a proprietary detergent solution after extensive washing with water. A small amount of a chelator, such as EDTA, can be added if contamination by metal ions is thought to be a problem. For tubes of 5 mm diameter, these operations can be difficult to perform, due to the high surface tension of aqueous solutions. However, devices are available which overcome this problem by using an applied vacuum to direct a spray of cleaning or rinsing solution against the bottom of the NMR tube (see *Figure 2a*).

Organic solvents are not normally necessary for cleaning NMR tubes used for biological samples. Laboratory grade acetone or other solvents contain impurities which will be deposited on the inside surface of the tube. If these solvents must be used for particular cleaning applications, it should not be as the final step, and particular care should be exercised in rinsing out all final traces. The NMR tube washer shown in *Figure 2a* is particularly convenient for this.

NMR tubes should be stored, capped, in a dust-free environment where they will not become scratched. Scratches on the outside of the tube are just as detrimental to resolution as those on the inside.

3.3 Drying the NMR tube

It is impossible to get an NMR tube completely dry; the glass will always retain a layer of adsorbed water on its surface. The use of very high temperatures

11

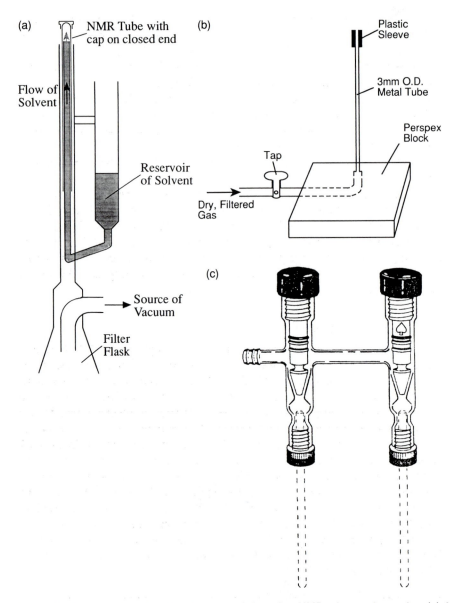

(a) NMR Tube with cap on closed end

Flow of Solvent

Reservoir of Solvent

(b) Plastic Sleeve

3mm O.D. Metal Tube

Perspex Block

Tap

Dry, Filtered Gas

(c)

Source of Vacuum

Filter Flask

Figure 2. Apparatus for cleaning, drying, and degasing NMR tubes and samples. (a) A design for an NMR tube cleaner which allows easy cleaning and rinsing of 5 mm diameter tubes. When the cap over the NMR tube is pushed down to make a seal, the applied vacuum sprays solvent from the reservoir against the very bottom of the inside of the tube, facilitating thorough cleaning. Cleaning solutions can be varied quickly and easily and a stream of air can be drawn through the tube in order to dry it. (b) Design of a simple apparatus for blowing dry, filtered N_2 gas into 5 mm NMR tubes. The NMR tube can be slid over the long, straight copper tube in order that gas may be directed right to the

to remove this last trace of moisture will cause the glass to creep, losing its cylindrical perfection and degrading its high specifications. Adequate drying can be accomplished by blowing a stream of filtered dry nitrogen gas against the bottom of the tube (see *Figure 2b*), by keeping the tube, inverted, in a dust-free oven at 50°C (no higher), or by applying a vacuum to the tube, either by storing it in a vacuum desiccator or lyophilizer chamber, or directly via a specially designed valve (see *Figure 2c*). A combination of these three techniques should suffice for most purposes. Where it is necessary to remove the residual bound water, say due to very low sample concentration and/or overlap of signals of interest with the water proton resonance, the tube can be pre-treated with deuterium oxide. Exhaustive pre-rinsing with ^2H_2O will substitute most of the bound water, or alternatively in-tube lyophilization is possible (4). With careful handling, it is possible to freeze a column of ^2H_2O inside a 5 mm tube and then lyophilize it in a large vacuum container or using a special fitting made to hold NMR tubes for degasing (see *Figure 2c*). It is possible that the advantage gained in removal of the last trace of water will outweigh the finite risk of breaking the tube during the freezing process.

3.4 Sample volume

As discussed in Section 2.1, the volume of the sample in the NMR tube is critical for good sensitivity, resolution, and lineshape. Since the amount of the sample is often limiting in biological NMR, we want to use the minimum volume of sample that we can, without compromising sensitivity. This volume is the amount required just to cover the receiver coils in the probe (see *Figure 1*), without giving rise to any 'end effects'. The change in magnetic suscepti-bility at the solvent/glass (bottom of the tube) and solvent/air (meniscus) interfaces will cause loss of resolution. If the sample volume is limiting, various devices can be employed to fill the remaining space within the coils— for example, thin walled glass spacers, which can be filled with buffer to minimize field distortions. However, even these tricks can not offset the loss of resolution caused by using a short sample in a 5 mm tube, and in this case it is probably just as effective to dilute the sample. For larger diameter tubes, where dilution of the sample is undesirable, the methods mentioned are probably worth considering.

---

bottom. A thin plastic sleeve at the top of the copper tube prevents scratching of the inside of the NMR tube. The weight of the NMR tube holds it down and allows the gas to be left flowing through it unattended. (c) A method for attaching an NMR tube to a source of vacuum. The open end of the tube is inserted into the valve by an O-ring sealed screw with axial bore. A source of vacuum and dry gas are available via the side hose connections. The NMR tube can then be evacuated and filled with dry gas repeatedly. Tubes can also be sealed with a flame whilst attached to this apparatus to protect highly oxygen sensitive samples. The apparatus depicted is available from the Aldrich Chemical Company.

4. Sample handling

This section deals with the techniques required to actually prepare your sample for the NMR tube. Given the appropriate amount of the macromolecule in a particular buffer, we generally require to reduce the sample volume to the minimum necessary for the NMR experiment, change the sample conditions (pH, ionic strength, buffer components) to those desired, and filter the final solution. Methods for performing each of these steps are described below.

4.1 Methods for concentration

One of the simplest methods for reducing the volume of the sample, is to dry it down and then to reconstitute it in the appropriate volume of solvent. This can be the last step in NMR sample preparation, if the sample components have been previously adjusted to take account of subsequent concentration changes.

Lyophilization can be used if the material can withstand the rigours associated with this technique, i.e. the increasing concentration of the macrosolute and buffer ions, and the extremes of pH occurring during the removal of the last traces of water. Not all proteins can be successfully lyophilized, especially those with surface thiol groups. Lyophilization of small volumes of biological samples is most conveniently performed in microcentrifuge vials, preferably those with a screw cap and rubber O-ring often used for cryo-storage. The efficiency of lyophilization is proportional to the surface area of the frozen sample, so freeze-drying in such vials can be slow. This is a problem where the sample has a high ionic strength and the rate of evaporation is not high enough to keep the sample frozen. Frothing can occur in these cases, which will lead to loss of some of the sample. Such samples are better handled in a vacuum centrifuge (see below). Samples with lower ionic strength should produce the final pellet in a small, narrow container, from which it can be efficiently removed, after dissolution in just the amount of solvent necessary for the NMR tube and no more. Sample loss is thus minimized in these cases. An additional advantage of using small tubes that can be centrifuged, is that, after addition of solvent and mixing/vortexing, the solution can be spun to the bottom of the tube for removal by pipette or syringe.

One problem with lyophilization concerns the freezing of the sample, which—apart from causing denaturation in some cases—can lead to the adsorption of atmospheric water vapour on to the frozen surface of the sample, unless strict precautions are taken to prevent this occurring. This becomes a problem when a ^2H_2O sample is being prepared and strict exclusion of water is necessary. Handling of the sample within a dry atmosphere can overcome this to some extent and the frozen sample vial can be contained within a manifold whilst on the lyophilizer. When the sample is dry, the manifold is isolated and removed to a dry environment for opening and subsequent sample preparation.

The danger of the sample 'bumping' whilst on the lyophilizer can be reduced by ensuring that other users do not put large volume flasks on at the same time. A lyophilizer dedicated to NMR samples is a good idea. Vacuum centrifuges (such as the Savant Speed-Vac) eliminate the possibility of 'bumping' completely and produce a final product that is dense and convenient to handle, without the difficulties of static charge associated with the fine powdery samples often produced by standard lyophilization. A slight problem of the vacuum centrifuge is that the lyophilized sample may become 'glassy' and difficult to redissolve. It is important to realize that lyophilization of a sample containing certain buffers may lead to large pH changes, even after reconstitution. This occurs because the neutral form of buffer ions such as ammonium, acetate, bicarbonate, and formate are volatile. Evaporation of the solvent at or above room temperature avoids some of the problems of freeze-drying, although this will only be successful for the more robust samples. It is especially useful for oligosaccharide samples and may also be used for oligo-deoxynucleotides and a limited number of protein samples. A heated centrifuge is the ideal piece of equipment for accomplishing this task, for the same reasons as stated above.

Simple concentration of the sample may not be the most appropriate method for reducing its volume to that required, if the initial sample conditions are far removed from those necessary for the NMR experiment, or if lyophilization or evaporation would damage the biomolecule. Under these conditions, ultrafiltration should be used instead. The stirred cell ultrafiltration device (see *Figure 3a*) has been used most widely in the past and is still the most convenient way of handling large initial volumes. The large surface area of the membrane and continual stirring ensure a smooth, continual, relatively fast filtration. A major drawback is the difficulty of handling smaller volumes in a large cell. It may be necessary to redilute the concentrated sample to ensure maximum recovery from the cell, followed by transfer to a smaller cell and refiltration; this is time-consuming and increases the amount of handling. The most serious drawback of the stirred cell is the possibility of filtering all of the solvent off and leaving an intractable and potentially denatured sample smeared over the surface of the membrane. The spun concentrator (see *Figure 3b*) is capable of handling small volumes efficiently and has a dead stop to prevent filtration to dryness. Centrifugal recovery of the concentrate minimizes losses at the membrane and on the reservoir wall.

Both types of concentrator are available with different membrane types and molecular weight cut-offs. Adsorptive losses can occur with these membranes and are dependent on the concentration of the solute, its hydrophobicity, the temperature, and the time of contact with the solution. Low starting concentrations lead, in general, to a lower recovery. Hydrophobic membranes tend to sequester a certain amount of material irreversibly, but this can be

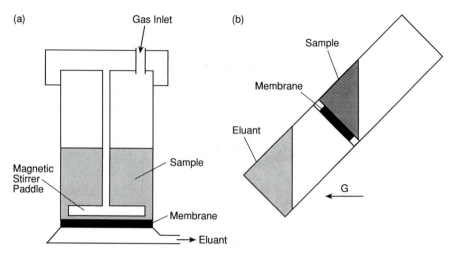

Figure 3. Apparatus for sample concentration, buffer exchange, and desalting. (a) The stirred cell ultrafiltration device (e.g. that available from Amicon). Gas under pressure forces the solute through the replaceable membrane at the bottom. Continual stirring by the internal magnetic flea prevents build up of the biomolecule at the surface of the membrane, which could cause blockage of the pores. Available in sizes from 3–500 ml. (b) The spun concentrator (e.g. Amicon's Centricon). Centrifugal forces drive solvents and low molecular weight solutes through the membrane. In a fixed angle rotor, the dense layer of macrosolute slides outward and gathers at the edge of the membrane. After the appropriate amount of time within the centrifuge for the level of concentration required, the sample is then recovered by inverting the device and spinning the concentrate back into the cap.

minimized by prior treatment with a solution of bovine serum albumin, followed by thorough washing. All ultrafiltration membranes retain macro-molecules according to their molecular size and shape, so the nominal mol-ecular weight cut-off may not hold true for all macrosolutes. A cut-off should be chosen to be considerably less than the weight of the macromolecule involved, despite the increase in concentration time that this implies. All the components of the concentrator should be pre-rinsed with buffer prior to use. It is especially important to wash a small amount of buffer through the membrane to remove any trace of preservative (e.g. glycerine, azide).

4.2 Desalting and buffer exchange

4.2.1 General techniques

Repeated passage of a solution of a biological macromolecule through an ultrafiltration cell or spun concentrator with intermediate redissolution in water will effectively desalt the sample. Two successive 10:1 concentration steps will remove 99% of all the low molecular weight species. The sample can be transferred into a solution of specific buffer, ionic strength, or pH by

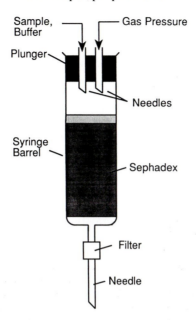

Figure 4. A simple home-made column for desalting or buffer exchange. The body of a disposable syringe is packed with Sephadex G-25 (or similar, depending on the biological macromolecule to be processed), fitted with a 0.45 μm syringe filter, and a Luerlock needle (preferably platinum, to prevent metal ion contamination of the sample). A source of dry, inert gas can be attached to the top of the column to facilitate passage of the sample through packing material. This can be conveniently provided through a needle pushed through the original rubber plunger tip. The sample and subsequent buffer can be loaded via another needle.

repeated concentration/dilution with the ultimate solution. This is the best way of preparing the NMR sample in the first instance.

Changing the buffer conditions of an existing NMR sample is more difficult if dilution is to be avoided. The sample may be passed through a short column packed with a size exclusion gel which has been previously equilibrated with the new buffer. Small pre-packed disposable columns are commercially available, (e.g. Pharmacia's NAP columns), or these can be simply constructed from a disposable plastic syringe, filter, syringe needle, and the appropriate adsorbent (see *Figure 4*). A bed volume of 1–2 ml is required for a 400 μl sample. The protocol is the same with commercial or home-made columns.

Protocol 1. Gel filtration of an NMR sample

Materials
- 2 or 5 ml disposable plastic syringe reassembled as shown in *Figure 4*
- Sephadex G-25

Protocol 1. *Continued*

- some syringe needles (preferably platinum)
- 0.45 μm (or similar) syringe filter of the type used for pre-filtering of HPLC samples
- a source of inert gas

1. Make a slurry of the gel filtration matrix as described in the manufacturer's instructions, pour 1–2 ml of it carefully into the column, avoiding air bubbles, and wash the column through thoroughly with the final buffer, allowing the buffer to run out until the top of the column material is only just covered.

2. Layer the sample evenly over the top of the column and allow it to percolate on to the column by gravity, or by the application of dry, inert gas to the top of the column.

3. Layer on more of the final buffer and keep the top of the column covered by it subsequently.

4. Allow a pre-determined volume of solution to flow through the column— the dead volume. This can be determined exactly by following the above procedure using a solution of dextran blue. The volume eluted between the loading of the sample and its first appearance in the eluate is the dead volume.

5. Collect the fraction containing the biomolecule. This will be the solution eluting immediately after the dead volume. At least the same volume as was loaded on to the column must be collected to recover the biomolecule. Due to dilution effects, total recovery of the biomolecule will invariably require collection of a slightly greater volume.

Spun columns are also available for buffer exchange (e.g. Isolab Inc.'s Quik-Sep columns). These have the advantage that more of the macromolecule can be recovered, but less operator control is possible whilst dealing with the small volumes involved. All methods involving the use of size exclusion gels will dilute the sample to some extent.

4.2.2 H_2O to ^2H_2O exchange

^{1}H NMR spectra of biological macromolecules are frequently run in ^2H_2O solution. This is partly historical, in that methods for suppression of the 110 M proton resonance in water have not been as efficient in the past as they are currently (see Chapter 3). However, spectra run in ^2H_2O continue to have an important place in NMR studies of proteins (see Chapter 4). The resonance lines for all the exposed, labile (N-, O-, and S-bound) protons disappear as they are exchanged for deuterons. The spectrum of the remaining non-labile protons is simplified by the loss of spin-spin coupling to the labile protons.

Importantly, the dynamic range problem is effectively decreased by the reduction of the intensity of the largest signal, allowing observation of small signals above the noise level.

The intensity of the residual water line is, of course, dependent on the efficiency with which the labile protons are exchanged, and on the isotopic enrichment of the ^2H_2O used. Suppliers of ^2H_2O label their products by the percentage deuterium content, and it is important to appreciate exactly what these figures imply for the NMR experiment. 99.96% ^2H_2O has a water content equivalent to 44 mM protons, which means that for a normal biological sample made up in this solvent, the water resonance will still be easily the most intense line in the spectrum. Even in 99.996% ^2H_2O (4 mM protons), the water line will be at least of equivalent size to those of the macromolecule.

It is worth noting that the viscosity of ^2H_2O is around 1.2 times that of H_2O at room temperature; this will affect the correlation time of the macromolecule and lead to a proportionate broadening of the lines in the NMR spectrum. It is also important to remember that the freezing point of pure ^2H_2O is 3.8 °C. Samples which have been run close to their freezing point in H_2O solution may actually freeze whilst in ^2H_2O solution; simple observation of the deuterium lock level will determine whether your sample is still fluid or not.

The methods for deuterium exchange are basically those described in Section 4.2.1, but with additional safeguards where maximum exchange is required. Lyophilization or evaporation and redissolution in ^2H_2O is effective for oligonucleotides and oligosaccharides, since all of their labile protons will exchange extremely rapidly. This is especially true for oligonucleotides at acid pH, where the intrinsic exchange rate is greater than at basic pH (see *Figure 5*). Repeating the process twice should suffice to make the concentration of labile protons in the macromolecule the same as that in the solvent. For proteins, with their additional tertiary structure, not all of the labile protons are exposed to the solvent and some may have very slow exchange rates, due to their involvement in hydrogen bonds. In addition, the shell of water surrounding a protein molecule can be considered to be an integral part of the protein structure and is very difficult to remove without damaging the protein itself. For some proteins it has been necessary to warm the ^2H_2O solution to 40 °C or even 70 °C for periods from a few minutes to several hours to achieve complete exchange. The pH dependence of the exchange rates of amide and other labile nitrogen-bound protons is shown in *Figure 5*. From this, it can be seen that for maximum possible exchange with ^2H_2O, either extremely acidic (pH <2.5) or, more realistically, basic pH (>7.0) should be used. The stability of the protein of interest at different temperatures and pH values will need to be determined before the correct choice of conditions for deuterium exchange can be made.

Where lyophilization or evaporation is not possible, repeated concentration and dilution with ^2H_2O buffers or the use of size exclusion gels pre-equilibrated with ^2H_2O is recommended.

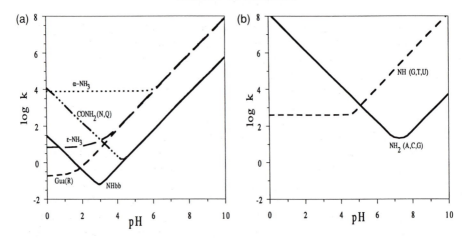

Figure 5. Plot of the logarithm of intrinsic exchange rate versus pH for various labile protons bound to nitrogen in polypeptides and polynucleotides at 25 °C. (a) Amide protons of the peptide backbone [NHbb] and of the side chains of asparagine and glutamine [$CONH_2(NQ)$]; amino protons of the N-terminus-[α-NH_3] and of the side chain of lysine [ε-NH_3]; guanido protons of arginine [Gua(R)]. Labile protons on oxygen and sulfur generally exchange too quickly to be observed. (b) Amino [NH_2] and imino [NH] protons of the bases in DNA and RNA which could become involved in hydrogen bonding in base pairing. (Adapted from Wüthrich and Wagner (5)—© Academic Press; and Wüthrich (6)—© John Wiley & Sons.)

The removal of the maximum amount of water is especially important for oligosaccharides. Not only are they are often only available in small quantities, but a large number of important resonances in their ^{1}H NMR spectra are close to that of water. Because of this, it may be necessary to work in a glove box, reducing 'proton contamination' from atmospheric water, and to repeatedly flash evaporate the oligosaccharide with ^2H_2O of the highest isotopic enrichment available. In extreme cases, it may be necessary to dry the sample cryogenically, e.g. by suspension in an evacuated sealed chamber immersed in liquid nitrogen.

4.3 Methods of filtration

After preparation of the NMR sample in a small vial, the next step is to transfer it to the NMR tube. Since high resolution NMR can be compromised by the presence of suspended dust or fibres, these contaminants have to be removed prior to recording the spectrum. Where the exclusion of water vapour or oxygen is vital, the following procedures can be performed in a dry box.

The vial containing the sample may be spun in a bench centrifuge to pellet the contaminants, which also suffices to remove any denatured material which may have been produced during the final steps of sample preparation. The clear supernatant may then be removed with a pipette or syringe and transferred to the NMR tube. Using a vial with a conical bottom minimizes the amount of solution that may be lost, which should not be more than 10–20 μl. Where the contaminants remain in suspension, or the pellet is easily disturbed, then the solution must be filtered. All filtering apparatus should be washed through with the NMR buffer and blown dry with a stream of dry, inert gas prior to use. This also serves to remove fibrous materials from the filters. In all cases the outflow from the filter can be directed straight into the NMR tube.

A plug of medical cotton wool or glass wool in the neck of a Pasteur pipette serves as a simple filter, although it is quite difficult to get the last drops of solution out and so the retentate is relatively large. Commercial glass fibre filters are available (e.g. Whatman GF/D) which overcome this problem. The best filters are those commonly used for HPLC samples. A 0.45 μm filter fitted on the end of a syringe has the advantages of high efficiency, a compact size leading to easy handling, and a negligible retention volume.

4.4 Degassing the sample—removal of dissolved oxygen

Oxygen has a dual deleterious effect on samples of biological macromolecules for NMR analysis, in that it is both a paramagnetic species and an oxidizing agent. The former property leads to the broadening of the lines of the NMR spectrum, whilst the latter can cause chemical damage to the macromolecule. The linewidths of resonances of most macromolecules are quite large in the first place, and so the paramagnetic broadening effect of the concentrations of oxygen found in buffer solutions in equilibrium with air is proportionally small and is usually ignored. This is not the case for oligosaccharides, however, which have much sharper lines, and for high resolution spectra of these compounds, appropriate precautions should be taken. On the other hand, oxidation is a particular problem for some protein samples, where cysteine and methionine residues are especially susceptible. Free thiol groups may be protected from oxidation by the addition of a small amount of DTE or DTT.

Freeze thaw cycles are one way to degas the sample. Given the difficulty of freezing a sample within the NMR tube, it is best to perform this in a small vial, such as those used for cryo-storage. After the requisite number of cycles, the sample can be transferred to the NMR tube, taking care to minimize the possibility of reintroducing air into the solution at that time. For maximum oxygen removal, the whole process can be performed in an inert atmosphere in a dry box. If the final step in the sample preparation process is a lyophilization, then it will probably be sufficient to redissolve the biopolymer in degassed sample buffer.

Vacuum inert gas cycles will also degas the sample. Using a valve such as the one illustrated in *Figure 2c*, the sample contained in the NMR tube can be alternately evacuated and then exposed to inert gas, such as dry argon, from which the residual oxygen has been removed by passage over a BTS catalyst.

The tube should be tilted during these operations, to maximize the surface area of the sample and facilitate gas exchange.

Long NMR tubes are available from all the manufacturers for sealing. Whilst attached to the valve and under an inert atmosphere, the tube can be tipped-off with a small flame. This may only be necessary for the most sensitive samples, since a snug-fitting plastic NMR tube cap routinely provides a reasonably gas-tight fit.

5. Sample parameters

The buffer constituents are a vital component in maximizing the information that can be obtained by NMR about a particular biopolymer sample. Correct choice of pH, ionic strength, buffer type, concentration, and temperature can make a very real difference to the quality of the NMR spectrum, particularly for protein samples. The signal-to-noise ratio of the spectrum can be increased by increasing the concentration of the biological macromolecule. However, aggregation of the macromolecule must also be minimized at the same time, so that production of a clear solution is not the sole criterion that must be met. The only way to tell whether the sample is going to provide a good NMR spectrum is to run that spectrum. The effect of variation of the sample parameters on both the intensity and the resolution of the NMR spectrum then provides information as to the optimal conditions to be used. A short time spent running simple 1D NMR spectra under varying sample conditions will enable the best quality multi-dimensional spectra to be obtained in the most efficient manner. It is important to realize that the best conditions worked out for, say, an enzyme assay, are not necessarily the best conditions under which to run the NMR spectrum. Given the significant effect which the sample conditions can have on resonance position, variation of these conditions (particularly pH and temperature) can also help to overcome overlap problems within the spectrum. The various components of the buffer are treated separately below, but it should be emphasized that changing one parameter may allow the spectrum to be further improved by the subsequent variation of one or more of the others.

5.1 Definition of pH in ^2H_2O—the isotope effect

For a sample in 100% ^2H_2O, the pH reading on the pH meter is 0.4 units lower than the true p^{2}H, due to an isotope effect on the glass electrode (7). The meter reading is normally reported uncorrected and designated pH*. If a sample made up in 90% H_2O/10% ^2H_2O has a measured pH of 7.0, then the same sample in 100% ^2H_2O will have a measured pH* of 6.6. This is an important point to remember if direct comparisons are to be made between the NMR spectra of H_2O and ^2H_2O solutions of the same biopolymer under otherwise identical sample conditions. A better estimate of the true p^{2}H is obtained by adding 0.4 units to the meter reading or by using the standard buffers calibrated for ^2H_2O for which recipes are available. The reason this is not commonly done, and the pH* value is used, is that there is also a deuterium isotope effect on the ionization equilibrium of the ionizable groups in the sample. For instance, most acids are between three to five-fold weaker in ^2H_2O than they are in H_2O (8), corresponding to a raising of their pKs by 0.5–0.7 units. This then tends to offset the error obtained from not correcting the pH meter reading to p^{2}H. Since, when comparing data in H_2O and ^2H_2O solution, one is usually more concerned that the ionizable groups of the molecule of interest should have the same charge state than that the free proton/deuteron concentrations should be the same, the use of the simple meter reading may be more useful. Care should however be taken to report this as pH*, not pH.

5.2 pH adjustment

For NMR tubes larger than 5 mm diameter, the adjustment of the pH of the sample is as simple as for any other solution in the laboratory. A long, narrow bore, combination pH electrode (such as that available from Aldrich) is inserted into the NMR tube. Small amounts of acid or base, as appropriate, are then added by micropipette or syringe. The solution is stirred throughout by a small magnetic stirrer bar (always remembering to remove this before returning the tube to the spectrometer magnet!). However, it is extremely difficult to adjust the pH of a sample conveniently whilst it is in a 5 mm NMR tube. Efficient mixing is a serious problem, and this can lead to local extremes of pH, which may be highly deleterious to the sample, causing precipitation in some cases. Although very thin pH electrodes are available, they are expensive and delicate and probably not suitable for routine use in most laboratories. pH electrodes which just fit inside the NMR tube (around 3.5 mm diameter), and are therefore more robust, tend to smear the sample all over the inside of the NMR tube during use. The simplest answer to these problems is to remove the sample from the NMR tube for pH adjustment, as detailed in *Protocol 2*.

Protocol 2. Adjustment of pH in a 400 μl NMR sample

Materials

- a long glass Pasteur pipette or syringe fitted with a long needle (preferably platinum)
- an Eppendorf tube or similar, with a screw cap and tapered bottom
- a vortex mixer
- a bench ultracentrifuge (microfuge or equivalent)
- a pH meter fitted with a pH electrode of about 4 mm diameter

Protocol 2. *Continued*

- solutions of acid and base of various concentrations (ten-fold serial dilutions of stock preferably)

1. Using the pipette or syringe transfer the sample from the NMR tube to the vial. To prevent unnecessary loss of the sample, put the tip of the pipette just under the surface and move it down smoothly as the solution is drawn up.

2. Cap the vial and spin the sample to the bottom of the vial by a few seconds centrifugation.

3. Measure the pH of the sample. On removal of the pH electrode from the sample, do not wash down the electrode; the short amount of time that the electrode is left dry should not harm it.

4. Holding the vial at a 45° angle, place a few microlitres of the pH adjustment solution at the top of the vial, above the surface of the sample solution. This will be held in place by surface tension. Cap the vial and vortex the main bulk of the solution through the droplet of acid or base. This will ensure swift and thorough mixing.

5. Spin the sample for a few seconds in the centrifuge.

6. Repeat steps **3–5** as necessary. At the point where the measured pH is close to that required, use lower concentration acid or base for the adjustment. If the pH change is slow, use higher concentration. By this means, the amount of added solution will be a few microlitres each time, and the sample dilution can be kept to a minimum.

7. Spin the sample for 5–10 min to pellet any dust, fibres, or denatured material.

8. Remove the sample from the vial in an analogous manner to step **1**, being careful not to disturb any sediment and transfer it back to the NMR tube. Place the tip of the pipette a few millimetres above the bottom of the NMR tube and slowly deliver the solution, moving the tip up smoothly to keep it just above the liquid surface. Try to keep the neck of the pipette away from the sides of the NMR tube. This will minimize the amount of fluid lost on to the outside of the pipette and up the walls of the NMR tube.

In situations where the sample being adjusted is in ^2H_2O, a few additional precautions need to be taken to prevent proton contamination. After referencing the pH electrode as normal (in H_2O buffers), rinse it off thoroughly with ^2H_2O and then soak it in a vial containing ^2H_2O. Before inserting the electrode in the sample, dry it off with a lint-free tissue. NaO^2H and ^2HCl solutions in ^2H_2O are commercially available, and a series of varying concentrations of these should be made by diluting the stocks with ^2H_2O. These solutions can be stored, tightly stoppered, until required. For NMR samples containing a small amount of water in the first place, following the protocol in the open laboratory leads to a small, but not unacceptable, increase in the size of the water resonance in the ^{1}H NMR spectrum. Where the intrinsic humidity is high, or where absolute exclusion of water is necessary, the NMR tube can be transferred straight from the magnet to a dry box for subsequent handling.

5.3 Choice of buffer

The most convenient buffers for ^{1}H NMR are those containing no non-exchangeable protons. Since we commonly require to have the buffer concentration at 10–50 mM to maintain the pH, any protons on the buffer are likely to give rise to ^{1}H NMR signals many times larger than those from the macromolecule of interest. These will obscure some regions of the spectrum and may give rise to a dynamic range problem. Some simple inorganic buffers can be exchanged very readily by simple lyophilization and redissolution in ^{2}H$_2$O, whilst others (particularly acetate and Tris) can now be purchased fairly cheaply with deuterium at non-labile positions. The most commonly used NMR buffers are shown in *Table 2*. If a buffer other than these is absolutely required, then it may be available in deuterated form from one of the isotope suppliers, though at a price. For NMR of nuclei other than protons, the choice of buffer is much less restricted. ^{13}C and ^{15}N NMR would normally be performed on isotopically labelled macromolecules, so that the effective concentrations from the natural abundance heavy isotope (<1%) of the buffer would be at most approximately equal to that of the atom(s) of interest and would be unlikely to provide a dynamic range problem. There are no common fluorinated buffers that might interfere in ^{19}F NMR. ^{31}P NMR should not generally be attempted in phosphate buffer.

Table 2. Buffers for ^{1}H NMR of biological samples (9,10)

Buffer	pK_a	Useful pH range	Notes
Acetate	4.76	3.7–5.6	a, b
Cacodylate	6.27	5.0–7.4	c
Phosphate	7.20 (2.15, 12.33)	5.8–8.0	
Tris	8.06	7.1–8.9	a, d
Borate	9.24 (12.74, 13.80)	8.0–10.2	
Glycine	2.35, 9.78	2.2–3.6, 8.6–10.6	a
(Bi)carbonate	6.35, 10.33	6.0–8.0, 9.2–10.8	b, e
Succinate	4.21, 5.64	3.5–6.0	a

[a] Readily available in deuterated form.
[b] Neutral species volatile to lyophilization.
[c] Poisonous, and has residual methyl proton resonance at 3.8 p.p.m.
[d] The pH of Tris solutions is very temperature sensitive.
[e] Requires closed system.

Inorganic buffers may appear to be the best for ^{1}H NMR work, but certain points should be borne in mind. Phosphate, cacodylate, borate, and bicarbonate can all interact with certain enzymes and their substrates. Borate can complex with saccharides and with the ribose moiety of oligonucleotides. Phosphate will, at least to some extent, inhibit kinases, dehydrogenases, and other enzymes which require phosphate esters as substrates. Bicarbonate requires a closed system, as it is in equilibrium with CO_2 gas. Cacodylate has found extensive use in ^{1}H NMR studies of oligonucleotides, despite its toxicity and the presence of non-exchangeable methyl protons, since it will buffer around neutrality and does not catalyse proton exchange in the way that phosphate does, making the observation of the base imino protons possible.

5.4 Ionic strength effects

Variation of the concentration of salts in the solution has far reaching consequences for the solubility of a particular biological macromolecule and therefore for the intensity of the NMR signal from it. The salt concentration also has bearing on the state of aggregation and therefore on the spectral resolution. The phenomenon of salting-in and salting-out of proteins is a well known one (11) and the effect is summarized in *Figure 6*.

As can be seen, the solubility of proteins in solutions of very low ionic

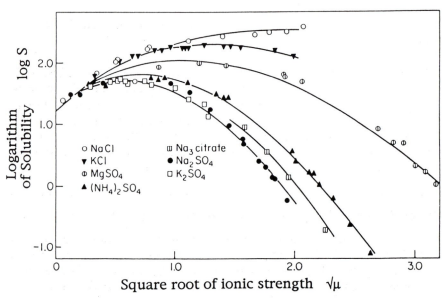

Figure 6. The solubility of carbonmonoxyhaemoglobin in various electrolyte solutions at 25°C. Ionic strength μ is given by $0.5 \sum c_i z_i^2$, where c_i is the number of a particular ion in the molecular formula, and z_i is the charge on that ion, (e.g. 1 M NaCl $\mu = 1$ M; 1 M CaCl$_2$ $\mu = 2$ M; 1 M MgSO$_4$ $\mu = 4$ M). (Adapted from Green (12)—© Journal of Biological Chemistry.)

strength is itself low, and this solubility steadily increases as the salt concentration is increased. This is due to the increase in the dielectric constant of the solution, which helps to shield the charges on individual protein molecules from each other. In this part of the solubility curve, there is little difference between one salt and another. As the salt concentration increases it starts to compete with the protein for the solvent molecules. Water is excluded from the immediate vicinity of the protein and it 'salts-out', as will be familiar to anyone who has ever done an ammonium sulfate precipitation. In this part of the solubility curve, the nature of the salt is very important. The effect is more dependent on the anion than the cation, and generally, multi-charged anions are least efficient at solubilizing proteins. But the effect is dependent on the size and charge distribution on the ion and on the protein, so the best salt for maximizing the solubility of each particular protein must be determined afresh. The Hofmeister series orders ions by their ability to 'salt-out' proteins, and may be used in this case as an approximate indication of their solubilizing power. However, care should be taken in using this, as ions at the left-hand side of each series may solubilize proteins at the expense of their native structure, (e.g. thiocyanate anion). Sodium chloride represents a good starting point for a systematic search for the correct salt to use in the NMR sample buffer.

The Hofmeister series
Anions: $SCN^- > NO_3^- > Cl^- > citrate > acetate > phosphate > SO_4^{2-}$
Cations: $Ca^{2+} > Mg^{2+} > Na^+ = K^+ > NH_4^+ > (CH_3)_4N^+$

Within the NMR experiment, variation in both ionic strength and ionic type can be used to maximize the solubility of the protein, in order to maximize the NMR signal. This is important, as the concentrations of protein necessary for NMR analysis ($\approx$1 mM) are close to the maximum solubilities that can be achieved for a large number of proteins. Slight variations in the sample conditions may make significant differences to the attainable solubility. In practical terms, one way of doing this is described in *Protocol 3*.

Protocol 3. Optimizing the NMR spectrum by varying the ionic strength

1. Make up the protein solution under some particular conditions of pH with an ionic strength of, say 50 mM (for some proteins, a lower ionic strength will be possible). The concentration of the protein should be the maximum possible that still produces a clear solution. This will vary from protein to protein.

2. Run a 1D ^1H NMR spectrum.

3. Add to the solution a small amount of a highly concentrated buffered salt solution (5 M NaCl is convenient) to raise the ionic strength by 50 mM.

Protocol 3. *Continued*

4. Rerun the 1D ^{1}H NMR spectrum and compare with the one run previously. Look for the sharpening of signals within the spectrum as evidence for decreased aggregation.

5. Continue adding small aliquots of the salt solution until there is no further evidence for improvement in the spectrum or until the spectrum starts to deteriorate. This can be taken as evidence for the beginning of aggregation and salting-out.

6. Make up a fresh solution of protein at the newly established optimal ionic strength and at the maximum concentration possible to obtain a clear solution.

7. Run a 1D ^{1}H NMR spectrum and compare with the one run at the same ionic strength but at lower protein concentration. The sensitivity will be greater, and if the resolution is comparable, then this solution can be used for subsequent NMR experiments. If the resolution is poorer, then the solution should be diluted with buffer until the resolution is acceptable.

8. For maximum resolution, and where protein stocks allow, the ion type can also be varied. Try repeating the above steps with a different salt.

9. If a specific component is required for enzyme function, e.g. Ca^{2+} is often an essential ion, try adding aliquots of a concentrated solution of this, and comparing NMR spectra as before.

---

pH and temperature can also be varied in concert with ionic strength; their effects are described elsewhere. High ionic strengths ($\geq$ 500 mM) do have some deleterious effects on the NMR experiment. Samples of high ionic strength require longer 90° pulses than those of lower ionic strength, and if these pulses become excessively long this may compromise a wide range of NMR experiments. In addition, the high dielectric that is produced leads to heating when a radiofrequency pulse is applied, and the temperature gradients that are set up within the sample decrease the resolution of the spectrum; this can be a serious problem if solvent presaturation or long isotropic mixing pulses are being used.

For oligonucleotides, the effect of ionic strength is not primarily on the macromolecular solubility, but on its conformation. The small oligonucleotides studied by NMR so far have tended to be inverted repeat sequences for reasons of spectral simplicity. At low ionic strength, these sequences can form hairpins, and in order to be sure of generating duplexes in solution, ionic strengths of 100–200 mM should be used. Mg^{2+} is an almost universal component of oligonucleotide sample buffers.

NMR studies of oligosaccharides are generally carried out in water. The effect of varying the ionic strength is negligible.

5.5 Organic solvents

The addition of organic solvents to solutions of biological macromolecules can increase solubility, decrease aggregation, and improve the resolution of the NMR spectrum, particularly for protein samples. Exactly which solvent and what concentration to use can be determined in a similar manner to that described for ionic strength in *Protocol 3*, that is: systematic variation of the solvent concentration by the addition of small aliquots whilst monitoring the NMR spectrum. The problem with the addition of organic solvent, which also applies to a lesser extent to high ionic strengths, is the perennial one of whether the data obtained is relevant to more physiological conditions. Generally, concentrations of organic solvent less than 10% are used and other techniques are available to determine whether a gross change in protein structure is occurring under these conditions (e.g. enzyme assay, CD spectroscopy). Organic solvents can bind directly to the protein via hydrophobic inter-actions, which at high concentrations will turn the protein inside out, and they can decrease the bulk dielectric, enhancing inter- and intramolecular electro-static interactions. Alcohols, dioxane, acetone, and DMSO can all help to stabilize a protein structure under very specific conditions. At 10°C, 0–20% ethanol will stabilize ribonuclease, whilst at 50°C it has a concentration-dependent denaturing effect (13). *Table 3* shows some important parameters for commonly used organic additives. Addition of non-aqueous solvents will elevate the p*K* of carboxylates and other neutral acids, but has little or no effect on amines, imidazoles, and other cationic acids.

The addition of organic solvents also permits solution NMR at tempera-tures below 0°C. Methanol and DMSO have been used as co-solvents to extend the fluid range as low as −50°C, for investigation of thermally un-stable molecules or where reaction intermediates can be 'frozen out'. The viscosity of the solutions at these temperatures seriously affects resolution, however. This problem essentially rules out the use of other cryo-solvents,

Table 3. Properties of some solvents used in macromolecular NMR

Solvent	Melting point °C	Boiling point °C	δ ^{1}H[a]	δ ^{13}C[a]
Acetone	−93.8[b]	55.5[b]	2.17	206.0, 29.8
DMSO	18.0	190.0	2.5	39.5
Dioxane	12.0	100.0	3.75	67.3
Ethanol	−130	78	1.17, 3.63	58.2, 17.6
Methanol	−98.0	65.4	3.31	49.0
Water	3.8[b]	101.4[b]	4.75	—

[a] Chemical shifts are approximate and depend on other sample conditions.
[b] Data for deuterated species.

such as ethylene glycol or glycerol, for NMR, and indeed presents difficulties in the use of glycerol, a common stabilizer, even at temperatures greater than 0 °C.

Spectra of oligosaccharides are often run in neat DMSO, allowing identification of alcohol and amine proton resonances in ¹H NMR. These are necessary for primary sequence determination and for obtaining information on intramolecular H-bonding (see Chapter 9).

5.6 Detergents

Whilst chaotropic agents are useful for the solubilization of proteins, this is at the expense of native structure, and there is, therefore, nothing to be gained in the addition of such compounds to cytosolic or extracellular proteins. Structural and functional studies on membrane bound proteins can however benefit from the use of detergents, since in these cases detergents can stabilize the native conformation. NMR studies of peptides and proteins incorporated into SDS micelles are thus possible (14). Both deuterated SDS and dodecyl-phosphocholine are commercially available for studies involving ¹H NMR.

5.7 Temperature

Whilst temperature is not strictly a sample preparation parameter, it is important to consider it in the same light as pH, ionic strength, and additives; the optimization of conditions for the NMR experiment requires variation of all the possible parameters. In general, the higher the temperature within the probe of the spectrometer, the better the resolution of the NMR spectrum, as a consequence of the decrease in the correlation time. This assumes, of course, that the sample is still stable at the elevated temperature over the time period of the experiment. The increase in resolution achieved may not be sufficient to justify the risk of losing the entire sample, since the thermal unfolding of some proteins is effectively irreversible, due to aggregation of the unfolded protein, at the concentrations used for NMR. For protein NMR, temperatures somewhat greater than room temperature, normally in the range 300–320 K, have become standard for structure determination work. Individual overlaps within a spectrum can often be resolved by slight changes in temperature. The differential temperature dependence of the amide proton chemical shifts within a protein have been used to estimate the extent of hydrogen bonding at each amide (15, 16), although the exchange rate is more commonly used for this purpose. A series of 1D spectra, run at slightly different temperatures, in a manner similar to that described in *Protocol 3*, will allow one to determine the optimum temperature for each particular sample.

If, as is often the case, exchange processes are taking place in the sample—such as conformational equilibria, ionization equilibria, or ligand binding—the appearance of the spectrum may be particularly sensitive to temperature. These effects are described more fully in Chapter 6.

5.8 Chemical shift references

Chemical shift referencing can be either external or internal, direct or indirect. External references are contained in a capillary within the NMR tube and do not contact the sample directly. The presence of an external capillary makes it much more difficult to attain good field homogeneity and resolution, and an external reference does not properly compensate for susceptibility changes in the sample solution. External referencing should be avoided if at all possible. A direct internal reference is dissolved in the sample buffer and should therefore be biochemically inert. This reference will correct properly for susceptibility effects, but its resonance may be affected by changes in the sample conditions. Indirect reference can be made by running a spectrum of the reference compound on its own, subsequent chemical shift measurements on the sample of interest then being made relative to the calibrated resonance(s) of the reference solution. In this case it is not always possible to duplicate the sample conditions for the reference solution, which may affect the exact chemical shift measured. After indirect referencing, it may be possible to use a strong, invariant line within the spectrum of the biopolymer itself for subsequent referencing—but one will often not know *a priori* which line(s) will be invariant.

For ^{1}H NMR, a number of different references may be used (*Table 4*). The residual water peak is often used and assigned a chemical shift around 4.75 p.p.m., but this is very temperature sensitive, varying by 0.005 p.p.m./°C.

Table 4. Common chemical shift references for biological NMR

Nucleus	Compound	Chemical shift[a]	Indirect or direct	Notes
^{1}H	3,3,3-Trimethylsilylpropionate (TSP)	0	Direct	b
^{1}H	5,5-Dimethylsilapentanesulfonate (DSS)	0	Direct	c
^{1}H	Acetone	2.22[d]	Direct	e
^{1}H	Dioxane	3.75[f]	Both	
^{1}H	Dimethylsulfoxide (DMSO)	2.5	Direct	e, g
^{13}C	Dioxane	67.3	Indirect	
^{13}C	Tetramethylsilane	0.0	Indirect	
^{15}N	2 M ^{15}NH$_4$NO$_3$ in 5 M HNO$_3$	0.0[h]	Indirect	
^{31}P	Trimethyl phosphate (TMP)	3.53[i]	Indirect	

[a] ^{1}H chemical shift relative to TSP at 0 p.p.m.
[b] Available deuterated at C-2 and C-3 (no residuals).
[c] Proton resonances also at ~2.9 p.p.m.
[d] At 300 K.
[e] Used for oligosaccharides.
[f] At 308 K.
[g] Residual ^{1}H's in ^{2}H$_6$-DMSO.
[h] Relative to liquid ammonia at −25.0 p.p.m.
[i] Relative to 85% phosphoric acid at 0 p.p.m.

Comparison of chemical shifts for resonances assigned under different conditions of pH, ionic strength, and temperature can be extremely difficult, especially if the data has been obtained in different laboratories. This is further complicated by the fact that the chemical shift given in the literature for the reference compound is often not consistent with the sample conditions actually used, and differs from the value that would have been obtained had it been measured relative to a second reference. The best that can be realistically hoped for is a self-consistent set of data referenced throughout using the same reference, the approximate chemical shifts of which are given in *Table 4*.

6. Preventing sample contamination

Standard laboratory practice should prevent most contamination by unwanted materials. Disposable gloves should be worn throughout, as fingertip contamination is otherwise very common. This is manifested by the presence of lactate, observable as a sharp doublet around 1.4 p.p.m. and a quartet around 4 p.p.m. in the ^{1}H NMR spectrum. There are some additional precautions that should be taken, particularly concerning the elimination of paramagnetic species and the prevention of microbial growth.

6.1 Paramagnetic contamination

The effect of the presence of metal ions such as Cu(II), Mn(II), Cr(III), high spin Fe(III), and low spin Co(II) is to broaden beyond detection the resonances of many nuclei in the neighbourhood of these paramagnetic centres. This can be a particular problem for proteins that bind divalent cations such as Ca^{2+} and Mg^{2+}, where the paramagnetic ion can substitute, and for phosphorylated proteins, or proteins with phosphate containing coenzymes, which can bind cations such as Mn^{2+} at the phosphate anion. The best way of dealing with this problem is not to introduce metal ions into the solution in the first place.

- avoid contact with metal, including so-called stainless steel
- where practicable, substitute non-contaminating materials—use FPLC instead of HPLC, platinum needles, PTFE coated spatulas
- do not use chromic acid to clean apparatus
- treat all glass and plastic ware with 0.1 M HCl and 1 mM EDTA prior to use, rinsing thoroughly with deionized water
- use commercially available 'low heavy metal' grades of inorganic salts throughout
- treat all buffer solutions with a metal chelating agent such as Chelex prior to use

The sample itself can be treated, prior to transfer into the NMR tube. Chelex is best for the removal of metal ions since its non-specific absorption is low. It is highly effective at neutral pH and in the absence of Mg^{2+} and Ca^{2+}, which compete for the chelation sites. Passage through a column of Chelex will lead to dilution of the sample and some slight loss of material, but it is just as effective to add a few particles of Chelex to the sample solution, stir gently for a few minutes and then decant, filter, or centrifuge.

Low concentrations of soluble chelating agents can be added to the sample during the NMR experiment. 5–10 μM EDTA will usually be sufficient. Fully deuterated EDTA is not commercially available, but at these concentrations its proton resonances should not interfere in ^1H NMR. If deemed necessary, the α-CH_2 protons can be exchanged for deuterium by boiling the Mg^{2+} salt of EDTA in ^2H_2O.

6.2 Microbial contamination

A concentrated solution of a biological macromolecule at high temperature and around neutral pH represents an ideal growth medium for algae, bacteria, and fungi. As well as consuming the compound of interest, the excretion of proteases and nucleases from the microbes will cause long-term damage. Algal growth can be eliminated by keeping the NMR sample out of the light whilst it is not in the magnet. Extremes of pH and temperature will retard microbial growth, but these are not generally useful conditions for NMR studies. Solutions containing greater than 80% ^2H_2O act as effective anti-microbial agents and this is possibly the simplest way of preventing microbial contamination. For water solutions, azide and fluoride at less than 50 μM are effective. However, azide is volatile at pH values less than 7 and fluoride can not be used in the presence of metal ions. The antibiotic chloramphenicol is highly effective at 10–50 μM, is chemically inert, and can be used with metal ions, but it does have non-exchangeable proton resonances which may inter-fere in ^1H NMR. Metal chelators (EDTA, EGTA) and rigorous exclusion of metal ions (Chelex) will also suppress microbial growth (see Section 6.1). The best anti-microbial agent to be used will depend on the exact nature of the sample involved.

General reading

Oppenheimer, N. J. (1989). In *Methods in enzymology* (ed. N. J. Oppenheimer and T. L. James), Vol. 176, pp. 78–89. Academic Press, London and New York.

Derome, A. E. (1987). In *Modern NMR techniques for chemistry research* (ed. J. E. Baldwin). Pergamon Press, Oxford.

References

1. Moody, P. C. E. and Wilkinson, A. J. (1990). *Protein engineering*. IRL Press, Oxford.

2. Gait, M. J. (ed.) (1984). *Oligonucleotide synthesis: a practical approach*. IRL Press, Oxford.
3. Chubb, R. T., Thanabel, V., Osborne, C., and Wagner, G. (1991). *Biochemistry,* **30,** 7718.
4. Oppenheimer, N. J. (1989). In *Methods in enzymology* (ed. N. J. Oppenheimer and T. L. James), Vol. 176, pp. 78–89. Academic Press, London and New York.
5. Wüthrich, K. and Wagner, G. (1979). *J. Mol. Biol.,* **130,** 1.
6. Wüthrich, K. (1986). In *NMR of proteins and nucleic acids*, p. 24. Wiley—Interscience, New York.
7. Glasoe, P. K. and Long, F. A. (1960). *J. Phys. Chem.,* **64,** 188.
8. Bunton, C. A. and Shiner, V. J. (1961). *J. Am. Chem. Soc.,* **83,** 42.
9. Blanchard, J. S. (1984). In *Methods in enzymology* (ed. W. B. Jakoby), Vol. 104, pp. 404–414. Academic Press, London and New York.
10. Dawson, R. M. C., Elliott, D. C., Elliott, W. H., and Jones, K. M. (1987). In *Data for biochemical research*, 3rd edn, p. 417. Oxford Science Publications.
11. Arakawa, T. and Timasheff, S. N. (1985). In *Methods in enzymology* (ed. H. W. Wyckoff, C. H. W. Hirs, and S. N. Timasheff), Vol. 114, pp. 49–77. Academic Press, London and New York.
12. Green, A. A. (1932). *J. Biol. Chem.,* **95,** 47.
13. Brandts, J. F. and Hunt, L. (1967). *J. Am. Chem. Soc.,* **89,** 4826.
14. Lee, K. H., Fitton, J. E., and Wüthrich, K. (1987). *Biochim. Biophys. Acta,* **911,** 144.
15. Kopple, K. D., Ohnishi, M., and Go, A. (1969). *J. Am. Chem. Soc.,* **91,** 4264.
16. Kopple, K. D., Ohnishi, M., and Go, A. (1969). *Biochemistry,* **8,** 4087.

3

Instrumentation and pulse sequences

T. A. FRENKIEL

1. Introduction

Despite having first been demonstrated over 40 years ago NMR is still in an exceptionally active state of development. Its range of applications is still widening and continual progress is being made in instrumentation and in experimental techniques. These technical advances have both fuelled and been fuelled by the growing interest in applying NMR to study biological macromolecules: as a result, the recent growth of biological NMR has been dramatic. Evidence of this growth and of the scope of biological NMR will be found in abundance in the following chapters.

The purpose of this chapter is two-fold. First, it is intended to provide descriptions of those NMR techniques which are of most importance in modern biological NMR. Without exception these are multiple-pulse techniques that depend on Fourier transformation to extract spectral information from data recorded in the time-domain. In particular, two-dimensional NMR methods have become the mainstay of biological NMR and this is reflected in the contents and emphasis of the chapter. Secondly, it is intended to help in putting these techniques into practice by highlighting the factors which need critical adjustment if good results are to be obtained. Essential to both of these is some understanding of the way in which an NMR spectrometer operates: this forms the subject matter of the next section. The remaining sections deal directly with the experimental techniques and their optimization. No attempt has been made to explain the theory behind any of these techniques. For this information the original literature cited for each technique, or the reviews listed in the introduction should be consulted; alternatively, a more general theoretical treatment can be found in the monograph by Ernst, Bodenhausen, and Wokaun (1). Nor are three-dimensional or four-dimensional methods covered explicitly: the general principles which are described here for two dimensions can be extended to three or four with differences only of detail.

2. NMR instrumentation

A number of hardware topics is covered in this section. Of these, the most important is a description, necessarily highly simplified, of the NMR spectrometer itself. This will be done from the viewpoint of the user rather than the designer, with most detail being given about those parts of the instrument which the user may need to adjust, or which determine how—or whether—particular types of experiment can be carried out. We shall then consider sample temperature control and some of the external factors which can influence spectral quality.

2.1 The NMR spectrometer

2.1.1 The magnet

Central to any spectrometer is the magnet itself. For biological NMR the use of helium-cooled persistent superconducting magnets is now almost universal, with the highest available (or affordable) field strength being preferred. At present it is generally regarded that the sensitivity and signal dispersion provided by magnets operating at proton frequencies of 500 MHz and 600 MHz are essential for work on biological macromolecules. Magnets operating at 600 MHz only became commercially available in 1987 and this field strength will have to suffice for the next few years at least: persistent superconducting magnets with significantly higher field strengths are not likely to be introduced until 1993 at the earliest. Nor is there any sign that the recent development of high temperature superconductors will result in a magnet suitable for NMR within a similar period.

2.1.2 The radiofrequency system

The principal radiofrequency components of a modern NMR spectrometer are shown in *Figure 1*. To make the description more tangible one particular instrument—the UNITY spectrometer manufactured by Varian Associates—has been chosen as an example. This is a highly flexible instrument, the design of which reflects developments in biological NMR that took place in the late 1980s. An alternative and similarly flexible design is discussed at the end of the section. The main functions of the subsystems shown in the figure are as follows:

(a) The transmitter. In conjunction with the power amplifiers (described in (c) below) this generates excitation pulses at the frequency of the nucleus to be observed. The exact frequency of the pulses can be set in steps of less than 1 Hz across a band many megahertz in width, covering all nuclei of biological interest. The output of the transmitter is fed through a programmable attenuator which allows the power output to be varied over a range that spans six orders of magnitude. In keeping with the requirements of multipulse experiments, the radiofrequency phase of the pulses can be set in steps of 0.5°: alteration of the phase, and also of the power

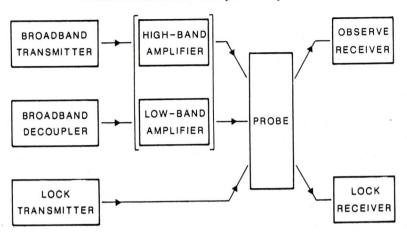

Figure 1. Block diagram showing the main rf components of a modern NMR spectrometer.

and frequency, can be accomplished under computer control during a pulse sequence with settling times of only a few microseconds.

(b) The decoupler. This provides a second frequency source for pulsing or irradiating the spins. Like the transmitter, this is fully broadbanded and the frequency, phase, and power of the decoupler output are adjustable in small steps under computer control.

(c) The power amplifiers. The system contains a pair of broadband power amplifiers which serve to boost the outputs of transmitter and decoupler to the relatively high levels (tens of watts) needed for non-selective pulses. The amplifiers operate at constant gain so that all variations in output power are accomplished by the programmable attenuators, as described in (a) above. Two amplifiers are needed because of the technical difficulty in designing a single one which would satisfactorily cover the necessary frequency range. Of the two, one covers the high gamma nuclei such as ^{1}H, ^{3}H, and ^{19}F, and the other the low gamma nuclei such as ^{31}P, ^{13}C, and ^{15}N. Routing of the attenuator outputs to the appropriate amplifier is carried out automatically. On this system both amplifiers can operate equally well with pulses and with continuous irradiation, i.e. they are CW amplifiers.

(d) The receiver. This serves to detect the free induction decay (FID), this being the time-domain signal which after Fourier transformation gives the NMR spectrum. To allow the spectrometer to detect signals from different nuclear species the receiver must be capable of operating over a wide frequency range, although of course only one nucleus will be of interest in any one experiment. As in a conventional radio receiver this wide frequency range is provided by using frequency-conversion techniques, such that most of the receiver circuitry operates at a fixed frequency and only a small part needs to work with the full range. The receiver can therefore be thought of as a series of stages in which the incoming signal is progressively amplified and reduced in frequency. The final function of the receiver is to carry out a real-time digitization of the free induction decay: from this point onwards all manipulations of the FID are done by the spectrometer's data processing system. In operation of the spectrometer careful consideration must be given to the strength of the NMR signal in relation to the signal-handling capabilities of the receiver. In an attempt to allow for the wide range of different NMR signal strengths the overall gain of the receiver is adjustable, although the available variation is not always sufficient to accommodate the strongest signals. Moreover, in most spectrometer designs the adjustment is applied only to the stage immediately preceding the analogue-to-digital converter (ADC), the intention being to avoid overload of the ADC. This arrangement has a serious drawback: strong signals may overload one of the *earlier* stages of the receiver, an occurrence known colloquially as front-end overload. Front-end overload is just as unacceptable as the more familiar ADC overload but its effects are not as obvious. Unfortunately, in some spectrometer designs front-end overload can occur without ADC overload, a situation which can be difficult to detect. Practical advice about setting the receiver gain and dealing with the strong signal case will be given in later sections.

(e) The lock transmitter and lock receiver. These make up the field-frequency lock system, the primary function of which is to ensure that the magnetic field experienced by the sample is constant. A secondary function is to ensure that any drift in the reference frequency of the spectrometer is compensated by a corresponding change in the magnetic field seen by the sample. The field corrections are applied by means of a solenoidal coil which is part of the room temperature shim assembly within the magnet bore. The ingenious feature of the lock system is that the NMR phenomenon itself is exploited to control the correction current. For this purpose a nuclear species other than the one being studied must be present in the sample; the use of deuterium as the locking nucleus is now almost universal. Once the spectrometer is 'locked' the lock system continuously monitors the deuterium NMR signal and varies the correction current such that the resonance position of the signal is kept constant. For a variety of technical reasons it is actually the zero-crossing of the dispersion-mode signal which is used in this way: the absorption-mode response is used to provide the operator with an indication of the *homogeneity* of the magnetic field. The lock transmitter and receiver broadly resemble their counterparts described above but are somewhat simpler. The transmitter needs no power amplifier since only low level excitation is required and no ADC is needed in the receiver.

The flexibility of this design of spectrometer arises in large part from the fact that the decoupler and transmitter are each capable of pulsing or irradiating any nucleus. Typically, in experiments which involve just a single nuclear species the transmitter alone would be used: only if a second species has to be pulsed or irradiated would the decoupler be employed. In biological NMR such heteronuclear experiments will almost invariably involve protons as one of the two nuclear species. Comparable flexibility can therefore be obtained by having one broadband channel and one dedicated proton channel, each of which can be used for either CW irradiation or pulses. This approach has been adopted by Bruker Spectrospin for their AMX spectrometers. With these modern spectrometers the distinction between the decoupler and the transmitter is less clear cut than it is in older instruments, in which the transmitter could only produce pulses and the decoupler could operate only at the proton frequency.

2.1.3 The probe

The probe is in many ways the most critical component in the spectrometer. It has two main functions:

(a) To convert the radiofrequency power from the transmitter and the decoupler into oscillating magnetic fields and to apply these fields to the sample.

(b) To convert the oscillating magnetic fields which are generated by the nuclear spins into an electrical signal which can be detected by the receiver.

The most efficient method of achieving the first of these is by means of a parallel tuned circuit which is resonant at the frequency of interest and has a coil that surrounds the sample. It can be shown that this arrangement also fulfils the second function since the two are reciprocal: modern probes use the same tuned circuit for transmitting and receiving at a given frequency.

A particular spectrometer system is likely to be equipped with several probes. An important physical characteristic of a probe is the sample size it is intended to accommodate. Most high resolution biological NMR is done using samples in 5 mm tubes, although 10 mm probes are available for both 500 and 600 MHz systems. Apart from their size, commonly available probes may be classified into three groups.

(a) Proton probes. As their name suggests these are intended purely for proton observation. They have a single coil, optimized for the proton resonant frequency and tunable only over a narrow range. In these probes a technique known as double-tuning is used to allow the proton coil to also operate at the deuterium frequency.

(b) Broadband X-nucleus probes. These are intended for observation of low gamma nuclei and are tunable over a broad frequency range: on most systems only one probe is needed to cover the frequency range spanned by ^{15}N and ^{31}P. In addition to this X-nucleus coil (the observe coil) these probes are invariably also equipped with a second coil which is tuned to the proton resonant frequency (the decoupler coil). In these probes the lock channel is usually provided by double-tuning on to the decoupler coil.

(c) Proton probes with X-nucleus decoupling coils (also known as inverse probes). These probes resemble those in (a) but with the addition of a decoupling coil which is tunable over much or all of the frequency range spanned by the low gamma nuclei. Typically, the addition of the decoupling coil reduces the proton sensitivity by about 15% relative to the simple proton probes of type (a).

The availability of the appropriate probe is usually a prerequisite for carrying out an NMR experiment. However, there are circumstances in which one probe can be used in place of another, albeit with less than optimum results. For example, coils intended to operate only at proton frequencies can often be tuned to ^{19}F or ^{3}H, and probes of type (b) could be used where ones of type (c) would be best (and vice versa).

2.1.4 The pulse programmer and data system

In order to carry out the many multiple-pulse experiments which are a fundamental part of modern NMR, all spectrometers contain a sophisticated pulse programmer. This controls all of the radiofrequency components of the spectrometer and is in turn programmed by the operator to produce the pulse sequence of interest. The flexibility of the pulse programmer is an important factor which affects the overall flexibility of the spectrometer. In the current generation of spectrometers the pulse programmer forms part of a specialized acquisition computer which is also responsible for co-adding incoming free induction decays before transferring them to the main data system for processing.

Although the processing requirements of one-dimensional NMR could be met reasonably well by the simple minicomputers which were part of the earliest Fourier transform NMR instruments, two-dimensional NMR makes much heavier demands on the data system. In addition to high processing speed, the system needs to have extensive disc storage capacity, and fast, high resolution, graphics. An efficient means of archiving data is essential since a typical 2D data set can occupy 16 Mbytes. At the time of writing automated analysis has made little inroad into 2D spectral interpretation and much use is still made of contour plots. For the congested spectra which typify biological samples the larger these plots can be the better: plotters of A1 size (594 × 840 mm) or larger are recommended.

2.2 Sample temperature control

The importance of ensuring accurate control of the temperature of an NMR sample can not be overstressed. Most biological samples have an explicit temperature dependence as a result of conformational equilibria or exchange phenomena. In many such systems the temperature dependence is itself an object of study, but, even if it is not meaningful, results can only be obtained if the temperature of the sample can be maintained at a constant and known value.

There is however a second reason for requiring good control of the sample temperature. This is because any sample which is run with the spectrometer locked to ^2H_2O acquires an apparent temperature dependence of about 0.01 p.p.m./°C. This has its origins in the temperature dependence of the ^2H_2O signal, which the lock circuit inexorably cancels by making temperature-dependent alterations to the magnetic field. The net effect is that the solute signals all appear to be temperature-dependent whereas the residual H_2O signal appears not to be. It is easy to see that this can result in broadened and distorted lineshapes in one-dimensional spectra, poor subtraction in difference experiments, and bad t_1-noise in two-dimensional spectra (as explained in Section 5.2). The seriousness of this problem depends on the linewidths of the sample being studied, with the narrowest lines requiring the greatest stability: temperature variations corresponding to frequency shifts of more than about one-tenth of a linewidth are the most that can be tolerated. In practical terms, for typical protein spectra short-term and long-term temperature stability to within about 0.2 K is desirable. It follows that some method of stabilizing the sample temperature should always be used, even if the desired temperature is equal to room temperature.

In an attempt to provide the required control, spectrometers have been equipped with variable temperature ('VT') facilities for many years. The remainder of this section will deal briefly with some practical aspects of variable temperature operation.

2.2.1 The VT system

Commercial VT systems make use of a stream of heated or cooled gas to control the sample temperature. The gas is fed into the bottom of a probe at a temperature below the desired sample temperature and is then passed over a heater within the probe before being routed around the sample. The heater is controlled by a thermocouple which senses the gas temperature near to the sample. Heat transfer between the gas stream and the lower part of the probe is minimized by containing the gas within a tubular dewar until it is within about 1 cm of the sample.

With this system the incoming gas must be pre-cooled when sample temperatures near or below room temperature are to be attained. The standard methods of doing this are either to use nitrogen gas obtained directly from liquid nitrogen, or to use nitrogen gas which has been passed through a heat-exchanger coil immersed in liquid nitrogen. Although these schemes are admirable for reaching temperatures well below 0°C neither is particularly appropriate for biological NMR, where temperatures below 0°C are rarely needed. A convenient and satisfactory alternative is to use dried air, cooled by means of a heat-exchanger surrounded by dry-ice. Air drying units which need little maintenance and remove water vapour down to a dew point of $-70\,°C$ are available at low capital cost. A relatively modest quantity of dry-ice (about 15 kg) suffices for a 24 hour experiment and can easily be replenished for longer runs. An added advantage of this system is that the dried air can be made to serve a second purpose, as described in Section 2.2.2.

Temperatures above room temperature are more straightforward. Compressed air may be used for most temperatures encountered in biological NMR, although the spectrometer manufacturers recommended the use of nitrogen gas to eliminate any risk that the heater may be damaged by becoming oxidized.

2.2.2 VT operation

Several practical points should be borne in mind whenever the VT system is being used.

(a) The incoming gas should be pre-cooled for sample temperatures which are below room temperature or less than about $5\,°C$ above room temperature.

(b) If the NMR sample is to be spun while it is below room temperature a supply of dry gas should be used to drive the spinner since water might otherwise condense from the spinner gas on to the probe or shim assembly. (A supply of dried gas will be readily available if dried air is used for the VT system, as described above.)

(c) When the temperature is changed there is a tendency for most systems to overshoot on upward changes and undershoot on downward ones. On some systems these errors can be as large as $5\,°C$, with obvious risks of denaturing or freezing samples. A procedure which is strongly recommended is to allow the temperature to stabilize for at least five minutes before introducing the sample.

(d) The time needed for a system to stabilize after a change in temperature varies widely from one system to another and would depend on probe and sample size, and the direction of the change. For 5 mm samples between 5 and 15 minutes is usually sufficient for the sample itself to reach the correct temperature. However, there may be much slower changes to the magnetic field homogeneity, which occur because some VT systems allow the temperature of the shim coils and related components in the magnet bore to change gradually towards the sample temperature. In bad cases an hour may be needed before acceptable stability is attained.

(e) Probe tuning may vary slightly after a change in sample temperature. It is therefore prudent to postpone carrying out any pulse calibrations until the system has stabilized.

2.2.3 VT calibration, accuracy, and stability

Wherever the sample temperature needs to be known accurately it should not simply be assumed that it is equal to the temperature indicated by the VT controller. There are three main reasons for this:

(a) The controller may be miscalibrated.

(b) The sensor which controls the system is not actually in the sample, but is of necessity some millimetres away in the gas flow upstream of the sample.

(c) Certain NMR experiments involve the application of substantial amounts of radiofrequency (rf) energy to the sample. In these cases the sample can be heated by the rf to a temperature above that of the gas flow. This self-heating is only likely to happen in spin-locking experiments and in experiments using heteronuclear decoupling.

When self-heating can be neglected the standard method of assessing the accuracy of the VT system is to use a so-called 'chemical shift thermometer'. These are samples which give spectra containing two peaks that have a shift-difference which is a well-characterized function of temperature. It is conventional to use methanol for temperatures below about 40°C and ethylene glycol for higher temperatures. Both of these give shift differences which have temperature coefficients of approximately 0.01 p.p.m./°C; full details are given in a number of places (2–4). The usual procedure is to construct a correction curve by making measurements over a range of temperatures. For this curve to be applicable to other samples care must be taken to ensure that the VT gas flow is kept unchanged.

As an aside it should be mentioned that an independent calibration method can be obtained by using platinum resistance thermometry (PRT). PRT sensors small enough to fit inside a 5 mm NMR tube can easily be obtained and a straightforward resistance measurement is then all that is needed for an accurate determination of the temperature within the tube.

Either of these methods can be used to estimate the stability of the temperature and the settling time following a temperature change. For these purposes relative temperature changes are all that need to be measured. Since all spectra obtained whilst locked to ^2H_2O have an apparent temperature dependence of 0.01 p.p.m./°C (as discussed above) it may be possible to use the sample of interest instead of a special calibration sample: reasonable precision can be obtained if there is at least one sharp line in the spectrum. This provides the basis for what is perhaps the only good method of estimating the self-heating effects referred to in (c): the spectrum should be compared before and immediately after running the suspect pulse sequence.

2.3 Environmental factors affecting spectral quality

Unfortunately, the quality of the spectra obtainable from an instrument can be significantly degraded by factors which are external to the spectrometer itself. In most cases there is little that the operator can do about these effects except be aware of their existence, although they should of course be taken into account if a new installation is being planned. The relative importance of these effects depends on their magnitude and on the quality of the spectrometer: as spectrometers have improved so environmental factors have been taken more seriously. It is now recognized that they can be a noticeable cause of t_1-noise in 2D spectra. There are three areas of concern:

(a) Floor vibration. These cause problems by making structures within the magnetic assembly vibrate, resulting in a modulation of the magnetic field. The net effect is that intense signals are accompanied by random-phase sidebands at low frequencies, usually less than 20 Hz. Until recently, very little isolation was provided between the magnet assembly and the laboratory floor, but it is now more common to use some form of vibration-damping mounting. Vibrations above about 10 Hz are relatively easy to reduce by standing the magnet on a set of pneumatically damped supports. Lower frequency vibrations are more difficult to eliminate, but relatively costly anti-vibration platforms are available if necessary. Either of these solutions can be applied to magnets at field.

(b) Variations in room temperature. These can have effects on the magnet, the sample temperature, and the radiofrequency electronics. As long as the sample temperature is controlled by a well-designed VT system the main problems are likely to arise from the magnet, the homogeneity of which will alter. Temperature gradients across the magnet are particularly harmful. If possible, the temperature in the vicinity of the magnet should be kept within a band of $\pm 1°C$ with minimal draughts.

(c) Magnetic fluctuations. Not surprisingly, moving a ferrous object within the stray field region of the magnet can have a significant effect on the field homogeneity at the sample. The size of the effect will depend upon the mass of the object: as a rule of thumb objects such as laboratory trolleys or refrigerator doors should not be moved within an area where the stray field is stronger than five gauss whilst an experiment is in progress. It should be remembered that the field extends vertically as well as horizontally: if there are rooms directly above or below a high field magnet they will undoubtedly be within the 5 G 'exclusion zone'. As an example, the 5 G field of a 600 MHz magnet extends 3.6 metres out from the magnet centre horizontally but 4.5 metres above and below the centre along the cylindrical axis.

3. Obtaining one-dimensional spectra

Many of the basic procedures involved in obtaining an NMR spectrum are identical for both one-dimensional and two-dimensional spectra and for both proton observation and X-nucleus observation. However, there are differences of detail and for proton spectra the nature of the solvent is important. The first part of this section describes these procedures assuming as an example that the proton spectrum of a sample dissolved in ^2H_2O (or another deuterated solvent) is to be obtained. X-nucleus spectra are treated briefly in Section 3.2. The special techniques which are needed for running proton spectra in solvents that are not fully deuterated are described in detail in Section 4.

3.1 Proton spectra

Setting up the instrument to obtain a simple proton spectrum of a sample in a deuterated solvent involves four fundamental operations: establishing the field/frequency lock, tuning the probe, optimizing the magnetic field homogeneity, and calibrating the radiofrequency pulse lengths. In addition, the acquisition parameters must be chosen and for many applications measures should be taken to optimize the spectral baseline.

3.1.1 Locking the sample

This consists of manually adjusting the magnetic field to bring the deuterium signal of the sample into resonance with the lock frequency of the spectrometer. Once lock is found it is important that the lock transmitter is not run at too high a power otherwise the field stabilization will be impaired and shimming will be unresponsive. The lock phase should also be adjusted to give the highest lock signal: this is essential for satisfactory shimming and also ensures that the stabilization circuit is working with a signal in pure dispersion, as discussed in Section 2.2.

Some deuterated solvents (such as $C^2H_3O^2H$) give more than one deuterium signal and in these cases much confusion can be avoided by locking on to the same signal consistently.

3.1.2 Probe tuning

With modern probes the observe coil tuning is strongly sample-dependent and can be affected by the sample volume and position as well as the more obvious factors such as solvent and ionic strength. Conversely, lock channel tuning is reasonably sample-independent and should not need routine adjustment. Accurate probe tuning is essential for good sensitivity and to get the strongest rf field from the available power. The phrase 'probe tuning' is slightly misleading since probes need to be both tuned to the correct resonant frequency and matched to the correct impedance, which is invariably 50 ohms. Most recent probes have adjusting rods which are supposed to provide independent optimization of the tuning and matching.

The best method of tuning an observe coil is to use the transmitter to produce a low level signal at the observe frequency. When the probe is tuned to this same frequency and matched to 50 ohms all of the rf power will be absorbed by the probe. In practice a meter is used to indicate the reflected power and this should be minimized by adjusting the tuning rods, allowing for the fact that the controls often interact to some extent. If a probe inadvertently becomes severely mistuned this method may be difficult to apply. One alternative is to adjust the probe tuning whilst monitoring the free induction decay obtained with a single scan; as the tuning improves the FID will increase in size. This is only applicable to protons: a more general method is to use a sweep generator and rf bridge, if these are available.

3.1.3 Shimming

The object of shimming is to improve the homogeneity of the magnetic field experienced by the sample. This is done by altering the currents in a set of correction coils, each of which is designed to counteract a particular type of field gradient. Shimming can either be carried out manually by the operator, or automatically under computer control: there seems to be little to choose between the two if the operator is reasonably expert. It is important to keep a sense of proportion when shimming: most biological samples have lines which are 5 or 10 Hz wide, which means that there is little point shimming until a solvent peak has a linewidth measured in fractions of a hertz. What is often more significant is the lineshape, particularly at the base of a solvent peak, since this controls how many solute signals are obscured by solvent. More will be said about this in Section 4.1.1 since it is of especial importance when samples are dissolved in H_2O.

Two methods are in common use for monitoring the field homogeneity whilst shimming is in progress. The first and most straightforward is to use the height of the lock signal. The second is to operate with the instrument continuously acquiring single scan proton free induction decays and to judge by the area bound by the envelope of the FID. These seemingly different criteria are in fact equivalent if the lock channel and the proton channel sense the same region of the sample. This is usually the case since in most probes the lock channel is double-tuned on to the proton coil, as described in Section 2.1.3. The choice between the two methods is therefore often dictated by personal preference and instrumental convenience. However, the FID method is only really applicable if the sample contains enough residual non-deuterated solvent (or some other sharp peak) to give a clearly visible contribution to the FID. An advantage when this is the case is that the shape of the decay gives an immediate guide to the lineshape; to emphasize this information relatively short acquisition times (0.1–0.2 sec) are recommended. With either method the spectrum itself should be checked at regular intervals to assess progress.

The conventional shimming procedure for an experiment that is to be carried out with the sample spinning is to adjust the axial shims whilst the sample is spinning and the transverse shims whilst it is stationary. If however the experiment is to be carried out non-spinning all shimming should also be carried out non-spinning. Finally, it should be realized that there are often significant interactions amongst the various gradients produced by the shim system. This is particularly likely to be true of the axial shims and means that if they are shimmed manually they should be adjusted interactively as a set to ensure that a global optimum is found.

3.1.4 Pulse calibration

Multiple-pulse experiments all depend on applying pulses of particular flip-angles to the spins and pulse calibrations are therefore an essential stage in setting-up an experiment. Calibrating a proton observe pulse is straight-forward when the sample gives a signal that can be seen with a small number of scans with no danger of receiver overload. The method is well known: the width of the pulse in a simple pulse acquire sequence is varied until either the first signal maximum, or the first or second null is found. These correspond to flip-angles of 90°, 180°, and 360° respectively and each has its devotees. In general, the null methods are to be preferred because they are more precise, and of these the 360° null is recommended once an approximate calibration has been made since it allows the scans to be repeated most rapidly. Which-ever is adopted, if accurate results are to be obtained signals close to the transmitter frequency should be used to assess the null or maximum. On most modern spectrometers the ratio of the 90°, 180°, and 360° pulse-widths is within a few per cent of the ideal ratio of 1:2:4. If this is not the case it should be taken into account when pulse-widths are set in multipulse experiments.

3.1.5 Acquisition parameters

The settings required for most acquisition parameters are widely understood and only a few need special consideration:

(a) The transmitter offset. If the spectrum contains one peak (such as a solvent peak) that is markedly more intense than the rest of the spectrum it is preferable to set the transmitter to be coincident with the intense peak. This minimizes spectral artefacts which arise from errors in digit-izing the strong signal: it is most important with low resolution ADCs (5).

(b) The spectral width. There is a common tendency to choose spectral widths which are only just sufficient to accommodate the spectrum of interest. This may have been attractive when NMR computers were slow and it was preferable to keep transform sizes to a minimum, but it can have adverse effects on the baseline and spectral widths at least 50% larger are recommended. (As long as the acquisition time is kept the same the signal-to-noise ratio of the spectrum will not change with changes in the spectral width.)

(c) The receiver gain. The general principle is that the gain should be set as high as possible subject to the absolute requirement that neither the receiver front-end nor the ADC be overloaded (see Section 2.1.2). ADC overload can best be detected by looking for clipping of a single scan

FID; this can be made clearer by first deliberately setting the gain to an excessive value to 'calibrate' the FID display. If the instrument's minimum gain setting still results in ADC overload then front-end overload must also be suspected and the remedies described below should be adopted. Front-end overload that occurs without ADC overload is more difficult to detect. Since it can cause lineshape distortions and the appearance of spurious signals one method is to compare two spectra, in one of which the signal strength has been temporarily much reduced, for example by using a very low pulse flip-angle or by introducing an attenuator early in the receiver chain. When front-end overload is a problem, or if the lowest gain setting is insufficient to prevent ADC overload, then one of the methods used to reduce the signal strength must be retained for the experiment. In multipulse experiments the flip-angles can not normally be altered so the attenuator method must be used.

3.1.6 Baseline optimization

This is a complex subject which can only be treated here in a superficial way. Unless special care is taken spectra are obtained which have some degree of baseline distortion, typically a DC offset, a slope, a curvature, or some combination of the three. This may or may not be important in 1D NMR depending on the application to which the spectrum is going to be put, but it is always of the utmost importance in 2D NMR—as will be shown in Section 5. Although DC offsets are relatively easy to eliminate from spectra after they have been obtained, the other forms of distortion are less tractable; routines are available but they are not particularly successful when applied to the overlapping spectra typical of biological macromolecules. It is therefore preferable to attack the problem at source.

There are three main causes of baseline distortion:

(a) The time-domain signal is delayed in its passage through the receiver, primarily by the audiofrequency filters. This will create a frequency-dependent phase-shift in the spectrum unless sampling starts immediately after the delay. Frequency-dependent phase-shifts give rise to baseline roll (6).

(b) The receiver, and in particular the audio filters, are not able to respond faithfully to the beginning of the FID since by definition this is a transient which must have frequency components outside the bandwidth of the filter. The resulting distortion of the time-domain data inevitably causes a baseline distortion (7).

(c) The real Fourier transform routine used by some spectrometers gives distorted baselines unless the spectrum prior to phase correction is in pure dispersion (8, 9).

The importance of each of these and the complexity of the measures needed to counteract them depends on the spectrometer and whether it uses simultaneous or sequential quadrature detection. Simultaneous quadrature instruments are easier to deal with: (c) does not apply and (b) gives rise mainly to a DC offset, leaving only (a) to be corrected at source. Good baselines can therefore be obtained in this case simply by adjusting the delay between the observe pulse and the start of acquisition to give a spectrum needing no first-order phase correction.

Conversely, all three causes need to be addressed on instruments that use sequential quadrature. Experience with Bruker AM spectrometers suggests that if (c) is satisfied the consequences of the filter distortion described in (b) are also minimized. As before, the first-order phase gradient should be eliminated by adjusting the delay between the observe pulse and the start of acquisition. Further improvements can be obtained by ensuring that the receiver is gated on at the beginning of this delay.

3.2 Observing nuclei other than protons

The general principles developed in Section 3.1 apply without exception to the less common case in which the nucleus to be detected is not ^{1}H. One additional consideration is that high power proton decoupling is often required and there may be some practical differences to the setting-up procedures if the X-nucleus signal is too weak to be seen without extensive time-averaging.

3.2.1 Instrument set-up

It will usually not be possible to shim using the X-nucleus signal as an indicator of the field homogeneity, so shimming must instead be done using either the lock or the proton FID. If a standard X-nucleus probe is being used both the lock signal and the proton FID will be detected by means of the decoupler coil. Since this is physically somewhat larger than the observe coil the shimming criterion is based on a larger sample volume than will be of relevance in the eventual X-observe experiment; the homogeneity may be adequate even though the shimming suggests otherwise. This effect is most marked with short samples and with 10 or 15 mm probes.

The proton and X-nucleus channels of the probe can both be tuned by the reflected power method described in Section 3.1.2. One possible complication is that in some probe designs the tuning of the two channels interacts to a significant extent. When this is the case the tuning of one channel should always be checked after the other has been altered.

The pulse calibration procedure may need modification if the X-nucleus signal is difficult to detect. In extreme cases it may be necessary to resort to using a different (more concentrated) sample for the calibration, taking care to match the solvent and the ionic strength. However, where a few minutes time-averaging would suffice it is better to use the sample of interest: the 360° null method is recommended since it allows faster repetition of the scans.

3.2.2 Broadband proton decoupling

When an X-nucleus such as ^{13}C, ^{15}N, or ^{19}F is being observed it is usually advantageous to use broadband proton decoupling to eliminate the splittings and broadening caused by the scalar couplings between the X-nuclei and the protons. However, broadband decoupling always has to be applied with care because of the long-standing problem of achieving satisfactory decoupling without causing excessive sample heating. In this regard the older noise-modulated and swept-square-wave decoupling methods are markedly inferior to the more modern composite-pulse methods and only the latter can be recommended (10). This is especially true for biological NMR, where the samples are often electrically conducting and temperature-sensitive.

The two most important properties of a decoupling method are the range of proton offsets that can be decoupled simultaneously—the bandwidth—and the quality of the decoupling within the bandwidth. With composite-pulse decoupling (CPD) poor quality decoupling results in the appearance of weak sideband signals and in the 'decoupled' peaks having residual fine structure, possibly unresolved. The bandwidth can be expressed more quantitatively and is approximately proportional to the strength of the rf field produced by the decoupler. (The field strength in frequency units is equal to the reciprocal of the time required for a 360° pulse, i.e. $B_2 = 1/\tau_{360}$.) Three CPD sequences are likely to be met with in practice, each with different attributes:

(a) GARP (11). This gives the widest bandwidth ($\pm$ 2.5 B_2) but at the expense of the poorest quality.

(b) WALTZ-16 (12). This gives better quality decoupling, but only over a narrower bandwidth ($\pm$ B_2). It is available pre-programmed or hard wired on most modern spectrometers.

(c) DIPSI-2 (13). Although this sequence has the poorest bandwidth ($\pm$ 0.6 B_2) it gives the best quality decoupling: unlike the other sequences it was designed taking ^1H-^1H couplings into account.

For most proton decoupling applications WALTZ-16 provides the best compromise. Its bandwidth is such that with current 5 mm probes only one to two watts of rf power are required to decouple the full proton bandwidth. Although GARP would allow lower powers to be used the resulting side-bands would be troublesome: conversely, for many biological applications the power requirements of DIPSI-2 are too high and the improvements in quality relative to WALTZ-16 unlikely to be noticeable. However, these sequences fulfil useful roles in other NMR experiments and will be mentioned again in Section 5.

When using broadband decoupling several things have to be borne in mind.

First, heating effects may not be negligible, even if composite-pulse methods are being used, and in critical cases estimates of the heating should be made using the method described in Section 2.3. Secondly, when the X-nucleus being observed is ^{19}F or ^{3}H special precautions may need to be taken because of the relative proximity of the observe and decoupler frequencies; usually all that is needed is an additional filter in each channel to eliminate the unwanted frequencies. Finally, since the CPD sequences depend upon the application of pulses of particular flip-angles to the spins, for best results a calibration of the decoupler field strength should be carried out. In practice, the sequences can tolerate flip-angle errors up to about ± 25% and it is usually satisfactory to use a rough estimate of the field strength, based on occasional calibrations made with convenient test samples. Several calibration methods are available of which the most appropriate is the traditional technique of coherent off-resonant decoupling. This is well described in several places (4, 14).

4. Working with samples dissolved in H_2O

In biological NMR there are many situations in which it is essential to obtain proton spectra from samples that are dissolved in H_2O—or rather more precisely in $H_2O/^2H_2O$ mixtures containing ~10% ^2H_2O. This requirement creates a number of problems, all of which stem from the very high proton concentration of the solvent and the potential it has to generate an NMR signal many times more intense than any of the solute signals. Although it is relatively easy to bring the solvent signal within the range of the receiver using the methods described in Section 3.1.5, the resulting spectrum would still be dominated by the solvent peak and any artefacts associated with it. More importantly, the signal-to-noise ratio of the solute signals would be far below what it would have been if the solvent had been ^2H_2O. Sections 4.2 and 4.3 concentrate on some better approaches to the problem: techniques for reducing the solvent signal whilst retaining high sensitivity to the solute. The presence of very strong solvent signals also makes shimming much more critical and necessitates some minor changes to the setting-up procedures described in Section 3.1, and these will be discussed first.

4.1 Setting-up the spectrometer

Two effects need to be taken into account. The first is that when the probe is tuned correctly the solvent signal obtained from a simple pulse-acquire experiment will overload the receiver unless the flip-angle of the pulse is less than a few degrees. The second is that the decay of the intense component of an FID can be dramatically accelerated by the phenomenon known as radiation damping, in which the current induced in the observe coil by the abundant protons generates a resonant rf field that tends to restore the magnetization towards its equilibrium position. This rf field is only resonant for the species that generated it, so only the solvent signal is affected by the damping: typically, a water signal may acquire an apparent linewidth of 20–30 Hz on a high field spectrometer. Closer inspection would show that the water lineshape is strongly dependent on the flip-angle of the pulse and on the tuning of the probe, and that it bears little resemblance to that of other signals from the sample.

These two effects have clear implications for shimming and for probe calibrations. Because of radiation damping, using the FID for shimming is extremely unproductive and can not be recommended, unless the probe is first detuned to eliminate the damping. (The lock signal is not affected by radiation damping so lock shimming can be used with no changes to the probe tuning, irrespective of the solvent.) Pulse calibrations are best carried out with presaturation (as described in Section 4.2) to reduce the solvent signal: the procedures described in Section 3.1 then apply without modification. In the absence of presaturation most trial pulse-widths will result in receiver overload and because of radiation damping flip-angles near to 180° could give particularly confusing results.

Acquiring spectra in H_2O using presaturation makes far more stringent demands on the homogeneity than any other type of experiment. The important factor is the width of the water peak at and below the height of the solute peaks, i.e. below about 0.01% of the height of the water peak. For reasons which are described later only those solvent molecules which resonate within a band 20–30 Hz wide can be suppressed by presaturation. Any which resonate at greater offsets will persist in the spectrum, possibly obscuring solute signals. In practice spectra obtained with presaturation often have solvent peaks with unsuppressed tails that can be anything from 50 Hz to 1000 Hz wide, the latter resulting in spectra that are unusable for most purposes. The presence of these tails implies that there are field gradients within the sample such that some water molecules experience magnetic fields that are very different from the field seen by the bulk of the sample. In this situation it is important to realize that the field gradients may well be generated by the observe coil itself (15), in which case little can be done to eliminate them by shimming alone. This should be suspected whenever the tails extend for more than about 100 Hz, and in particular if the probe in use was designed more than a few years ago. It appears that the problem often originates with sample which is near the leads of the coil and some improvement can often be had by successively raising the sample a few millimetres followed by careful reshimming. Although this is time-consuming once a good sample position has been found it should be applicable to all other samples. An alternative approach is to use an observe pulse which discriminates against regions where the rf field is weak (16): when the tails do arise from sample near the leads of the coil the static and rf fields will both be atypical. A suitable pulse is $90°_x\ 90°_y\ 90°_{-x}\ 90°_{-y}$ and even if this is not used routinely it is recommended as an aid in identifying the cause of the problem.

4.2 Solvent irradiation methods

Presaturation is the easiest and most effective solvent suppression technique. It is widely used in 1D and 2D experiments and it is the method of choice whenever exchange between the solvent and the sample can be discounted. The idea is shown in *Figure 2(a)*, in which a simple pulse acquire experiment has been taken as an example. The observe pulse of the sequence is preceded by a continuous, unmodulated period of low level irradiation applied exactly at the frequency of the solvent to be suppressed. Ideally the irradiation should be so weak that only the solvent is affected: field values of 20–30 Hz are commonly used.

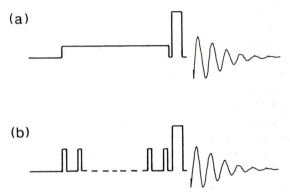

Figure 2. (a) Solvent suppression using CW presaturation. (b) Equivalent sequence using DANTE presaturation.

Although presaturation is a name of long-standing it is somewhat misleading (17): the process taking place during the irradiation is a combination of saturation (in the sense of reducing the solvent z magnetization to zero), dephasing in the plane perpendicular to the irradiating field, and spin-locking along the irradiating field. Incomplete dephasing and spin-locking lead to there being a small component of the solvent magnetization in the transverse plane at the end of the irradiation period. Part of this transverse component will appear in dispersion in the spectrum, the amount depending on the phase difference between the irradiation and the observe pulse.

More repeatable and much improved suppression can be obtained by taking these effects into account:

(a) To improve the repeatability the presaturation should be strictly coherent with the transmitter, i.e. the two must be at identical frequencies and have a constant phase difference (18).

(b) To improve suppression the absolute phase difference between the presaturation and the pulse should alternate between $+90°$ and $-90°$ on successive scans.

(b′) An alternative to (b) is to use a four-step cycle in which the phase of the presaturation varies in 90° steps, without worrying about the absolute phase difference. In comparison with (b) this gives the same suppression after four scans but worse suppression on individual scans.

If these suggestions are ignored presaturation can be carried out on any spectrometer, using the decoupler to irradiate the solvent and the transmitter for the observe pulse. The improved methods are more demanding since they need absolute coherence between the saturation and the pulse and this can only really be guaranteed if the two are derived from the same source. This in turn means that it must be possible to switch the power between the very low value needed for the saturation and the high value needed for the pulse, and that it must be possible to couple power at both levels to the probe. Although this is within the capabilities of most modern spectrometers (as described in Section 2.1) it is beyond older instruments. An alternative which offers the same advantages but makes lesser demands on the spectrometer is shown in *Figure 2(b)*: it uses a train of pulses of intermediate strength—a DANTE sequence (19)—in place of the CW irradiation of *Figure 2(a)*. This obviates the need for the very low power level and so also avoids the problem of coupling the low level to the probe. Setting-up the DANTE sequence is straightforward. Firstly the pulse repetition rate should be chosen so that it is at least as great as the offset between the solvent signal and the edge of the spectrum. The power and/or width of the pulses can then be adjusted to give the required effective strength for the saturating field, either empirically or by noting that it is approximately equal to the strength of the constituent pulses scaled by the on-off ratio. Typically, a DANTE train for this purpose might consist of 5 μsec pulses of 1 kHz strength repeated every 250 μsec. It should be noted that with some spectrometers pulse-widths of less than 1 μsec can not be recommended.

Whichever method is adopted care should be taken to keep the saturation power to the minimum. It is tempting but misguided to use higher powers in an effort to saturate broad components of the solvent signal that arise from field inhomogeneities (Section 4.1). This is because an inevitable consequence of presaturation is that solute signals close to the solvent resonance will also be saturated. Although relatively few signals might be affected directly, in large molecules cross-relaxation can result in a partial saturation of all the remaining solute signals. As the saturation power is increased more and more solute signals are affected directly and the general signal loss becomes more severe. Setting the saturation power is therefore a compromise and several values should be checked before a choice is made.

Finally it should be mentioned that methods are available for restoring magnetization to solute signals that have been directly affected by the solvent saturation; i.e. the signals that are 'under' the water. Although these techniques are not applicable to 1D NMR (the restored signals would be obscured by unsuppressed solvent) they are extremely useful in 2D NMR, since the restored signals can give rise to clearly visible cross-peaks. Two different methods have been devised: in general the first works best with larger molecules (molecular weight $> 15\,000$) and the second with smaller ones.

(a) SCUBA (20). This consists primarily of turning off the presaturation 50–80 msec before the first rf pulse. In this delay cross-relaxation partially restores intensity to the missing signals. To minimize the effects of solvent recovery a non-selective 180° pulse is inserted at the centre of the delay.

(b) Pre-TOCSY (21). In contrast to SCUBA this makes use of scalar couplings to redistribute magnetization. An isotropic mixing sequence (Section 5) lasting 30–50 msec is inserted after the first rf pulse. Missing signals that have significant couplings to signals that were unaffected by the presaturation will be restored to some extent.

4.3 Solvent non-excitation methods

Although presaturation is the easiest way of obtaining spectra from samples in H_2O, it is not suitable for all applications. In particular, alternative methods are needed when the protons of interest are in exchange with the solvent at a rate that is comparable to or faster than their longitudinal relaxation rates. In this situation what is required is a method of turning z magnetization into transverse magnetization for the signals of interest whilst generating as little transverse magnetization from the solvent as possible. Over the years a great deal of effort has been expended on this problem and dozens of different solutions have been proposed (22–25). They differ in several factors: the quality and reproducibility of the suppression, the usable excitation bandwidth, and the phase properties of the signals that are excited. The relative importance of these attributes depends on the application. If the method is to be used as part of a 2D sequence highly reproducible suppression is advantageous, and methods which give spectra with neither first nor higher-order phase gradients are preferred since they give superior spectral baselines.

To conclude this section two methods which have the latter property are described briefly.

(a) The jump-return sequence (24). This is one of the simplest members of a large class of sequences which use non-selective pulses combined with delays to give non-uniform excitation. It consists simply of two non-selective 90° pulses of opposite phase separated by a delay τ, as shown in *Figure 3(a)*. It gives an excitation profile that is a cyclic function of offset, with a periodicity of $1/\tau$. The principal null is at the transmitter frequency and the first excitation maximum is at an offset of $1/4\tau$. It is usual to set the transmitter frequency exactly on to the solvent resonance and to choose the delay τ to centre the first maximum on the signals of interest.

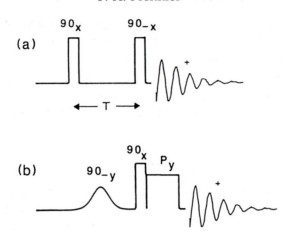

Figure 3. Sequences for observing spectra without exciting a solvent resonance: (a) Jump-return and (b) Bax–Sklenar. In (b) improved solvent suppression can be obtained by independently alternating the phase of the selective pulse and of the purge pulse P.

No y magnetization should be produced and the signals should all be in phase, apart from a change in sign between signals on one side of the transmitter and the other. In practice some empirical adjustment of the length of the pulses may be necessary to get reasonable solvent suppression.

(b) The Sklenar–Bax sequence (25). This is a more recent method which was designed expressly for 2D applications. As can be seen from *Figure 3(b)* the sequence starts with a shaped $90°_{-y}$ pulse which should be selective for the solvent. This is immediately followed by a non-selective $90°_x$ pulse and a 2–3 msec CW purging pulse. The idea is that after the first two pulses the solvent magnetization is along the x axis whereas magnetization from the solute is along the y axis. The spin-locking pulse then dephases the solvent magnetization but leaves the solute signals unaffected. The result should be a solute spectrum with no phase gradient and essentially uniform intensity, with the exception of a narrow gap in the region of the solvent. In practice, the solvent suppression on a single scan is not terribly good, since it is limited by the accuracy of the phase difference between the first and second pulses and by the amount of solvent magnetization left on the z axis by the selective pulse. A simple four-step phase-cycle (described in the figure caption) improves the suppression but the second of these two limitations remains. However, for many biological applications the poor suppression is outweighed by the advantages of uniform excitation and constant phase, and the method has the added advantage of being highly reproducible. Implementing the sequence as shown in the figure requires quite flexible hardware. If pulse-shaping and power-switching facilities are not available an alternative is to use the same power level for all three pulses, and to make a DANTE sequence serve as the selective pulse.

It should be appreciated that incorporating any of the solvent non-excitation methods into 2D experiments may be far from straightforward. Solvent presaturation is always easier and should be used in preference wherever it is applicable.

5. Two-dimensional NMR

Since its inception 20 years ago two-dimensional NMR has grown rapidly in importance, most notably in the last ten years. At present almost all NMR studies of biological macromolecules depend heavily upon 2D methods. The subject is treated here from a practical point of view: more theoretical treatments can be found in several places (1,26,27). The section starts with some general material that applies to all 2D experiments and then concludes with subsections devoted to ·specific techniques. Very little phase-cycling information is given beyond that needed to select the correct signals: the original literature should be consulted for further details. It may appear that the choice of symbols and 2D parameters in Sections 5.1 and 5.2 is unconventional or even perverse. This is quite deliberate and is intended to reduce confusion by avoiding spectrometer-specific terminology.

5.1 Introduction to 2D NMR

A characteristic feature of all 2D pulse sequences is that they contain an incrementable delay, conventionally called the evolution time and given the symbol t_1. When a 2D experiment is carried out the sequence is run with a range of values of the incrementable delay, usually evenly spaced from near zero to a value t_{1max}. With each t_1 value an FID is recorded in the usual way, with any necessary phase-cycling and time-averaging. In this context the real-time variable that defines the FID itself is given the symbol t_2 and the period during which the FID is recorded is called the detection period. The pulse sequences are designed so that the signals detected in t_2 are modulated in amplitude or phase as a function of the incrementable delay t_1. This means that the frequencies of the signals present during t_1 can be measured indirectly through their effect on the real-time signals detected directly in t_2. From this standpoint the various 2D experiments differ in the type of signal that is present during t_1 and in the interaction employed to transmit information from t_1 to t_2. Once the complete data set has been acquired it is Fourier transformed with respect to t_1 and t_2 to give a two-dimensional spectrum that is a function of two frequency variables F_1 and F_2. In this description the term 'signals' must be interpreted rather broadly as far as t_1 is concerned, since both allowed and forbidden transitions can be detected indirectly with equal ease.

It should be clear from this description that the FIDs acquired with successive t_1 values should vary only by virtue of the intended modulation. Any other variation will give artefacts in the 2D spectrum with genuine F_2 frequencies but spurious F_1 frequencies. Instrument instability is a potent and notorious source of these artefacts, which in this case are known as t_1-noise since they appear as bands of noise spanning the entire F_1 spectral width at every signal-bearing F_2 frequency. Minimizing t_1-noise is an important aim since it is almost always the dominant source of noise in proton experiments. Unfortunately, many of the causes of t_1-noise are outside the control of the operator since they include the environmental factors described in Section 2.3, the temperature stability of the sample (Section 2.2), and the intrinsic stability of the spectrometer. Only a few factors are under the operator's control: 2D experiments should always be run non-spinning, and if solvent presaturation is used it should be by means of one of the coherent methods (Section 4.2).

The final topic to be addressed is that of the 2D lineshape and the closely connected problem of providing quadrature detection in the F_1 dimension. In common with 1D NMR, the absorption-mode lineshape is strongly preferred because of its superior resolution. There are two general methods which allow 2D spectra to be obtained with peaks that have absorption-mode lineshapes in both dimensions, and sequences using these methods should always be selected instead of older sequences that need magnitude-mode displays. The two methods are closely related but in setting-up experiments it is worth being aware of the basic differences:

- The TPPI method (28) takes one acquisition per t_1 value and to get an F_1 spectral width of w_1 the increment between successive t_1 values must be set to $1/2w_1$.

- In the hypercomplex method (29) two acquisitions are taken for each t_1 value and for the same F_1 spectral width the increments must be $1/w_1$.

In either case the phase of the pulse (or pulses) that controls the phase of the magnetization present in t_1 is changed by 90° on successive acquisitions: this is quite distinct from the normal phase-cycling that takes place on successive scans. An excellent comparison of the two methods has been given by Keeler and Neuhaus (30).

5.2 General considerations in setting-up 2D experiments

Most of the procedures involved in setting-up 2D experiments have already been described. Shimming, tuning, and pulse calibration are all important and were discussed in Section 3. The benefits of operating without spinning the sample and of using coherent methods of solvent saturation were mentioned in the previous section. Guidelines for choosing some of the acquisition parameters were given in Section 3 and are entirely applicable to 2D NMR.

However, setting the receiver gain may require more care: the procedure described before is a good starting point, but in 2D experiments it is possible that the signal strength might increase as t_1 increases, leading to overload in the later stages of the experiment. This is only likely if the experiment involves solvent suppression: if so, the large values of t_1 should be taken into account in the setting-up procedure. A good way of doing this is to use the 2D pulse sequence to run trial 1D experiments with a few t_1 values chosen from across the range from zero to t_{1max}. Optimizing the baselines of the individual F_2 spectra is also important since baselines which differ from one spectrum to the next are a potential source of artefacts in the 2D spectrum. The suggestions made in Section 3.1.6 are all relevant, although it should be mentioned that this should be one of the last things to be tackled since the necessary corrections might depend on the spectral width and the receiver gain. It is usually best to optimize the baseline using the 2D sequence in a 1D mode.

The remainder of this section deals with the additional factors that arise by virtue of the experiments being two-dimensional. The most important of these are the F_1 spectral width w_1 and the number of acquisitions n. The two are related and in setting them it is helpful to appreciate that they determine the maximum value of the incrementable delay, since $t_{1max} = n/2w_1$. The significance of t_{1max} is that it controls the trade-off between the sensitivity and the attainable F_1 resolution: if t_{1max} is too short the resolution will be impaired and if it is too long the sensitivity will suffer. In biological work t_{1max} values of between 20 and 80 msec are generally used, depending on the application.

(a) The F_1 spectral width. Many of the comments made in Section 3.1.5 about the real-time spectral width also apply to w_1. It is commonplace to use the minimum possible value, in the belief that this gives the best sensitivity and in the interests of attaining the desired t_{1max} with the minimum possible number of acquisitions. This certainty minimizes the size of the data matrix that needs to be processed but the sensitivity argument is fallacious (31). This is important since better F_1 baselines can often be obtained by making w_1 at least 50% larger than the minimum, in which case the same sensitivity can be attained by scaling up the number of acquisitions n (to give the same t_{1max}) with a corresponding decrease in the number of scans per acquisition (to give the same total experimental time).

(b) The number of acquisitions. If the above procedure is followed this is simply set to give the desired t_{1max} with the chosen F_1 spectral width. Between 500 and 1000 acquisitions are usually used.

(c) The repetition rate. If the delay between scans is too short artefacts are likely to arise as a result of the spin system not being in the same state at the start of each scan. Conversely, if the delay is too long the sensitivity will be poor. A good compromise is to allow a delay between scans which is around one to one and a half times the T_1 of the protons of interest, and to precede the delay with a purging sequence which dephases all

59

transverse and z magnetization, thus eliminating the variation between scans. Suitable purging sequences include:

(*i*) a pair of field-gradient pulses separated by a 90° rf pulse, and

(*ii*) a pair of spin-lock pulses with a 90° phase-shift between them.

The spectrometer hardware will determine which of these is applicable.

(d) The number of scans per acquisition. This must be set taking the number of acquisitions into account so that the total number of scans in the experiment will give an adequate signal-to-noise ratio. The only constraint is that since 2D pulse sequences invariably use phase-cycling the number of scans must be set in multiples of the cycle size (or possibly in multiples of the size of a subcycle, depending on the sequence).

(e) F_1 baseline optimization. Two of the three causes of baseline distortion described in Section 3.1.6 also apply to the F_1 dimension, the exception being that signal distortions produced by the receiver are of direct significant for F_2 but not for F_1. The remedies proposed in Section 3.1.6 depended on whether the data were acquired using simultaneous or sequential quadrature detection, and the corresponding distinction for F_1 is between the hypercomplex and TPPI methods of carrying out 2D acquisition. The F_1 baselines of hypercomplex experiments can be improved by acquiring the data such that the resulting 2D spectrum needs no first-order phase corrections in F_1. This is also important for TPPI experiments, which in addition should be acquired so that a 90° zero-order phase correction is needed in F_1 to give absorption lines—in other words the data should be sine-modulated (8). These proposals mirror those made previously for F_2. Ensuring that TPPI experiments yield sine-modulated data needs little more than minor adjustments to the phase-cycling: the sequences given later in this section are shown with phases which give cosine modulation, in keeping with convention. Eliminating the F_1 phase gradients is more intricate, since these are caused by the effective evolution times all exceeding the desired values by a fixed amount. This timing error is a result of evolution occurring during the pulses that flank the t_1 delay in the pulse sequence, and the magnitude of the error depends on the lengths of these pulses. It can be corrected by making a corresponding reduction in the explicitly programmed t_1 delay (32): this is easy on some spectrometers but difficult on others. The first t_1 value should be set to the minimum realizable value: any resulting DC errors are usually easy to correct.

5.3 Simple ^{1}H-^{1}H experiments

The sequences shown in *Figures 4* and *5* represent some of the oldest 2D experiments and also some of the first to be used extensively for biological applications. They are still used widely, although in some situations they are being superseded by the rotating-frame experiments described in Section 5.4.

One of the advantages of these experiments is that they are easy to implement and can be carried out on most spectrometers. No special procedures need to be followed in setting them up beyond those described in the previous sections.

5.3.1 The COSY sequence and its derivatives

COSY (26, 27, 33) is the simplest 2D experiment that can be used to identify pairs of protons that have a mutual scalar coupling. It has two inherent drawbacks that limit its usefulness with macromolecules:

- the fine-structure components that make up each cross-peak appear in antiphase,
- the diagonal peaks and the cross-peaks are 90° out of phase in both dimensions.

The first of these means that sensitivity is poor unless the coupling responsible for the cross-peak is at least as large as the linewidth. The second means that extremely harsh weighting functions must be used to give spectra with usable lineshapes, and this in turn gives a further reduction in sensitivity.

More favourable lineshapes can be obtained by using the double-quantum filtered COSY sequence (34) shown in *Figure 4(b)*. In this experiment the diagonal peaks and the cross-peaks all show the same antiphase absorption-mode fine-structure. Although the double-quantum filtering attenuates the signals to some extent this is more than offset by the gain resulting from not

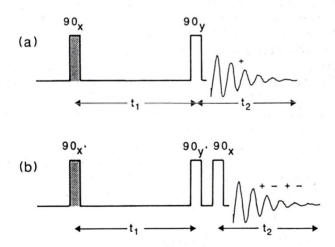

Figure 4. Pulse sequences for (a) COSY and (b) double-quantum filtered COSY. A four-step phase-cycle is needed in (b): x' = x y −x −y. This cycle provides a high degree of axial peak suppression; in (a) axial peaks can be suppressed by alternating the phase of the first pulse together with the receiver phase. In this and the following figures the pulses whose phases must be altered to give F_1 quadrature are indicated by stippling.

having to use resolution enhancement to impose acceptable lineshapes on the peaks. This experiment still suffers from the generic COSY problem of giving poor sensitivity unless the relevant splitting is resolvable.

Several other COSY variants may be encountered in practice, the most important of which is relayed-COSY (35, 36). This uses an additional coherence-transfer step to give extra connectivity information. For example, a three-spin system A-B-C which has no coupling between A and C would only give A-B and B-C cross-peaks in a normal COSY spectrum, whereas with relayed-COSY a cross-peak would also be detected between A and C. This could be crucial in proving that all three spins were part of the same network, particularly if B appears in a crowded region of the spectrum. Relayed-COSY has the same disadvantages as COSY but the extra information is obtained at the price of even poorer sensitivity: where TOCSY experiments are viable alternatives they are much preferred (Section 5.4).

5.3.2 NOESY

The NOESY experiment (37, 38) is usually used to identify pairs of protons that are undergoing cross-relaxation, i.e. protons that would show an NOE in 1D experiments. The characteristic feature of the NOESY pulse sequence—shown in *Figure 5*—is the mixing time τ. The cross-peaks are generated by magnetization transfer that takes place during the mixing time so the length of this delay must be chosen according to the rate of transfer process. In contrast to the COSY experiment, both cross-peaks and diagonal peaks in NOESY spectra can be obtained with absorption-mode lineshapes using the methods described in Section 5.1, and if any fine structure is resolvable in the NOESY cross-peaks all the multiplet components will also be in phase. The experiment is therefore eminently suitable for studying macromolecules. If presaturation is used in a NOESY experiment it is usually also necessary to irradiate the solvent during the mixing time.

A complicating factor is that the NOESY pulse sequence can not distinguish between magnetization transfer caused by cross-relaxation and magnetization transfer caused by chemical exchange (see Chapter 6). In macromolecules both give cross-peaks of the same sign as the diagonal.

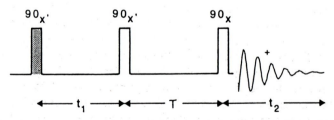

Figure 5. Pulse sequence for the NOESY experiment. A four-step phase-cycle is usually used: x' = x y −x −y. In contrast to *Figure 4* the receiver phase remains constant. This cycle provides good axial peak suppression.

5.4 Rotating-frame experiments

The somewhat arcane title of this section refers to two techniques in which cross-peak formation depends upon the behaviour of the spins in the presence of a radiofrequency field. The two techniques provide alternatives to the COSY and NOESY experiments in that one is used primarily to give information about scalar coupling and the other about cross-relaxation. The hardware requirements of the two are rather similar and may well be beyond the capabilities of older spectrometers. Another feature the two have in common is that both are known by two different names, reflecting the fact that each has been developed by more than one group. For the purposes of this section no distinction need be made between TOCSY (39) and HOHAHA (40), or between ROESY (41) and CAMELSPIN (42), and the first name from each pair is used to cover the other.

5.4.1 TOCSY

The idea underlying the TOCSY experiment is that coupled spins can be made to exchange magnetization by applying a pulse train which eliminates the effects of the chemical shift but retains the scalar coupling. This phenomenon is known as isotropic mixing and the advantage of it is that to a good approximation all the multiplet components in the resulting cross-peaks have the same phase; as a consequence, the cross-peaks do not suffer 'self-cancellation' if the multiplet structure is poorly resolved. A second advantage is that the cross-peaks and diagonal peaks all have absorption-mode line-shapes, again to a good approximation. These two factors make the experiment considerably more sensitive than COSY, despite the fact that there is some signal loss as a result of relaxation during the isotropic mixing sequence.

The crux of the TOCSY experiment is the way in which the isotropic mixing is achieved. A number of suitable sequences are available, most of which involve applying continuous but phase-modulated irradiation to the spins. In this context isotropic mixing and broadband decoupling have a certain amount in common and the sequences in current use are all based on ones introduced for broadband decoupling. The rf field strength required to give good mixing across the full proton bandwidth depends somewhat on the mixing sequence but values in the range from 6 to 10 kHz are usually used. In a simple two-spin system the magnetization transfer in cyclic and at a maximum after a time 1/2J, where J is the coupling between the two spins. In more complicated spin systems multi-step transfers occur and there is no general optimum mixing time (43). In practice the mixing time is usually chosen on a fairly rough-and-ready basis, with values of 25–50 msec being used to favour large couplings and single-step transfers, and values from 50 to 100 msec for smaller couplings and multi-step transfers. It should be remembered that there is a risk that excessive sample heating or even probe damage could be caused by high levels of rf power being applied at relatively high duty

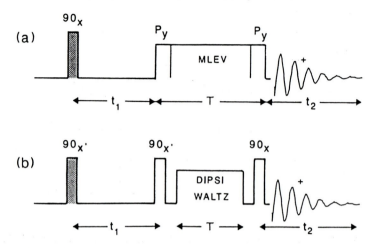

Figure 6. Pulse sequences for carrying out TOCSY experiments. Isotropic mixing is generally by means of MLEV in (a) but WALTZ or DIPSI in (b). The phase-cycling in (b) is as described for NOESY, and in (a) as for COSY.

cycles. With some probes or samples this imposes constraints on the allowable rf field strengths and/or mixing times.

TOCSY experiments are usually carried out in one of two principal ways (40, 44, 45), as illustrated in *Figure 6(a)* and *(b)*. The main difference between them is that in the first the isotropic mixing operates with transverse magnetization but in the second with z magnetization. Superior results are claimed for the second method with DIPSI-2 as the mixing sequence (45). The method shown in *Figure 6(a)* is quite difficult to implement since the frequency source used for the mixing sequence must be coherent with that used for the pulse; in addition, the pulse should ideally be at a higher rf level than the mixing sequence. On modern spectrometers with fast power switching and CW amplifiers these requirements can easily be met, but on other spectrometers it may be necessary to use the same power level throughout with the proton decoupler providing both pulses and spin-locking. The alternative method shown in *Figure 6(b)* is less demanding since the spin-locking does not need to be coherent with the pulses: the decoupler can be used for one and the transmitter for the other (46).

5.4.2 ROESY

The ROESY experiment may be thought of as the rotating-frame analogue of NOESY, although there are a number of differences between the two. The most important of these for work with macromolecules is that cross-peaks arising from spin-diffusion are weaker and easier to identify. For smaller molecules a second advantage arises from the fact that the rotating-frame cross-relaxation rate has the same sign for all correlation times. This means

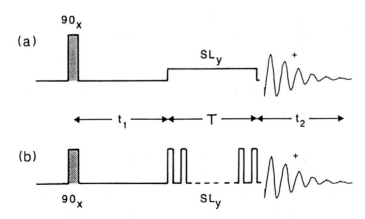

Figure 7. ROESY pulse sequences using (a) continuous spin-locking and (b) DANTE spin-locking. In both sequences axial peaks can be suppressed by alternating the phase of the first pulse together with the receiver phase.

that the ROESY experiment can be applied to all molecules irrespective of correlation time, in contrast to NOESY. It should be noted that the ROESY cross-peaks are opposite in phase to the diagonal.

The pulse sequence for the simplest type of ROESY experiment is shown in *Figure 7(a)*. Cross-peaks develop during the mixing time τ, which must be chosen according to the cross-relaxation rate. During the mixing time an unmodulated low level rf field is applied to spins: field strengths of 1–2 kHz are commonly used. The effect of this irradiation is to select just that component of the magnetization which is aligned along the irradiation axis, a process known as spin-locking. The sequence has rather similar hardware requirements to the TOCSY sequence shown in *Figure 6(a)*, but since the spin-locking must be very much weaker than the initial pulse it is not possible to use a single compromise power level for both. For older spectrometers a more amenable form of the experiment is shown in *Figure 7(b)*. The advantage of this sequence is that it uses a DANTE pulse train for the spin-locking and so allows a single frequency source at a fixed power level to suffice for both the initial pulse and the spin-locking (47). This mirrors the use of the DANTE sequence for presaturation as described in Section 4.2. The setting-up procedure given there applies in principle but for the ROESY experiment the flip-angle of the individual pulses should be kept close to 30° (to minimize isotropic mixing) and the duty-cycle should be adjusted to give the desired effective spin-locking field.

5.4.3 TOCSY and ROESY: concluding remarks

It should be appreciated that the distinction between the two experiments is not an absolute one and that in each case cross-peaks can arise by mechanisms other than the intended one. In particular, cross-peaks in TOCSY spectra can be attenuated by ROESY effects, possibly to the point that weak cross-peaks might become undetectable. A modified isotropic mixing scheme which avoids this problem has been described (48). TOCSY-type cross-peaks can appear in ROESY spectra between coupled spins that have nearly equal, or equal and opposite, offsets: a solution to this has also been proposed (49). In addition, chemical exchange can give rise to cross-peaks in both types of experiment. When more than one of these mechanisms is operating simultaneously spectra must be interpreted with caution (49, 50).

5.5 Heteronuclear correlation experiments

This is a large subject which can only be treated superficially. In biological NMR the heteronuclear experiments of interest are those which identify the chemical shifts of protons and X-nuclei that are directly coupled. One obvious use of this correlation information is to allow assignments already made for one nuclear species to be transferred to the other. A more important but less obvious use is in overcoming problems caused by overlap in the proton spectrum: in the heteronuclear 2D spectrum the proton resonances are spread out according to the shifts of the heteronuclei to which they are coupled. This factor has motivated a recent resurgence of interest in heteronuclear methods and the HMQC technique to be described below is an integral part of many 3D and 4D experiments.

There are many different ways in which the basic correlation experiment can be carried out. The principal distinction between them lies in which nucleus is detected directly and which indirectly (1). Proton-observe methods are usually the most sensitive but are instrumentally more demanding: for historical reasons they are widely known as inverse experiments. For the potential sensitivity to be achieved they should be run using inverse probes and broadband decoupling of the X-nuclei should be used during acquisition. If the X-nucleus is a rare isotope at natural abundance severe t_1-noise may be caused by protons that are *not* coupled to the X-nucleus. Methods that rely on direct observation of the X-nucleus are less sensitive but can be carried out on older spectrometers (51). At the time of writing the main biological application of heteronuclear NMR is in studies of proteins that have been uniformly labelled with ^{13}C or ^{15}N. In this situation the greater sensitivity of the proton-observe method is crucial and only this will be described further. In particular, the widely used HMQC experiment will be discussed in some detail but other experiments that serve a similar purpose will be omitted (52, 53).

5.5.1 HMQC

The HMQC experiment shown in *Figure 8* is a simple and effective proton-observe correlation experiment (51, 54). As mentioned above, it would typically be used to study proteins labelled with ^{13}C or ^{15}N. An important feature of the experiment is that the proton magnetization which is detected during t_2

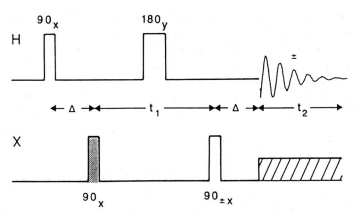

Figure 8. Pulse sequence for the HMQC experiment. Heteronuclear broadband decoupling is applied during t_2 using a sequence such as GARP. The basic two-step phase-cycle shown in the figure suppresses axial peaks and selects signals from protons coupled to X-nuclei.

originated as proton magnetization at the start of the sequence. This contributes to the good sensitivity of the sequence and means that the repetition rate of the scans can be set according to the T_1 of the protons and not of the X-nuclei.

In setting-up this experiment a few factors that have not arisen earlier need to be considered:

(a) X-nucleus transmitter offset. This can be determined from a 1D X-observe experiment, as described in Section 3.2. An alternative is to carry out a trial 2D HMQC experiment using low resolution and a larger-than-necessary F_1 spectral width: this may well be the quicker of the two.

(b) X-nucleus pulse calibration. A good method is to use the HMQC sequence to carry out 1D experiments, using the phase-cycling shown in *Figure 8* and with t_1 set to zero. If imperfections are neglected the proton signal intensity should vary as $\sin^2\alpha$, where α is the flip-angle of the X-nucleus pulse. In practice it is best to look for the first signal minimum, which corresponds to a flip-angle of 180°; for greatest accuracy signals with a low X-nucleus offset should be used to assess the null. (A disadvantage of this method is that the signal has the same sign for flip-angles on either side of the null. Alternative methods (55) avoid this by giving a $\sin\alpha$ dependence but since they need a well-resolved proton resonance a separate test sample must usually be used.)

(c) X-nucleus decoupling. The points made in Section 3.2.2 in connection with proton decoupling are all applicable to the X-nucleus case. In particular it will be necessary to use a composite-pulse decoupling sequence such as GARP or WALTZ, and even with these the possibility that sample heating could be excessive should be borne in mind. Experience with 5 mm inverse probes suggests that 2–3 W of decoupling power is sufficient for most biological applications involving ^{15}N or ^{13}C. The rf field strength can be measured either by means of coherent off-resonant decoupling as described in Section 3.2.2, or by means of the pulse methods mentioned above: in either case it is preferable to use a labelled test sample with well-resolved X and H resonances. When X-nucleus decoupling is used for the first time it is important to check that it is not perturbing the lock or introducing extra noise into the observe channel. If there are any adverse effects they can usually be remedied by inserting some or all of the following filters into the cables which connect to the probe: a high-pass filter in the proton line; a deuterium band-pass filter in the lock line; and a low-pass filter and a deuterium notch filter in the X-nucleus line. If filters are needed they should be in place before any pulse calibrations are carried out.

(d) The Δ delay. This is governed by the X-H coupling constant and if relaxation were negligible the optimum setting would be $1/2J_{XH}$. In practice values somewhat shorter than this are usually used and 5 msec and 3 msec are common choices for HMQC experiments using ^{15}N and ^{13}C respectively.

The HMQC experiment is a fitting place at which to conclude this section and the chapter. A perennial aim in biological NMR is the development of new techniques to allow systems with ever increasing molecular weights to be studied. The techniques described in this chapter have all played a part in this process, and it now appears that further significant advances will be made by means of three- and four-dimensional experiments formed by amalgamating some of the existing techniques. Heteronuclear methods, and in particular the HMQC sequence, are an important constituent of many of these developments.

References

1. Ernst, R. R., Bodenhausen, G., and Wokaun, A. (1987). *Principles of nuclear magnetic resonance in one and two dimensions.* Clarendon Press, Oxford.
2. Van Geet, A. L. (1968). *Anal. Chem.,* **40,** 2227.
3. Van Geet, A. L. (1970). *Anal. Chem.,* **42,** 679.
4. Martin, M. L., Martin, G. J., and Delpuech, J.-J. (1980). *Practical NMR spectroscopy.* Heyden, London.
5. Lindon, J. C. and Ferrige, A. G. (1980). *Prog. NMR Spectrosc.,* **14,** 27.
6. Plateau, P., Dumas, C., and Guéron, M. (1983). *J. Magn. Reson.,* **54,** 46.
7. Hoult, D. I., Chen, C.-N., Eden, H., and Eden, M. (1983). *J. Magn. Reson.,* **51,** 110.
8. Otting, G., Widmer, H., Wagner, G., and Wüthrich, K. (1986). *J. Magn. Reson.,* **66,** 187.

9. Marion, D. and Bax, A. (1988). *J. Magn. Reson.*, **79**, 352.

10. Shaka, A. J. and Keeler, J. (1987). *Prog. NMR Spectrosc.*, **19**, 47.

11. Shaka, A. J., Barker, P. B., and Freeman, R. (1985). *J. Magn. Reson.*, **64**, 547.

12. Shaka, A. J., Keeler, J., and Freeman, R. (1983). *J. Magn. Reson.*, **53**, 313.

13. Shaka, A. J., Lee, C. J., and Pines, A. (1988). *J. Magn. Reson.*, **77**, 274.

14. Pachler, K. G. R. (1972). *J. Magn. Reson.*, **7**, 442.

15. Dykstra, R. W. (1987). *J. Magn. Reson.*, **72**, 162.

16. Bax, A. (1985). *J. Magn. Reson.*, **65**, 142.

17. Hoult, D. I. (1976). *J. Magn. Reson.*, **21**, 337.

18. Zuiderweg, E. R. P., Hallenga, K., and Olejniczak, E. T. (1986). *J. Magn. Reson.*, **70**, 336.

19. Morris, G. A. and Freeman, R. (1978). *J. Magn. Reson.*, **29**, 433.

20. Brown, S. C., Weber, P. L., and Mueller, L. (1988). *J. Magn. Reson.*, **77**, 166.

21. Otting, G. and Wüthrich, K. (1987). *J. Magn. Reson.*, **75**, 546.

22. Redfield, A. G. (1978). In *Methods in enzymology* (ed. C. H. W. Hirs and S. N. Timasheff), Vol. 49, pp. 253–70. Academic Press, New York.

23. Hore, P. J. (1989). In *Methods in enzymology* (ed. N. J. Oppenheimer and T. L. James), Vol. 176, pp. 64–77. Academic Press, San Diego.

24. Plateau, P. and Guéron, M. (1982). *J. Am. Chem. Soc.*, **104**, 7310.

25. Sklenar, V. and Bax, A. (1987). *J. Magn. Reson.*, **75**, 378.

26. Aue, W. P., Bartholdi, E., and Ernst, R. R. (1976). *J. Chem. Phys.*, **64**, 2229.

27. Bax, A. (1982). *Two-dimensional nuclear magnetic resonance in liquids*. Delft University Press, D. Reidel Publishing Company, Dordrecht.

28. Marion, D. and Wüthrich, K. (1983). *Biochem. Biophys. Res. Commun.*, **113**, 967.

29. States, D. J., Haberkorn, R. A., and Ruben, D. J. (1982). *J. Magn. Reson.*, **48**, 286.

30. Keeler, J. and Neuhaus, D. (1985). *J. Magn. Reson.*, **63**, 454.

31. Levitt, M. H., Bodenhausen, G., and Ernst, R. R. (1984). *J. Magn. Reson.*, **58**, 462.

32. Marion, D. and Bax, A. (1989). *J. Magn. Reson.*, **83**, 205.

33. Nagayama, K., Kumar, A., Wüthrich, K., and Ernst, R. R. (1980). *J. Magn. Reson.*, **40**, 321.

34. Rance, M., Sorensen, O. W., Bodenhausen, G., Wagner, G., Ernst, R. R., and Wüthrich, K. (1983). *Biochem. Biophys. Res. Commun.*, **117**, 479.

35. Eich, G., Bodenhausen, G., and Ernst, R. R. (1982). *J. Am. Chem. Soc.*, **104**, 3731.

36. Wagner, G. (1983). *J. Magn. Reson.*, **55**, 151.

37. Jeener, J., Meier, B. H., Bachmann, P., and Ernst, R. R. (1979). *J. Chem. Phys.*, **71**, 4546.

38. Kumar, A., Ernst, R. R., and Wüthrich, K. (1980). *Biochem. Biophys. Res. Commun.*, **95**, 1.

39. Braunschweiler, L. and Ernst, R. R. (1983). *J. Magn. Reson.*, **53**, 521.

40. Davis, D. G. and Bax, A. (1985). *J. Am. Chem. Soc.*, **107**, 2820.

41. Bax, A. and Davis, D. G. (1985). *J. Magn. Reson.*, **63**, 207.

42. Bothner-By, A. A., Stephens, R. L., Lee, J.-M., Warren, C. D., and Jeanloz, R. W. (1984). *J. Am. Chem. Soc.*, **106**, 811.

43. Cavanagh, J., Chazin, W. J., and Rance, M. (1990). *J. Magn. Reson.*, **87**, 110.

44. Bax, A. and Davis, D. G. (1985). *J. Magn. Reson.*, **65**, 355.
45. Rucker, S. P. and Shaka, A. J. (1989). *Mol. Phys.*, **68**, 509.
46. Rance, M. (1987). *J. Magn. Reson.*, **74**, 557.
47. Kessler, H., Griesinger, C., Kerssebaum, R., Wagner, K., and Ernst, R. R. (1987). *J. Am. Chem. Soc.*, **109**, 607.
48. Griesinger, C., Otting, G., Wüthrich, K., and Ernst, R. R. (1988). *J. Am. Chem. Soc.*, **110**, 7870.
49. Cavanagh, J. and Keeler, J. (1988). *J. Magn. Reson.*, **80**, 186.
50. Feeney, J., Bauer, C. J., Frenkiel, T. A., Birdsall, B., Carr, M. D., Roberts, G. C. K., and Arnold, J. R. P. (1991). *J. Magn. Reson.*, **91**, 607.
51. Freeman, R. and Morris, G. A. (1978). *J. Chem. Soc. Chem. Comm.*, 684.
52. Bodenhausen, G. and Ruben, D. J. (1980). *Chem. Phys. Lett.*, **69**, 185.
53. Bendall, M. R., Pegg, D. T., and Doddrell, D. M. (1983). *J. Magn. Reson.*, **52**, 81.
54. Bax, A., Griffey, R. H., and Hawkins, B. L. (1983). *J. Magn. Reson.*, **55**, 301.
55. Bax, A. (1983). *J. Magn. Reson.*, **52**, 76.

Resonance assignment strategies for small proteins

CHRISTINA REDFIELD

1. Introduction

Prior to 1982, assignment of resonances in the complex NMR spectra of proteins was achieved, using 1D NMR, and was based, for the most part, on the assumption that the structure of the protein in solution was the same as in the available X-ray structure. The introduction of 2D NMR techniques such as COSY and NOESY in the early 1980s dramatically increased the resolution of protein NMR spectra and as a result the amount of information which could be obtained from these spectra also increased. In 1982 Wüthrich and co-workers introduced a systematic method for the full assignment of the 2D NMR spectra of proteins which relied only on information about the amino acid sequence of the protein; this method is the sequential assignment procedure (1, 2). This approach is now used by most researchers who apply NMR spectroscopy to the study of proteins.

In this chapter the elements of the sequential assignment strategy for the ^{1}H NMR spectra of proteins will be outlined. This method is particularly useful for the assignment of the spectra of small proteins containing less than 75 residues but it can be used to assign the spectra of proteins twice that size. It will be shown in Chapter 5 that isotopic labelling with ^{15}N and ^{13}C is a powerful addition to NMR techniques for the study of proteins of more than 100 residues and can permit extension of sequential assignment to larger molecules. However, if labelled protein is not available and the spectrum of the protein is of high enough quality then assignment by ^{1}H methods alone should certainly be attempted for these larger proteins.

The sequential assignment method consists of two-stages. The first involves the identification of the systems of spin-spin coupled resonances which belong to a particular amino acid residue. This is achieved using COSY, RELAY, double-RELAY, and HOHAHA experiments. During this process the spin systems of, for example, all the alanine residues may be identified but there is nothing to distinguish one alanine spin system from another. The second-stage involves the assignment of each spin system to a particular residue in the amino acid sequence. This can not be achieved using the COSY-type experiments because there is no resolved spin-spin coupling across the peptide bond which connects neighbouring residues. Instead assignments are deduced from the through-space dipole-dipole information found in NOESY spectra. The procedures used in this two-stage assignment method are described in this chapter.

2. Assessing the quality of 2D spectra

Before attempting the sometimes lengthy process of sequential assignment it is sensible to determine whether the protein of interest will give 2D NMR spectra of the required quality. In order to do this a COSY and NOESY spectrum of the protein in H_2O should be collected. Ideally these spectra should be collected using a pre-TOCSY, SCUBA or $1\bar{1}$ method (see Chapter 3) so that loss of cross-peaks due to saturation of H^α resonances can be avoided.

The coupled H^N and H^α resonances of each residue should give rise to a cross-peak in the so-called 'fingerprint' region of the COSY spectrum ($F_2 \simeq$ 10.5 to 6.5 p.p.m., $F_1 \simeq$ 6.0 to 2.5 p.p.m.). In addition cross-peaks arising from the side chain NH's of Arg and Lys may be observed in this region of the spectrum ($F_2 \simeq$ 7.5 to 6.5 p.p.m., $F_1 \simeq$ 3.5 to 2.5 p.p.m.). Assign an arbitrary number to each H^N-H^α cross-peak observed in the COSY spectrum. If more than about 10% of the expected cross-peaks are missing then the steps outlined in *Protocol 1* should be taken in order to determine the reason for the absence of peaks and to overcome the problem before continuing with the assignment process.

Protocol 1. Reasons for loss of H^N-H^α cross-peaks

1. Cross-peaks are missing because of saturation of H^α resonances beneath the water resonance. This can be overcome using the pre-TOCSY COSY or SCUBA methods. If these experiments are not available then collect the spectrum at 10°C above or below the previous temperature; the H_2O resonance shifts significantly with temperature whereas most H^α resonances do not.

2. H^N intensity is decreased due to cross saturation from the H_2O signal because the H^N is in fairly rapid exchange with the H_2O. This problem can be identified by comparing the intensity of H^N's in 1D $1\bar{3}3\bar{1}$ or $1\bar{1}$ spectra collected with and without presaturation of the water. If the intensity of the H^N's differs in these experiments then cross saturation is the problem. H^N exchange rates can be decreased by adjusting the sample pH so that it is closer to the pH minimum for hydrogen exchange (pH ~ 4) or by decreasing the sample temperature.

3. Cross-peaks may be below the signal-to-noise level because of a small H^N-H^α coupling constant (J) or a large linewidth (LW). Cancellation of cross-peak intensity occurs because of overlap of antiphase components when J is less than or equal to the linewidth. Collect a HOHAHA spectrum with a short isotropic mixing period to ensure that only direct connectivities are observed. Compare the number of H^N-H^α cross-peaks in the COSY and HOHAHA spectra. If there are additional peaks in the HOHAHA spectrum then these may correspond to residues where J ≤ LW. Linewidths can be decreased by decreasing sample viscosity; this can be achieved by increasing the temperature or by decreasing protein concentration. If either of these is feasible then collect another COSY spectrum and look for additional peaks.

The NOESY spectrum should contain a large number of H^N-H^N or H^α-H^N NOE effects; count up the number of these cross-peaks observed. If there are many fewer peaks than residues then sequential assignment, which depends on these NOE effects, will be difficult. The mixing time selected for the NOE experiment may be too short; try collecting a NOESY spectrum with a longer mixing time. If too many peaks are observed then spin diffusion may be the problem; the experiment should be repeated with a shorter mixing time.

If experimental conditions in which a significant number of COSY and NOESY peaks are observed can not be found then the protein is not a good candidate for sequential assignment using ^{1}H 2D techniques alone and alternative methods should be considered. If the COSY and NOESY spectra are of acceptable quality then the first-stage of the sequential assignment process can be started.

3. Stage 1: spin system identification

The first stage of assignment involves the identification of systems of spin-spin coupled resonances. Because there is no spin-spin coupling across the peptide bond each of these spin systems corresponds to the resonances of an individual amino acid residue. The random coil chemical shift values for resonances in the common amino acids are shown in *Figure 1* (3). Many of the amino acids have unique spin system topologies and will give rise to unique patterns of cross-peaks in a COSY spectrum (4). If there is no overlap in the spectrum then spin systems can readily be identified from the COSY spectrum alone. However, in most proteins there is some degree of overlap and this overlap is likely to increase as the size of the protein increases. When overlap occurs unambiguous spin system assignments are not usually possible on the basis of the COSY spectrum alone and other 2D data sets are required. The experiments that should be collected for the first-stage of assignment are shown in *Table 1*.

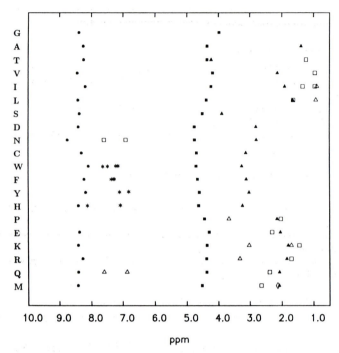

Figure 1. Random coil chemical shifts for the 20 common amino acids (3). The *filled circle, filled square, filled triangle, open square, open triangle, open diamond,* and *star* represent the H^N, H^α, H^β, H^γ, H^δ,H^ϵ, and aromatic CH, respectively.

It will be shown later that the analysis of NOE data for the second-stage of sequential assignment relies most heavily on the H^N and H^α resonances. Therefore, the approach to spin system identification described here emphasizes the COSY H^N-H^α peak. Information about the type of residue from which a particular H^N-H^α cross-peak arises is collected using the various 2D experiments. For many residues, such as alanine and threonine, complete spin system identification is quite straightforward even for large proteins. For other residues, including arginine and lysine, complete spin system identification is much more difficult. However, the lack of complete spin systems for these residues does not hamper the sequential assignment process.

Spin system information can be obtained from two types of experiment. The first is the HOHAHA experiment which is favoured by many researchers for several reasons. First, the in-phase nature of the cross-peaks means that signal-to-noise ratios may be better than would be observed in COSY spectra where cancellation of intensity occurs in the antiphase cross-peaks. Secondly, by adjusting the isotropic mixing period used in the experiment long-range connectivities in the spin system can be observed; in favourable cases complete spin systems can be identified even for Arg and Lys residues. The

---

Table 1. Suggested experiments for spin system identification

H_2O sample—all H^N's present

- H_2O pre-TOCSY COSY (512 complex t_1 values of 1 K complex points)
- H_2O relayed coherence transfer (RELAY, $2\tau = 30$–36 msec, 512 complex t_1 values of 1 K complex points)
- H_2O double relayed coherence transfer (double-RELAY, $2\tau = 30$–36 msec, 512 complex t_1 values of 1 K complex points)
- H_2O HOHAHA (several isotropic mixing periods 15–75 msec, 256 complex t_1 values of 1 K complex points)

^2H_2O sample—fully exchanged sample, no H^N's present

- ^2H_2O DQF-COSY (512 complex t_1 values of 2 K complex points)
- ^2H_2O relayed coherence transfer (RELAY, $2\tau = 30$–36 msec, 512 complex t_1 values of 1 K complex points)
- ^2H_2O double relayed coherence transfer (double-RELAY, $2\tau = 30$–36 msec, 512 complex t_1 values of 1 K complex points)
- ^2H_2O HOHAHA (several isotropic mixing periods 15–75 msec, 256 complex t_1 values of 1 K complex points)

If overlap is severe repeat at a different temperature or pH.

---

second type of experiment includes COSY, DQF-COSY, RELAY, and double-RELAY. The cross-peaks in these spectra are characterized by a pattern of antiphase fine-structure which reflects the active and passive coupling constants of the spin system. This pattern of fine-structure can provide useful information for assignment which is usually not available from HOHAHA spectra. Long-range four- and five-bond connectivities in spin systems can be identified in RELAY and double-RELAY spectra. The method for spin system identification described below relies on the COSY-type spectra which are then augmented by the HOHAHA spectra. However, as the molecular weight of the protein studied increases resonance linewidths will also increase and the method described below may not give satisfactory results because of poor quality COSY, RELAY, or double-RELAY spectra. If this occurs then most of the information needed can be obtained from HOHAHA spectra alone.

Before attempting to identify the side chain resonances associated with each H^N-H^α cross-peak it is useful to analyse the information available in the fingerprint region of the COSY spectrum. Part of the fingerprint region of COSY spectra of human lysozyme collected with 512 and 256 t_1 values are compared in *Figure 2*. The cross-peaks shown in *Figure 2a* have various shapes and cross-peak fine-structures. The cross-peaks composed of eight antiphase components are characteristic of the glycine spin system which has two H^α resonances coupled to the H^N. It can be shown by spectral simulations that the square four-component cross-peaks arise from spin systems in which

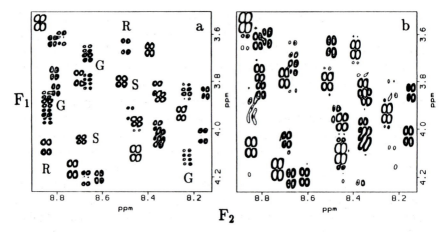

Figure 2. Part of the fingerprint region of the COSY spectrum of human lysozyme in H_2O. Spectrum (a) was collected with 512 complex t_1 values whereas spectrum (b) was collected with only 256 complex t_1 values. Pairs of glycine cross-peaks are labelled *G*. Square and rectangular cross-peaks are labelled *S* and *R*, respectively.

both H^α-H^β coupling constants or the single H^α-H^β coupling constant are small ($\leqslant 5$ Hz) whereas the rectangular four-component peaks arise from spin systems in which there is at least one large H^α-H^β coupling constant ($\geqslant 9$ Hz) or two intermediate H^α-H^β coupling constants ($\geqslant 6.5$ Hz) (5). Thus, the H^N-H^α cross-peak shape can provide information about coupling constants (see also Chapter 10). This information can not readily be obtained from the COSY spectrum shown in *Figure 2b*; the loss of information results from the poor digital resolution in F_1.

3.1 Glycine spin system identification

As discussed above the three coupled protons of the glycine spin system form an AMX spin system which gives H^N-H^α cross-peaks with a characteristic 8-component shape. As a result glycine spin systems are usually easy to identify as shown in *Protocol 2*.

Protocol 2. Glycine spin system identification

1. COSY (H_2O): identify a pair of cross-peaks, with the characteristic 8-component peak shape, which correlates an H^N chemical shift with two H^α chemical shifts.
   DQF-COSY (^2H_2O): confirm that these two H^α chemical shifts belong to a glycine by identifying a 4-component square cross-peak correlating the two H^α resonances.
2. COSY (H_2O): remaining single 8-component H^N-H^α cross-peaks may correspond to glycines in which the second H^N-H^α cross-peak is too weak to observe due to a small H^N-H^α coupling constant.

RELAY (H_2O): look for a strong H^N-H^α peak, not seen in the COSY spectrum, which has the same H^N shift as the 8-component COSY peak. DQF-COSY (^2H_2O): look for a 4-component square H^α-H^α cross-peak correlating the H^α resonance identified in the H_2O cosy spectrum and the H^α resonance identified in the H_2O RELAY spectrum.

3. DQF-COSY (^2H_2O): look for any remaining H^α-H^α cross-peaks with the characteristic 4-component square shape. If peaks are found then the absence of a corresponding pair of H^N-H^α peaks may indicate that the H^N resonance is very broad or the proton is exchanging rapidly with H_2O.

4. DQF-COSY (^2H_2O): if there are no more H^α-H^α cross-peaks resolved from the diagonal this could indicate that the two coupled H^α resonances of remaining glycines have the same chemical shift. In this case the H^N-H^α peak would have only 4-components and look like the peaks arising from other residues. Wait until the stage 2 NOE analysis to identify these glycines.

3.2 Alanine and threonine spin system identification

The spin systems of alanine and threonine residues can also be identified in a relatively straightforward manner. Inspection of *Figure 1* shows that the H^α-H^β cross-peaks of Ala and the H^β-H^γ cross-peaks of Thr will fall in the same region of the DQF-COSY spectrum. These cross-peaks can be distinguished from the other H^α-H^β cross-peaks in this region of the spectrum, arising from Arg, Lys, Pro, and Leu, on the basis of their strong intensity and narrow shape in the methyl dimension as illustrated in *Figure 3*. Identification of the Ala and Thr fingerprint region cross-peaks relies on establishing an unambiguous connectivity between the H^N or H^α and the methyl resonances. The detailed procedure for identifying the Ala and Thr spin systems is described in *Protocol 3*.

Protocol 3. Alanine and threonine spin system identification

1. DQF-COSY (^2H_2O): count up intense cross-peaks in the Ala and Thr CH-CH_3 region of the spectrum; number of peaks should equal the sum of Ala and Thr residues.

2. RELAY (^2H_2O): look for RELAY peaks from H^α resonances to methyl resonances identified in COSY. These RELAY peaks correspond to Thr H^α-H^γ; H^α can be upfield or downfield of H^β depending on the type of secondary structure. The intensity of the RELAY peak will depend on the H^α-H^β coupling constant. If the number of RELAY peaks equals the number of Thr then the remaining COSY cross-peaks (step **1**) correspond to Ala. If not then remaining peaks could be Ala or Thr.

Protocol 3. *Continued*

3. DQF-COSY (^2H_2O): look for confirmatory cross-peaks between Thr H^α and H^β identified in RELAY and COSY.

4. COSY (H_2O): look for all possible H^N-H^α cross-peaks with correct Ala or Thr H^α chemical shift.

5. RELAY (H_2O): look for RELAY peaks from H^N to Ala H^β. Ala RELAY peaks can be distinguished from other H^N-H^β RELAY peaks by the narrow peak shape in the methyl dimension as illustrated in *Figure 4a*. The intensity of the RELAY peaks depends on the H^N-H^α coupling constant but peaks are usually observed for all Ala. If RELAY peaks for all Ala are observed then remaining methyl groups in COSY spectrum (step **1**) belong to Thr. Look for RELAY peaks from Thr H^N to H^β. These are most likely to arise from the same Thr spin systems that gave H^α-H^γ RELAYS.

6. Double-RELAY (H_2O): look for Thr H^N to H^γ double-RELAY peaks as illustrated in *Figure 4b*. The intensity of the peaks will depend on both the H^N-H^α and H^α-H^β coupling constants. Only spin systems which give H^N-H^β and H^α-H^γ RELAYs are likely to give double-RELAY peaks.

7. Spin systems not seen in steps **1–6** are probably characterized by small H^N-H^α or H^α-H^β coupling constants. Try to complete these spin systems by locating H^N-H^α and H^α-H^β COSY peaks with the correct H^α or H^β chemical shifts.

3.3 Valine, isoleucine, and leucine spin system identification

Val, Ile, and Leu are the remaining residues which give cross-peaks involving methyl resonances in COSY spectra. Inspection of *Figure 1* shows that the Val H^β-H^γ, Ile H^β-$H^{\gamma2}$, Ile $H^{\gamma1}$-H^δ, and Leu H^γ-H^δ cross-peaks occur in the same region of the COSY spectrum but are well separated from the Ala H^α-H^β and Thr H^β-H^γ cross-peaks described above. The Leu H^γ-H^δ peaks and the Val and Ile H^β-H^γ peaks are narrow in the methyl dimension and more extended in the methine direction as a result of the complex coupling pattern of the H^β or H^γ. The δ-methyl group of Ile is coupled to a pair of γ_1-protons and is a triplet in 1D spectra; the pair of $H^{\gamma1}$-H^δ cross-peaks are of lower intensity and wider in the methyl direction than the other peaks described above because of the cancellation of the central components of the cross-peak. Connectivities between the H^N and the γ-methyl resonances of Val and Ile are fairly straightforward to identify but connectivities between the H^N and the δ-methyl groups of Leu and Ile are far more difficult to establish. The procedure for identifying these spin systems is described in *Protocol 4*.

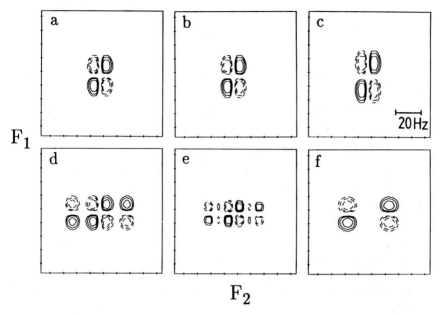

Figure 3. Simulated cross-peaks for some of the spin systems which give rise to peaks in the region $F_1 \simeq 6.0$–2.5 p.p.m. and $F_2 \simeq 2.2$–1.2 p.p.m. (a) Ala H^α-H^β. (b) Thr H^β-H^γ $^3J_{\alpha\beta}$ = 3.5 Hz. (c) Thr H^β-H^γ $^3J_{\alpha\beta}$ = 9.5 Hz. (d) Type-U residue H^α-H^β $^3J_{\alpha\beta}$ = 12.4 Hz, $^3J_{\beta\gamma}$ = 3.4, 12.4 Hz. (e) Type-U residues H^α-H^β $^3J_{\alpha\beta}$ = 3.4 Hz, $^3J_{\beta\gamma}$ = 3.4, 12.4 Hz. (f) Type-U residue H^α–H^β $^3J_{\alpha\beta}$ = 6.5 Hz, $^3J_{\beta\gamma}$ = 6.5, 6.5 Hz. The digital resolution before zero-filling was 11.77 Hz/pt in F_1 and 5.88 Hz/pt in F_2; after zero-filling the digital resolution was 2.94 Hz/pt in both dimensions.

Protocol 4. Valine, isoleucine, and leucine spin system identification

1. DQF-COSY (^2H_2O): count up pairs of peaks which correlate a common CH shift to two CH_3 shifts; the number of pairs of peaks should equal the sum of Val and Leu residues. Count up single peaks which correlate CH-CH_3; the number of peaks should equal the number of Ile residues. Count up pairs of peaks which correlate a common CH_3 shift with a pair of CH shifts; the number of pairs of peaks should equal the number of Ile residues.

2. RELAY (^2H_2O): look for pairs of RELAY peaks which correlate a common H^α shift with two methyl shifts; these peaks arise from Val. If the number of pairs of RELAY peaks equals the number of Val residues then the remaining pairs of CH-CH_3 COSY peaks (step **1**) arise from Leu. Look for single RELAY peaks which correlate an H^α and an Ile $H^{\gamma 2}$ shift. Look for RELAY peaks which correlate a pair of CH_3 chemical shifts; the pair of CH_3 resonances belongs to a Leu or Val residue. This RELAY

Protocol 4. *Continued*

    effect is useful if there are several Val, Leu, or Ile CH resonances with an overlapping chemical shift.

3. DQF-COSY (^2H_2O): look for H^α-H^β peaks to confirm Val and Ile assignments made in step **2**.

4. COSY (H_2O): look for all possible H^N-H^α peaks with correct Val or Ile H^α chemical shifts.

5. RELAY (H_2O): look for H^N-H^β RELAYs for Val and Ile residues identified in step **2**.

6. Double-RELAY (H_2O): look for pairs of H^N-H^γ peaks for Val, single H^N-$H^{\gamma2}$ peaks for Ile as illustrated in *Figure 4b*. Only Val and Ile residues which give H^N-H^β and H^α-H^γ RELAYs are likely to give double-RELAY peaks.

7. Val or Ile spin systems not seen in steps **1–6** are probably characterized by small H^N-H^α or H^α-H^β coupling constants. Try to complete spin systems by locating H^N-H^α and H^α-H^β COSY peaks with the correct H^α or H^β chemical shifts.

8. HOHAHA (^2H_2O): look for pairs of Leu H^α-H^δ peaks in spectra collected with long isotropic mixing times; these are sometimes seen for the smaller proteins. If these are not observed then the H^α-H^δ and H^N-H^δ connectivities are established later using NOE effects.

9. The connectivity between the Ile $H^{\gamma1}$-H^δ and the Ile H^N-H^α-H^β-$H^{\gamma2}$ subsystems can sometimes be made on the basis of $H^{\gamma1}$-$H^{\gamma2}$ and H^β-H^δ RELAY or HOHAHA peaks. If these are not seen then the connectivity can be established later using NOE effects.

3.4 Type-J spin system identification

The remaining amino acids can be divided into those with protons at the γ position and those without. The latter group includes Ser, Asp, Asn, Cys, Trp, Phe, Tyr, and His. These residues have H^β resonances with chemical shifts usually downfield of ~ 2.5 p.p.m. (*Figure 1*) and will be denoted as type-J.[a] These amino acids all have identical H^N-H^α-H^β spin subsystems because there is no resolved coupling between the β-protons and the protons beyond the γ position. In ^2H_2O the H^α and H^β protons of these eight residues form an AMX system in which there is resolved coupling between all protons. The H^α-H^β coupling constants depend on the torsion angle χ_1 whereas the geminal H^β-H^β coupling constants are fixed. The H^α-H^β coupling constants determine the cross-peak fine-structure observed for the H^α-H^β and H^β-H^β

[a] The letters J, U, and X are chosen to denote certain amino acid categories because these letters are not used in the one-letter amino acid code.

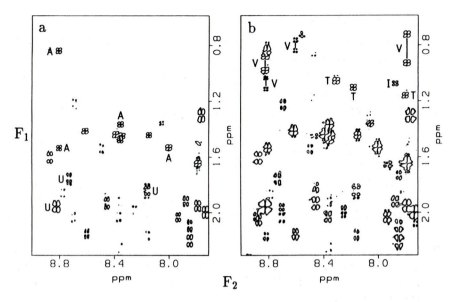

Figure 4. Part of the (a) RELAY and (b) double-RELAY spectra of human lysozyme in H$_2$O. RELAY peaks corresponding to some Ala and type-U residues are labelled in (a). Double-RELAY peaks corresponding to some valine, isoleucine, and threonine residues are labelled in (b).

cross-peaks of these residues as well as influencing the shape of the H^N-H$^\alpha$ cross-peak. An analysis of the various possible peak shapes can aid in the assignment process. Examples of the types of H$^\alpha$-H$^\beta$ and H$^\beta$-H$^\beta$ cross-peak patterns generally observed for these eight residues are shown in *Figure 5*. The procedure for identifying these spin systems is summarized in *Protocol 5*.

Protocol 5. Type-J spin system identification

1. DQF-COSY (^{2}H$_2$O): look for pairs of H$^\alpha$-H$^\beta$ cross-peaks with a common H$^\alpha$ shift and H$^\beta$ shifts downfield of ~ 2.5 p.p.m. The fine-structure of the cross-peaks should be similar to one of the examples shown in *Figure 5*. Confirm that the pair of H$^\alpha$-H$^\beta$ peaks arises from the same residue by locating the H$^\beta$-H$^\beta$ cross-peak; check that the peak fine-structure as well as chemical shift position is correct.

2. DQF-COSY (^{2}H$_2$O): look for single H$^\alpha$-H$^\beta$ cross-peaks with a H$^\beta$ shift downfield of ~ 2.5 p.p.m. If all Gly, Thr, Ile, and Val cross-peaks have been identified then these single peaks are likely to arise from one of the eight type-J residues. This peak is most likely to correspond to the strong cross-peak (large H$^\alpha$-H$^\beta$ coupling constant) in *Figure 5a*; the weak peak (small coupling constant) has disappeared into the noise.

Protocol 5. *Continued*

RELAY (^2H_2O): look for a strong H^α-H^β RELAY to the missing H^β position. DQF-COSY (^2H_2O): look for a H^β-H^β peak cross-peak of the correct fine-structure which lines up with the two H^β positions.

3. COSY, RELAY (H_2O): for each spin system identified in ^2H_2O identify all the possible H^N-H^α peaks with the correct H^α chemical shift. Try to identify the correct H^N-H^α peak using peak shape and RELAY effects.

  (a) If the spin system was characterized by one large and one small H^α-H^β coupling constant then the correct H^N-H^α cross-peak should have a rectangular shape. An H^N-H^β RELAY peak should be observed to the H^β resonance with the large H^α-H^β coupling constant. A double-RELAY peak may be observed to the H^β resonance with the small H^α-H^β coupling constant.

  (b) If the spin system was characterized by two small H^α-H^β coupling constants then the correct H^N-H^α cross-peak should have a square shape. Eliminate peaks which are rectangular. It is unlikely that strong H^N-H^β RELAY or HOHAHA peaks will be observed because of the small H^α-H^β coupling constants.

  (c) If the spin system was characterized by two intermediate coupling constants then the correct H^N-H^α cross-peak should have a rectangular shape. RELAY peaks may be observed to both H^β positions.

Some of the type-J residues have protons beyond the γ position. The resonances arising from the four aromatic type-J amino acids all fall in the spectral region downfield of the H_2O signal. Each aromatic amino acid gives a unique pattern of cross-peaks as shown in *Figure 6*. These aromatic spin systems can be identified using the methods outlined in *Protocol 6*.

Protocol 6. Aromatic spin system identification

1. DQF-COSY (^2H_2O): count up number of cross-peaks seen between 8.5 and 5.5 p.p.m. The total should equal one per His, one per Tyr, two per Phe, and three per Trp. If there are more than the expected number this may indicate slow flipping of one or more Tyr or Phe ring.

  (a) His: the weak cross-peaks between $H^{\delta2}$-$H^{\epsilon1}$ will shift significantly with pH close to the pK_α value. $H^{\epsilon1}$ is almost always downfield of $H^{\delta2}$.

  (b) Tyr: the intense square H^δ-H^ϵ cross-peaks should have the peak shape shown in *Figure 6b*. In the case of slow flipping a pair of peaks will be seen for each Tyr but it is not possible to determine which two peaks correspond to the same Tyr.

  (c) Phe: the H^δ-H^ϵ peak should be more intense than the H^ϵ-H^ζ peak; both should have the peak shapes shown in *Figure 6c*. Look for pairs of H^δ-H^ϵ and H^ϵ-H^ζ peaks that share a common H^ϵ position. In the case of slow flipping a total of four peaks will be observed ($H^{\delta 1}$-$H^{\epsilon 1}$, $H^{\epsilon 1}$-H^{ζ}, $H^{\epsilon 2}$-H^{ζ}, $H^{\delta 2}$-$H^{\epsilon 2}$).

(d) Trp: the $H^{\epsilon 3}$-$H^{\zeta 3}$ and $H^{\zeta 2}$-$H^{\eta 2}$ peaks should be more intense than the $H^{\zeta 3}$-$H^{\eta 2}$ peak; the three peaks should have the shapes shown in *Figure 6d*.

2. RELAY (^2H_2O): no additional peaks are observed for His or Tyr.
   (a) Phe: H^{δ}-H^{ζ} peak indicated by a square in *Figure 6c*. ($H^{\epsilon 1}$-$H^{\epsilon 2}$ and two H^{δ}-H^{ζ} peaks if ring is slowly flipping).
   (b) Trp: $H^{\epsilon 3}$-$H^{\eta 2}$ and $H^{\zeta 3}$-$H^{\zeta 2}$ peaks indicated by a square in *Figure 6d*.

3. Double-RELAY (^2H_2O): no additional peaks are observed for His, Tyr, or rapidly flipping Phe.
   (a) Phe: $H^{\delta 1}$-$H^{\epsilon 2}$ and $H^{\delta 2}$-$H^{\epsilon 1}$ peaks if ring is slowly flipping.
   (b) Trp: $H^{\epsilon 3}$-$H^{\zeta 2}$ peak indicated by a triangle in *Figure 6d*.

4. COSY (H_2O): a weak $H^{\epsilon 1}$-$H^{\delta 1}$ cross-peak is observed for Trp.

5. NOESY (H_2O): $H^{\epsilon 1}$ of Trp gives intraresidue NOE effects to $H^{\delta 1}$ (strong) and $H^{\zeta 2}$ (medium). These NOE effects are the only way to correlate the protons on the 5- and 6-membered rings of Trp.

The resonances beyond the γ position of type-J residues can only be correlated with the H^N-H^α-H^β subsystem on the basis of NOE effects because no resolved long-range coupling exists between the H^β and H^δ protons. The NOE effects that can be used to correlate the type-J side chain groups to the backbone spin systems are summarized in *Protocol 7*.

Protocol 7. NOE effects which correlate type-J side chain resonances to the H^N-H^α-H^β subsystem

1. Asn: COSY (H_2O): weak cross-peak between the pair of side chain amide ^1H resonances.
   NOESY (H_2O): very strong NOE peak between the pair of side chain amide ^1H resonances. Weaker NOE effects between the side chain amide resonances and the H^β resonances.

2. Ser: the hydroxyl proton is usually in fast exchange with H_2O and, therefore, not observed. However, if a separate OH resonance is observed then COSY and NOESY cross-peaks between the OH and H^β will be observed in H_2O.

3. Tyr and Phe: NOESY (H_2O): an NOE should always be observed between H^δ and H^β resonances. NOE effects between the H^N, H^α, and the H^δ-resonances may also be observed depending on the residue conformation (ϕ, χ_1, χ_2).

4. Trp: NOESY (H_2O): an NOE effect should always be observed between either $H^{\delta 1}$ or $H^{\epsilon 3}$ and the H^β resonances.

Protocol 7. *Continued*

5. His: NOESY (H$_2$O): an NOE effect is observed between the H$^{\delta 2}$ and H$^\beta$ resonances for some values of χ_2. The H$^\beta$ resonances should titrate with the same pK_a value as the ring protons, H$^{\delta 2}$ and H$^{\epsilon 1}$.

If NOE effects are observed between these side chain groups and the backbone protons then the spin system can be assigned specifically to one of the eight type-J residues. However, care must be taken not to confuse intra- and

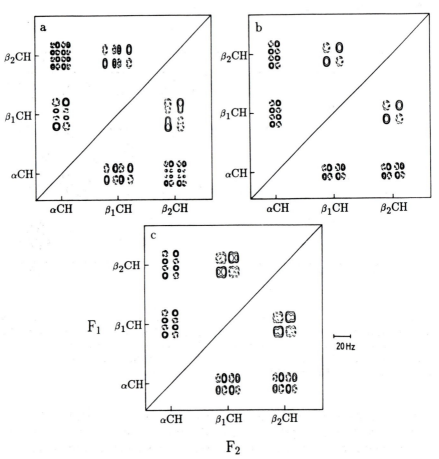

Figure 5. Simulated cross-peaks for the type-J H$^\alpha$-H$^\beta$ spin system. (a) $\chi_1 = -60°$ or $180°$, $^3J_{\alpha\beta1} = 12.4$ Hz, $^3J_{\alpha\beta2} = 3.4$ Hz, $^3J_{\beta1\beta2} = 16.0$ Hz. The H$^\alpha$-H$^{\beta2}$ peak will be of lower intensity than the H$^\alpha$-H$^{\beta1}$ because of the smaller active coupling constant. (b) $\chi_1 = 60°$, $^3J_{\alpha\beta1} = 3.4$ Hz, $^3J_{\alpha\beta2} = 3.4$ Hz, $^3J_{\beta1\beta2} = 16.0$ Hz. Both H$^\alpha$-H$^\beta$ peaks will be of low intensity because of the small active coupling constants. (c) χ_1 is averaged, $^3J_{\alpha\beta1} = 6.5$ Hz, $^3J_{\alpha\beta2} = 6.5$ Hz, $^3J_{\beta1\beta2} = 16.0$ Hz. The digital resolution before zero-filling was 10.56 Hz/pt in F$_1$ and 2.64 Hz/pt in F$_2$; after zero-filling the digital resolution was 1.32 Hz/pt in both dimensions.

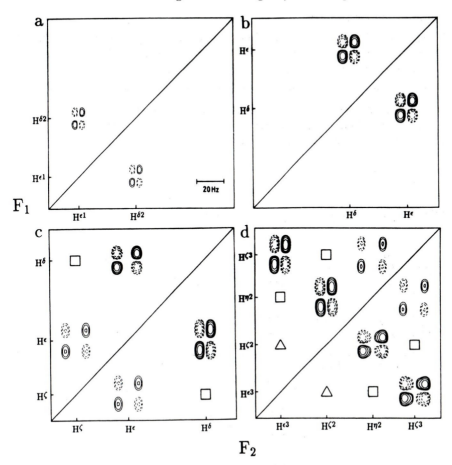

Figure 6. Simulated cross-peaks for the aromatic protons of (a) His, (b) Tyr, (c) Phe, and (d) Trp. The digital resolution before zero-filling was 10.56 Hz/pt in F_1 and 2.64 Hz/pt in F_2; after zero-filling the digital resolution was 1.32 Hz/pt in both dimensions. The *open squares* represent the positions of RELAY peaks and the *open triangles* represent the positions of double-RELAY peaks.

interresidue NOE effects. If such unambiguous NOE effects are not observed then the backbone spin system is assigned to the more general type-J category and the identification of side chain protons can be made after the sequential assignments are completed.

3.5 Type-U spin system identification

The final group of amino acid residues includes Lys, Arg, Met, Gln, Glu, and Pro. All of these residues have two protons at the γ position which are coupled to the H^β protons. These residues have H^β resonances with chemical shifts upfield of ~2.2 p.p.m. (*Figure 1*) and are designated type-U. The backbone protons of leucine residues for which H^N to H^δ or H^α to H^δ connectivities have not been established will also fall into this category. As a result of the additional coupling of the H^β protons the H^α-H^β cross-peak fine-structure is expected to be far more complicated than for the type-J peaks. However, because of large linewidths and limits in digital resolution the H^α-H^β peaks usually have only a 4- or 8-component fine-structure pattern which is elongated in the H^β dimension as illustrated in *Figure 3*. The H^α-H^β coupling constants for these residues depend on the torsion angle χ_1 in the same way that those of the type-J residues do and the shape of the H^N-H^α cross-peak will be influenced by these coupling constants. Residues within the type-U category can sometimes be further divided on the basis of the chemical shifts of the γ-protons as shown in *Figure 1*; Met, Gln, and Glu have H^γ resonances downfield of the H^β positions (type-U_1) whereas Arg, Lys, Pro, and Leu have H^γ resonances of the same chemical shift as or upfield of the H^β resonances (type-U_2). The procedure for identifying the type-U spin systems is summarized in *Protocol 8*.

Protocol 8. Type-U spin system identification

1. DQF-COSY (^2H_2O): look for H^α-H^β cross-peaks with H^β upfield of ~2.2 p.p.m.; peaks should be elongated along the H^β dimension. The side chains of these residues are often on the surface of the protein and consequently are more likely to undergo averaging about χ_1 and χ_2. Therefore, two strong H^α-H^β cross-peaks with coupling constants of ~6.5 Hz are often observed.

2. COSY (H_2O): for each spin system identified in ^2H_2O identify all possible H^N-H^α peaks with the correct H^α chemical shifts. Spin systems which give one or two strong H^α-H^β COSY peaks are unlikely to have an H^N-H^α peak with the square shape. Proline residues will correspond to H^α-H^β spin systems for which no H^N-H^α peak is identified. The H^α-H^β COSY cross-peaks of the proline residues should be identical in H_2O and ^2H_2O.

3. RELAY (H_2O): look for H^N-H^β RELAY peaks for each of the possible fingerprint region peaks. These RELAY peaks should be elongated in the H^β dimension compared to Ala H^N-H^β RELAY peaks as illustrated in *Figure 4*.

4. HOHAHA or RELAY (^2H_2O): look for pairs of H^α-H^γ cross-peaks where H^γ resonances are downfield of H^β; these cross-peaks are most likely to arise from the three type-U_1 residues, Glu, Gln, and Met. To distinguish between these residues look for the following effects in COSY and NOESY spectra.
   (a) Gln: δ-amide group gives a weak cross-peak in COSY (H_2O). The pair of δ-amide protons can be correlated to the rest of the spin system by identifying $H^{N\delta}$-H^γ NOE effects in NOESY(H_2O).

(b) Met: the ε-methyl group gives a strong singlet resonance. This resonance often gives NOE effects to the H^β and H^γ protons. However, be
careful not to confuse intra- and interresidue NOEs in proteins containing several Met residues.

5. HOHAHA or RELAY (^2H_2O): look for pairs of H^α-H^γ cross-peaks where
H^γ resonances are upfield of H^β; these cross-peaks are most likely to arise
from the four type-U_2 residues, Lys, Arg, Leu, and Pro. To distinguish
between these residues look for the following effects in COSY and
NOESY spectra.
(a) Arg: $H^{N\epsilon}$ gives cross-peaks to H^δ resonances in COSY(H_2O). RELAY
peaks to the γ-protons may be observed in H_2O. Correlate these H^γ
chemical shifts with those identified above from H^α-H^γ RELAY or
HOHAHA peaks.
(b) Lys: $H^{N\zeta}$ gives cross-peaks to H^ϵ resonances in COSY(H_2O). Look for
long-range $H^{N\zeta}$-H^δ, $H^{N\zeta}$-H^γ connectivities in HOHAHA spectra and
correlate these with the chemical shifts identified above.
(c) Pro: H^δ resonances give a strong COSY peak at ~3.6 p.p.m. Look
for H^γ-H^δ and H^β-H^δ peaks in COSY, RELAY, and HOHAHA
spectra and correlate these with H^α-H^β and H^α-H^γ peaks identified
above.

If resonances beyond the γ position can be identified then the spin system can
be assigned specifically to one of the seven type-U residues. Otherwise, the
backbone spin system is assigned to the more general type-U_1, type-U_2, or
type-U categories and the identification of side chain protons can be made
after the sequential assignments are completed.

3.6 Type-X cross-peak category

Once *Protocols 1–8* have been carried out a large number of spin systems
should have been identified. At this stage it should be possible to label each
peak in the fingerprint region with a one-letter amino acid identifier. For
cross-peaks corresponding to glycine, alanine or threonine this identifier will
be the specific one-letter amino acid code, G, A, or T. For other cross-peaks
corresponding to residues such as serine, arginine, or methionine this identifier will correspond to a more general group of amino acids such as type-J or
type-U. If all the spin systems have been identified then each fingerprint
region peak will be labelled. However, it is more likely that there will be some
H^N-H^α cross-peaks without an identifier at this point, and these peaks can be
labelled type-X. Most of these peaks will have the square shape characteristic
of small H^α-H^β coupling constants; it is these small J values which lead to the
absence of observable COSY and RELAY peaks. The type-X H^N-H^α peaks
can arise from any amino acid spin system. However, if all H^N-H^α peaks of a

particular type, for example, glycine, have been identified then this amino acid type can be excluded from the type-X category.

4. Stage 2: sequence-specific assignment

The second stage of assignment involves the assignment of an amino acid spin system identified in stage 1 to a specific residue in the protein sequence. This sequence-specific assignment is achieved by correlating one amino acid spin system with the spin systems of its neighbouring residues in the sequence. There is no resolvable spin-spin coupling between protons of adjacent residues and, therefore, COSY-type spectra can not be used to delineate the sequential connectivities. Instead, this stage of assignment relies on the short-range through-space connectivities observed in NOESY spectra.

In the early 1980s Wüthrich and co-workers showed that, for all sterically allowed values of ϕ, ψ, and χ_1 at least one of the distances between H^N, H^α, and H^β protons of adjacent residues was short enough to give rise to an observable NOE effect (1, 2, 4). This conclusion was based on the analysis of 19 high resolution protein crystal structures. The most useful NOE effects for sequential assignment were found to involve the H^α of residue i and the H^N of residue $i + 1$, $d_{\alpha N}$ $(i, i + 1)$, the H^N's of residues i and $i + 1$, d_{NN} $(i, i + 1)$, and the H^β of residue i and the H^N of residue $i + 1$, $d_{\beta N}$ $(i, i + 1)$. The intensities of these three NOE effects depend on the torsion angles ψ, ϕ and ψ, and χ_1 and ψ, respectively. The torsion angle dependence of the inter-residue distances means that specific types of secondary structure are characterized by specific sequential NOE effects. In the extended backbone structure, characteristic of β-sheet the distance $d_{\alpha N}$ $(i, i + 1)$ is very short, 2.2 Å, whereas the distance d_{NN} $(i, i + 1)$ is rather long, 4.3 Å. In helical secondary structure, on the other hand, the distance d_{NN} $(i, i + 1)$ is short, 2.8 Å, whereas the distance $d_{\alpha N}$ $(i, i + 1)$ is longer, 3.5 Å (4).

NOE effects between H^N, H^α, and H^β resonances are not, however, restricted to adjacent residues of the sequence. It will be shown in Chapters 10 and 11 that long-range NOE effects involving these resonances are used to identify regions of secondary structure in a qualitative manner. It is, therefore, important to assess the reliability of these NOE effects in making sequential assignments. In 1982, when this assignment method was proposed, the upper distance for the observation of an NOE effect was about 3.0–3.5 Å. Wüthrich and co-workers showed that 88% of $d_{\alpha N}$ NOE's, 76% of $d_{\beta N}$ NOE's, and 88% of d_{NN} NOE's with an interproton distance of ≤ 3.0 Å involve adjacent residues (1, 2, 4). With the improvement in spectrometer sensitivity achieved during the 1980s the upper limit for the observation of an NOE effect is now around 5.0 Å. At this greater distance a smaller percentage of the NOE effects observed will involve protons on adjacent residues. However, if only medium and strong intensity NOE effects are interpreted for assignment purposes then sequential assignments can be made with confidence. Wüthrich and co-workers also analysed the probabilities of observing two of the three NOE effects for a particular pair of residues. If both a $d_{\alpha N}$ and a d_{NN} NOE or a $d_{\alpha N}$ and a $d_{\beta N}$ NOE effect are observed between residues i and j then in $\sim 93\%$ of cases residues i and j are adjacent in the sequence if an upper limit of 3.8 Å is used for the observation of an NOE effect. Thus, the identification of two of the three NOE effects is a more reliable criterion for sequential assignment (1, 2, 4).

From the discussion above one would conclude that the second stage of assignment would simply involve the identification of sequential NOE effects beginning at the N-terminal residue and continuing to the C-terminal residue. In practice a full set of NOE connectivities from the N- to C-termini is usually not observed. Breaks in the sequential assignment occur for several reasons. First, a sequential d_{NN} ($i, i + 1$) NOE will not be resolved from the diagonal if the two H^N resonances involved have very similar H^N chemical shifts. Secondly, a break may occur if the COSY H^N-H^α cross-peak of a residue does not appear for one of the reasons listed in *Protocol 1*. Thirdly, a break in the sequential assignment can occur at each proline residue. The H^α and H^δ resonances of proline are often involved in the same sequential NOE's as the H^α and H^N resonances of other amino acids (4). However, continuity in the sequential assignment will only be possible if the full proline spin system has been identified in stage 1, that is if the proline H^α and H^δ resonances have been correlated. For these reasons the sequential assignment process is carried out in shorter peptide segments within the protein sequence.

If a strong $d_{\alpha N}$ NOE effect is observed between two amino acid spin systems then these residues are likely to be adjacent in the sequence. If the resonances of the spin systems have been assigned to a specific amino acid type, such as Gly and Ala, then sequential assignment may be possible from this single NOE effect because the pair G-A may occur only once in the sequence. However, as the size of the protein increases the occurrence of unique pairs of residues in the sequence decreases. Thus, it may be necessary to identify the residue adjacent to the glycine or alanine before a specific assignment can be made. If the resonances of the spin systems have been assigned to a more general class of amino acids such as type-J or type-U or if no information about the spin system is known, type-X, then sequential assignment of the basis of the single NOE is very unlikely because many pairs such as J-U or X-J are likely to be found in the sequence. Additional NOE's to the adjacent residues will always be needed before an assignment can be made. In larger proteins, where many fingerprint region cross-peaks have been identified as type-J, U, or X, the assignment strategy must rely, initially, on the fingerprint region cross-peaks which have been assigned a specific amino acid type such as glycine or alanine. The procedure for carrying out the second stage of assignment for proteins of this kind follows.

The only part of the NOESY spectrum which needs to be considered for sequential assignment is the section with H^N chemical shifts in F_2 and the entire chemical shift range in F_1. The first step in the analysis involves the identification of NOE effects involving aromatic resonances and of intra-residue NOE effects in the NOESY spectrum. The procedure for identifying these NOE effects is listed in *Protocol 9*.

Protocol 9. Identification of aromatic NOE effects and intraresidue NOE effects involving H^N's

1. Overlay the NOESY spectrum of fully exchanged protein in ^2H_2O and the spectrum of the protein in H_2O. The peaks between 10.5 and 6.5 p.p.m. along F_2 which occur in both spectra involve aromatic protons rather than H^N's. These peaks should be circled and labelled in the H_2O spectrum. They are not used in sequential assignment.

2. Identify intraresidue NOE effects which involve side chain H^N's identified in *Protocols 7* and *8*. These NOE effects are not used in sequential assignment.

3. Overlay the fingerprint region of the COSY and NOESY spectra in H_2O. For each H^N-H^α cross-peak in the COSY label the corresponding $d_{\alpha N}$ (i, i) cross-peak in the NOESY. The intraresidue H^N-H^α distance is always less than ~3.5 Å so an NOE should be observed for each residue. If a large number of peaks are missing then repeat the NOESY experiment with a longer mixing time.

4. For each residue label peaks in the NOESY spectrum which correspond to intraresidue H^N to H^β and other side chain protons of the spin system.

5. Look for sets of NOE effects at frequencies corresponding to any H^N chemical shift which does not give rise to a fingerprint region COSY peak. These NOE effects may arise from amides for which H^N-H^α cross-peaks are missing due to a small coupling constant or a large linewidth.

The unlabelled NOE peaks which remain in the H^N-H^N and H^α-H^N regions of the NOESY spectrum after this protocol has been followed will correspond to interresidue NOE effects. Further analysis of the NOESY spectrum should not require overlaying of the NOESY and COSY spectra

---

Figure 7. Schematic representation of the sequential NOESY connectivities described in *Protocol 10*. Symbols in the *left-hand boxes* represent d_{NN} NOE peaks, those in the *centre boxes* represent $d_{\alpha N}$ NOE peaks, and those in the *right-hand boxes* represent $d_{\beta N}$ NOE peaks. The *filled circles* and *filled squares* represent intra- and interresidue NOE effects, respectively. The size of the circle or square represents the intensity of the NOE effect. (a) A strong $d_{\alpha N}$ ($i - 1$, i) NOE peak and a strong $d_{\alpha N}$ (i, $i + 1$) NOE peak are observed (steps **2–4**). (b) A medium d_{NN} ($i - 1$, i) NOE peak and a medium d_{NN} (i, $i + 1$) NOE peak are observed (steps **5–6**). (c) A strong $d_{\alpha N}$ ($i - 1$, i) NOE peak and a medium d_{NN} (i, $i + 1$) NOE peak are observed (steps **2, 7,** and **8**). (d) A medium d_{NN} ($i - 1$, i) NOE peak and a strong $d_{\alpha N}$ (i, $i + 1$) NOE peak are observed (steps **3, 9,** and **10**).

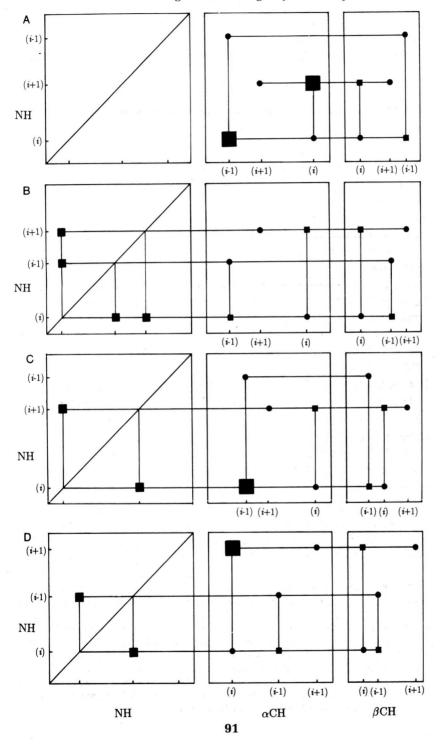

because the intraresidue NOE effects have now been labelled in the NOESY spectrum.

The next stage in the analysis involves the identification of the sequential NOE effects between residue i and its adjacent residues $i - 1$ and $i + 1$. Depending on the nature of the secondary structure present these adjacent residues can be identified on the basis of $d_{\alpha N}$ or d_{NN} NOE connectivities. The procedure for identification of adjacent residues is illustrated in *Figure 7* and described in detail in *Protocol 10*.

Protocol 10. Identification of adjacent residues from d_{NN}, $d_{\alpha N}$, and $d_{\beta N}$ NOE effects

1. Start with the most downfield H^N in the spectrum and locate its $d_{\alpha N}$ (i, i) peak in the NOESY spectrum.

2. Look along the H^N (i) shift for a strong $d_{\alpha N}$ $(i - 1, i)$ interresidue NOE peak. If a peak is found then look along the H^α shift of this peak $(i - 1)$ for an intraresidue $d_{\alpha N}$ $(i - 1, i - 1)$ NOE.

3. Look along the H^α (i) shift for a strong $d_{\alpha N}$ $(i, i + 1)$ interresidue NOE peak. If a peak is found then look along the H^N shift of this peak $(i + 1)$ for an intraresidue $d_{\alpha N}$ $(i + 1, i + 1)$ NOE.

4. Confirm that the H^N-H^α peaks of residues $i - 1$, i, and $i + 1$, identified in steps **2** and **3**, arise from adjacent residues by identifying $d_{\beta N}$ $(i - 1, i)$ and $d_{\beta N}$ $(i, i + 1)$ NOE effects if possible.

5. If a strong $d_{\alpha N}$ NOE effect other than $d_{\alpha N}$ (i, i) is not observed for this H^N then look along the H^N (i) shift for two medium intensity d_{NN} NOE peaks. If two peaks are observed then these are likely to correspond to residues $i - 1$ and $i + 1$. Look along the H^N $(i - 1)$ and H^N $(i + 1)$ shifts for weak intraresidue $d_{\alpha N}$ NOE peaks. In order to distinguish the H^N-H^α peak of residue $i - 1$ from that of $i + 1$ look along the H^N (i) shift for a weak $d_{\alpha N}$ $(i - 1, i)$ interresidue NOE peak to one of the two possible H^α shifts. The H^α shift observed corresponds to residue $i - 1$. Look along the remaining H^N $(i + 1)$ shift for a weak $d_{\alpha N}$ $(i, i + 1)$ interresidue NOE effect to the H^α of residue i

6. Confirm that the H^N-H^α peaks $i - 1$, i, and $i + 1$, identified in step **5**, arise from adjacent residues by identifying $d_{\beta N}$ $(i - 1, i)$ and $d_{\beta N}$ $(i, i + 1)$ NOE effects.

7. If a strong $d_{\alpha N}$ $(i - 1, i)$ NOE effect was observed in step **2** but no strong $d_{\alpha N}$ $(i, i + 1)$ effect was observed in step **3** then look along the H^N (i) shift for a d_{NN} $(i, i + 1)$ NOE peak. Look along the H^N $(i + 1)$ shift for a weak intraresidue $d_{\alpha N}$ $(i + 1, i + 1)$ NOE peak and a weak interresidue $d_{\alpha N}$ $(i, i + 1)$ NOE peak to the H^α of residue i.

8. Confirm that the H^N-H^α peaks $i - 1$, i, and $i + 1$, identified in steps **2** and **7**, arise from adjacent residues by identifying $d_{\beta N}$ $(i - 1, i)$ and $d_{\beta N}$ $(i, i + 1)$ NOE effects.

9. If a strong $d_{\alpha N}$ $(i, i + 1)$ NOE was observed in step **3** but no strong $d_{\alpha N}$ $(i - 1, i)$ effect was observed in step **2** then look along the H^N (i) shift for a d_{NN} $(i - 1, i)$ NOE peak. Look along the H^N $(i - 1)$ shift for a weak intraresidue $d_{\alpha N}$ $(i - 1, i - 1)$ NOE peak. Look along H^N (i) for a weak interresidue $d_{\alpha N}$ $(i - 1, i)$ NOE peak to the H^α of residue $i - 1$.

10. Confirm that the H^N-H^α peaks $i - 1$, i, and $i + 1$, identified in steps **3** and **9**, arise from adjacent residues by identifying $d_{\beta N}$ $(i - 1, i)$ and $d_{\beta N}$ $(i, i + 1)$ NOE effects.

11. Repeat steps **2** to **10** for the next intraresidue $d_{\alpha N}$ (i, i) peak in the NOESY spectrum.

12. Make a list of each H^N-H^α COSY peak and the H^N-H^α COSY peaks which correspond to the two adjacent residues. Also record the amino acid identifier for each H^N-H^α peak.

A complete list of adjacent NOE effects is fairly straightforward to compile if there is no overlap in the H^N and H^α chemical shifts. However, as the size of the protein increases overlap of H^N and H^α chemical shifts becomes increasingly likely. When overlap occurs the list of adjacent NOE effects for residue i may contain several possibilities for residues $i - 1$ and $i + 1$ instead of single values. There are several ways in which to overcome the problem of overlap. The overlap in H^N chemical shifts can be overcome by collecting the COSY and NOESY spectra in H_2O at 10°C above or below the original temperature; the H^N resonances will shift relative to each other as a result of differing temperature coefficients. This set of COSY and NOESY spectra can then be analysed using *Protocols 9* and *10*. The resulting lists of adjacent residues obtained at two different temperatures can be compared in order to resolve ambiguities. A particularly powerful method for reducing the problem of overlap involves reducing the number of peaks in the spectrum on the basis of differences in hydrogen exchange rates. This method can be useful if samples of the protein can be prepared which contain different subsets of amide hydrogens. This procedure for simplification of COSY and NOESY spectra is described in *Protocol 11*.

Protocol 11. Exploiting hydrogen exchange rates for overcoming overlap

1. Prepare two samples containing subsets of amides as follows.
   (a) Dissolve protein in ^2H_2O. This sample will contain only the slowly exchanging amides and is called partly exchanged.

Protocol 11. *Continued*

  (b) Dissolve protein in ^2H_2O and remove all amide protons by heating sample to denaturation temperature or by raising or lowering the pH, and incubating for an appropriate time. Freeze-dry the sample and redissolve in H_2O. This sample will contain only the rapidly exchanging amides and is called reverse exchanged.

2. Collect a COSY and NOESY spectrum for each sample.

3. Compare each COSY subspectrum with the COSY spectrum in H_2O containing all H^N-H^α peaks. Label the COSY peaks appearing in each COSY subspectrum.

4. Identify intraresidue NOE effects appearing in each NOESY subspectrum as described in *Protocol 9*.

5. Identify adjacent spin systems in each NOESY subspectrum as described in *Protocol 10*. It is important to keep in mind the following rules. A $d_{\alpha N}$ (i, $i + 1$) or $d_{\beta N}$ (i, $i + 1$) NOE effect will only appear in a NOESY subspectrum if the H^N-H^α cross-peak of residue $i + 1$ appears in the COSY subspectrum. A d_{NN} (i, $i + 1$) NOE effect will only appear in a NOESY subspectrum if the H^N-H^α cross-peaks of *both* residues i and $i + 1$ are present in the COSY subspectrum.

6. Make a list of the adjacent spin systems for each of the two hydrogen exchange subspectra and use these to resolve ambiguities in the complete list.

Once *Protocols 9* and *10* have been carried out and procedures for overcoming overlap problems have been applied a list of adjacent spin systems is available. This list is now used as the basis for making sequence-specific assignments. If the spin system identification was carried out as described above then each fingerprint region peak will have an identifier such as G, A, T, J, U, or X. If very complete spin system identification was achieved then specific identifiers such as Y, N, R, and Q will also be found. The spin systems which have been assigned to one specific type of amino acid residue are now used as the basis for assignment because short sequences containing these residues are more likely to yield unique assignments. The procedure used for sequence-specific assignment is described in *Protocol 12*

Protocol 12. Sequence-specific assignment

1. Choose an amino acid type which occurs frequently in the sequence as the starting point.

2. From the adjacent residue list (*Protocol 10*) obtain the triplets of sequential residues ($i - 1$, i, $i + 1$) which have the chosen amino acid type at their centre (i). Compare the amino acid identifiers of these triplets with the known sequence of the protein and look for assignments for the triplet which are consistent with the sequence. If a triplet occurs only once in the sequence then a unique assignment can be made.

3. Once a triplet has been assigned look for spin systems adjacent to residues $i - 1$ and $i + 1$, that is, obtain from the adjacent residue list triplets with peaks $i - 1$ and $i + 1$ at their centre. Extend the segment in both directions until no further sequential connectivities are found. Check that the spin system identifiers of these residues are consistent with the protein sequence.

4. If more than one possible assignment exists for a particular triplet then the residues adjacent to $i - 1$ and $i + 1$ should be used. Compare the amino acid identifiers of the quintet of sequential residues ($i - 2, i - 1, i, i + 1, i + 2$) with the known sequence of the protein and look for assignments which would be consistent. If a quintet occurs only once in the sequence then a unique assignment can be made.

5. Continue to extend segments at both ends (step **4**) until a unique assignment is found for each residue selected in step **1**.

6. Extend assigned segments in both directions until a break is reached (step **3**) and check that the spin system identifiers are consistent with the protein sequence.

7. Choose a second amino acid type as the starting point and repeat steps **2–6**.

8. When all the cross-peaks with specific amino acid identifiers have been assigned then only fairly short segments in the sequence will remain unassigned. At this point assignments can be made unambiguously on the basis of the J, U, and X classifications alone.

An example of the application of *Protocol 12* to the assignment of the spectrum of human lysozyme is shown in *Figure 8*. Human lysozyme contains 14 alanine residues scattered throughout the sequence; alanine is chosen as the starting point for sequential assignment. The steps leading to the assignment of two of the triplets containing Ala are summarized in *Figure 8a*. One of these triplets has the sequence U-A-T. Inspection of the human lysozyme sequence shown in *Figure 8b* indicates that this triplet occurs only once in the sequence and, therefore, the cross-peaks corresponding to U-A-T can be assigned to R41-A42-T43. Assignments for the nine residues between N39 and A47 result from the U-A-T triplet. The second triplet shown in *Figure 8a* has the sequence X-A-U. This sequence is not unique and can be assigned to four different groups of residues in human lysozyme. Extension of the triplet to include residue $i + 2$ leads to a unique assignment of residues L31-A32-K33-W34. Assignments for the eight residues between L31 and Y38 result from the X-A-U triplet. The peptide segments identified in the complete sequential assignment of human lysozyme are summarized in *Figure 8b*.

By following the procedure outlined in *Protocol 12* it should be possible to

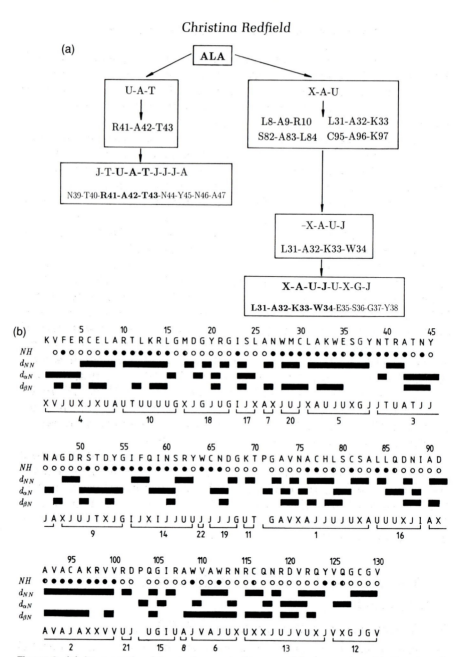

Figure 8. (a) An example of the assignment of two peptide segments using the steps in *Protocol 12*. (b) Amino acid sequence, summary of hydrogen exchange rates, and summary of the NOE connectivities used in the sequential assignment of human lysozyme. *Filled, half-filled,* and *empty circles* indicate amides with slow, intermediate, and rapid exchange rates, respectively. Peptide segments identified in the assignment procedure and the amino acid identifiers of the residues in these segments are shown. The *numbers* beneath these segments indicate the order in which they were assigned.

obtain sequence-specific assignments for the majority of spin systems identi-fied in stage 1. The assignments obtained should be entirely self-consistent. If there are any discrepancies in the data then the assignments should be reviewed carefully for possible errors. Once all the spin systems have been assigned it is possible to go back and determine whether the breaks that occur between the sequential segments can be justified. For example, do breaks occur at proline spin systems? If so can the appropriate $d_{\alpha\delta}$ and $d_{\delta N}$ connecti-vities be identified now that the spin system is known to belong to proline? Do breaks in the sequential assignment occur because the difference in the H^N chemical shifts of the adjacent residues is too small for a d_{NN} NOE to be resolved from the diagonal? A good way to check that the assignments are correct is to identify all the sequential connectivities in a NOESY spectrum collected under different conditions of temperature or pH.

Once the fingerprint region H^N-H^α COSY peaks of the type-J, U, and X residues have been assigned to specific amino acids it is possible to look for intraresidue and sequential NOE effects and RELAY, double-RELAY, and HOHAHA peaks that may lead to the identification of side chain resonances not previously recognized. For example, once a cross-peak has been assigned specifically to Met instead of the more general type-U category it is possible to search the NOESY spectrum for H^β-H^ε and H^γ-H^ε NOE effects and the HOHAHA spectra for H^N-H^γ connectivities. The H^γ-H^δ subsystem of Leu residues identified as type-U can be correlated to the backbone H^N-H^α-H^β subsystem by searching the NOESY spectrum for a strong NOE effect be-tween Leu H^α and either $H^{\delta 1}$ or $H^{\delta 2}$. In this way a more complete level of resonance assignment can be achieved for the protein.

This procedure for spin system identification and sequential assignment has been used for the assignment of the spectra of spinach plastocyanin, hen and human lysozymes. These are three of the largest proteins for which a high level of assignment has been achieved using only ^{1}H two-dimensional tech-niques. Refer to the corresponding assignment papers for more examples of the procedures used (6–8).

5. Alternatives to the sequential assignment approach

5.1 Mainchain directed approach (MCD)

The mainchain directed approach, which is an alternative to the sequential analysis of NOESY data, was developed by Englander and Wand in 1987 (9). This strategy does not rely on linear NOE connectivity patterns but instead relies on cyclic patterns which are characteristic of particular units of second-ary structure. The cyclic MCD pattern suggested for the identification of residues in helices involves the identification of three NOE effects: d_{NN} $(i, i + 1)$, $d_{\beta N}$ (i, i), and $d_{\beta N}$ $(i, i + 1)$. These NOE's are all intraresidue or sequential and would be identified in exactly the same manner in the sequential assignment approach outlined above. The cyclic MCD patterns suggested for the antiparallel and parallel β-sheet structures differ substantially from the sequential assignment patterns because they involve residues on adjacent strands of the sheet which are consequently separated in the sequence. The two MCD patterns for the antiparallel β-sheet are designated as the inner and outer loops. The inner loop involves five NOE effects: $d_{\alpha N}$ $(i, i + 1)$, $d_{\alpha N}$ $(i, j + 1)$, $d_{\alpha N}$ $(j, j + 1)$, $d_{\alpha N}$ $(j, i + 1)$, and $d_{\alpha \alpha}$ (i, j) where i and j are residues on opposite strands. The outer loop involves two COSY peaks (or intraresidue NOE's) and three NOE effects: H^N (i)-H^α (i), H^N $(i - 1)$-H^α $(i - 1)$, $d_{\alpha N}$ $(i - 1, i)$, $d_{\alpha N}$ $(i, j + 1)$, and d_{NN} $(j + 1, i - 1)$. The MCD pattern for parallel β-sheet involves two COSY peaks (or intraresidue NOE's) and four NOE effects: H^N (i)-H^α (i), H^N $(i + 1)$-H^α $(i + 1)$, $d_{\alpha N}$ $(i, i + 1)$, $d_{\alpha N}$ $(i + 1, j + 1)$, $d_{\alpha N}$ $(j, j + 1)$, and $d_{\alpha N}$ (j, i). Once the major elements of secondary structure have been identified from the MCD patterns specific assignments are made on the basis of spin system information available for easily identified residues such as glycine and alanine.

An advantage of the MCD approach is that the cyclic nature of the NOE patterns can be used to overcome problems of ambiguity and can provide a check for internal consistency at each step. A disadvantage of the approach is that it relies on the presence of regular secondary structure and is, therefore, unlikely to work for proteins with large loops or other less regular structures.

5.2 Use of X-ray diffraction data

In many studies the sequential assignment of the protein spectrum is the first-step in the determination of the three-dimensional structure of the protein in solution as discussed in Chapters 10 and 11. However, in other studies resonance assignments are required in order to probe the interaction of the protein with substrates and inhibitors or to study the folding and stability of the protein using hydrogen exchange methods. A high resolution crystal structure is sometimes available for the protein of interest and this can be used to aid in assignment if there is sufficient evidence that the solution and X-ray structures are very similar.

The X-ray structure can be used to generate a list of all interproton distances within a cut-off of ~ 5.0 Å. This list can then be used to predict whether the sequential connectivity between any two residues i and $i + 1$ will be a strong $d_{\alpha N}$ $(i, i + 1)$ NOE or a medium d_{NN} $(i, i + 1)$ NOE. This information will simplify the analysis of the NOESY spectrum to some extent but it does not provide any firm assignments to use as starting points for the analysis.

The list of short interproton distances will also contain pairs of protons from residues which are separated in the sequence. Many of these long-range NOE effects will involve aromatic and methyl protons. Analysis of the

NOESY spectrum in ^2H_2O in conjunction with this list of predicted long-range NOE effects can lead to a significant number of assignments for aromatic and methyl resonances (10). If these residues are located in or near the active site of the protein then they can be used as probes of the interaction of the protein with substrates and inhibitors. If a more complete level of assignment is desired then the aromatic and methyl assignments can be used as starting points in the sequential analysis if a connectivity to the backbone H^N or H^α can be established.

The use of information derived from an X-ray structure can lead to incorrect assignments if significant differences exist between the local or global solution and X-ray structures; this possibility must always be considered when using this approach. However, solution NMR studies of proteins such as lysozyme, cytochrome c, basic pancreatic trypsin inhibitor, and dihydrofolate reductase have shown that the structures of these proteins in solution closely resemble the structures obtained by X-ray diffraction. For example, assignments for a large number of resonances of human lysozyme obtained using an analysis of NOE data based on the X-ray structure have subsequently been confirmed in the 2D sequential assignment study of this protein (8, 10).

References

1. Wüthrich, K., Wider, G., Wagner, G., and Braun, W. (1982). *J. Mol. Biol.*, **155**, 311.
2. Billeter, M., Braun, W., and Wüthrich, K. (1982). *J. Mol. Biol.*, **155**, 321.
3. Bundi, A. and Wüthrich, K. (1979). *Biopolymers*, **18**, 285.
4. Wüthrich, K. (1986). *NMR of proteins and nucleic acids*. Wiley, New York.
5. Redfield, C. (1990). In *Frontiers of NMR in molecular biology*, (ed. D. Live, I. M. Armitage, and D. Patel), pp. 155–66. Wiley-Liss, New York.
6. Driscoll, P. C., Hill, H. A. O., and Redfield, C. (1987). *Eur. J. Biochem.*, **170**, 279.
7. Redfield, C. and Dobson, C. M. (1988). *Biochemistry*, **27**, 122.
8. Redfield, C. and Dobson, C. M. (1990). *Biochemistry*, **29**, 7201.
9. Englander, S. W. and Wand, A. J. (1987). *Biochemistry*, **26**, 5953.
10. Boyd, J., Dobson, C. M., and Redfield, C. (1985). *Eur. J. Biochem.*, **153**, 383.

5

Stable isotope labelling and resonance assignments in larger proteins

JOHN L. MARKLEY and MASATSUNE KAINOSHO

1. Introduction

This chapter introduces the general principles of isotope-assisted NMR spectroscopy of proteins and presents descriptions of several specific procedures. The rationale and methodology of protein labelling are discussed, and NMR experiments that exploit various patterns of labelling are surveyed.

NMR spectroscopy is widely used in protein structure–function investigations. Recent developments in NMR methodology (1) have made it possible to determine full structures of small proteins in solution at levels of resolution comparable to those of moderately-refined single crystal X-ray structures (2) (see Chapters 10 and 11). NMR spectroscopy can provide a wealth of additional data about proteins in solution. Such data, which are complementary to a molecular structure, include the kinetics and thermodynamics of protein folding or protein–ligand interactions (see Chapter 7), enzyme mechanisms, and the physical properties of individual nuclei or protein functional groups. With small proteins (M_r under 10000) much of this information can be gained from NMR studies of proteins at their natural isotopic abundance. With larger proteins, however, it is necessary to employ stable isotopic labelling in order to resolve and assign the signals of interest. [For reviews see references (3–11).] Even with smaller proteins, isotopic labelling is advantageous since additional information can be extracted from the chemical shift, spin coupling, and relaxation behaviour of the labelled nuclei. Stable isotope labelling is not necessarily expensive. Provided that the protein of interest can be produced in good yield from simple labelled precursors, the cost may be surprisingly low. The expense is often easily justified by simplification of the spectrum and its interpretation or by the extra information provided.

2. Labelling strategies

2.1 Isotopes of interest

The atoms whose isotopic compositions are most often altered for isotope-assisted NMR studies are hydrogen ($^1H/^2H$), carbon ($^{12}C/^{13}C$), and nitrogen ($^{14}N/^{15}N$). ^1H, ^{13}C, and ^{15}N are spin = 1/2 nuclei whose signals are observable in protein spectra. The ^{12}C nucleus has spin = 0 and is NMR inactive; ^2H and ^{14}N are quadrupolar nuclei (spin > 1/2), whose rapidly relaxing signals normally are not observed in solution state protein NMR experiments; these three isotopes can be used to block out the observable signals from their companion spin 1/2 nuclei. The properties of these and additional nuclei of potential utility for protein labelling investigations are listed in the Introduction.

Through-bond interactions in NMR occur by spin-spin coupling. The size of the coupling constant and the relaxation properties of the nuclei determine whether a particular coupling interaction can be exploited in multinuclear experiments of proteins. *Table 1* summarizes the magnitudes of coupling constants found in peptides and proteins. It is seen that the largest couplings (the ones that will be applicable, for example, to proteins larger than about M_r 30000) are the one bond ^1H-^{15}N, ^1H-^{13}C, and ^{13}C-^{13}C splittings. The smaller single bond ^{13}C-^{15}N coupling may prove to be a major limitation in

Table 1. Magnitudes of NMR spin-spin coupling in peptides and proteins

Atoms coupled	Coupling constant range (Hz)[a]
$^2J_{HH}$	12–18
$^3J_{HH}$	0–10
$^1J_{NH}$	85–95
$^2J_{NH}$	0–4
$^3J_{NH}$	0–10
$^1J_{CH}$	
Aliphatic	120–130
Aromatic	170–220
$^2J_{CH}$	0–8
$^3J_{CH}$	0–14
$^1J_{CC}$	
Aliphatic	35–50
Aromatic	60–70
$^{13}C^{\alpha}$-$^{13}C'$	50–60
$^1J_{C'N}$	12–15
$^1J_{C\alpha N}$	4–6

[a] Absolute values taken from: Bystrov, V. F. (1976). *Prog. NMR Spectrosc.*, **10**, 41.

multidimensional experiments with proteins larger than M_r 30 000; this coupling has been used, however, in one-dimensional NMR studies of a protein as large as M_r 150 000.

2.2 Applications of protein labelling

Stable isotope labelling is employed in protein NMR studies in several useful ways:

(a) Spectral simplification. In the additive strategy, one introduces nuclei that normally are at low abundance. For example, since the natural abundance of ^{15}N is low (0.37%), when a protein is labelled with this nucleus ^{15}N NMR signals are observed only from nuclei that have been inserted. Alternatively, one can use a subtractive strategy. For example, one can label a protein completely with ^{2}H, then bind a small ligand at natural ^{1}H abundance (99.98%); the ^{1}H NMR spectra of this protein–ligand complex will exhibit resonances only from the ligand.

(b) Spectral assignment. Depending on the labelling pattern used, selective labelling can provide:

- identifications of residue type (e.g. His rather than some other residue)

- assignments to the residue position in the protein sequence (e.g. Tyr31)

- atom position within an amino acid residue (e.g. the $H^{\beta 3}$ of Ser).

(c) Provision of coherence pathways. Stable isotope labelling provides coherence pathways for multi-dimensional experiments that can either be selected for or filtered out. Whenever possible, it is advantageous to observe less sensitive nuclei (such as ^{13}C and ^{15}N) indirectly through coherences with ^{1}H (the most sensitive stable isotope).

(d) Alteration of cross-relaxation pathways. Labelling can be employed to select for particular NOEs (nuclear Overhauser enhancements), which provide structural information, or to block (or minimize) unwanted spin diffusion pathways, which degrade the quality of structural information.

(e) Observation of the NMR spectral properties of less abundant, less sensitive nuclei. Labelling makes it easier to obtain information about carbon and nitrogen sites in proteins.

(f) Isotope exchange reactions. Hydrogen exchange can be followed by ^{1}H NMR since the proton signal intensity increases or diminishes, respectively, during deuterium out- or in-exchange. If the protein is labelled with ^{15}N, one can increase the resolution of N-H exchange experiments by collecting 2D ^{1}H-^{15}N data so that the exchanging peaks are spread out by the chemical shifts of both the proton ^{1}H^N and directly bound ^{15}N. Hydrogen exchange also can be followed indirectly by detecting isotope effects on the chemical shifts of other nuclei (e.g. ^{13}C or ^{15}N). For investigations of protein functional groups that participate in chemical reactions, labelling with ^{2}H, ^{13}C, or ^{17}O may provide kinetic or mechanistic information.

2.3 Labelling patterns

An almost limitless number of labelling patterns is achievable with proteins. These can be described systematically as follows.

(a) Complete labelling. All atoms of a given type within the protein (e.g. ^{13}C) are fully labelled to the practical limit with a particular isotope (e.g. 99% U-^{13}C). More than one atom type can be labelled; an example of dual complete labelling would be a protein labelled with 98% U-^{15}N and 96% U-^{13}C. Complete deuteration is termed 'perdeuteration'.

(b) Amino acid-selective complete labelling. Complete labelling can be restricted to one or more kinds of amino acid residue.

(c) Stochastic labelling. All atoms of a given type within the protein (e.g. ^{13}C) are labelled randomly (stochastically) to a given fractional level of isotope (e.g. 25% U-^{13}C). Stochastic deuteration has been termed 'random fractional deuteration' (e.g. 50% U-^{2}H).

(d) Amino acid-selective stochastic labelling. Stochastic labelling also can be restricted to one or more kinds of amino acid (e.g. a protein whose alanine residues contain 45% U-^{2}H).

(e) Single site labelling. One atom within an amino acid is labelled with a particular nucleus, (e.g. histidine is labelled at $N^{\epsilon 2}$ with 99% ^{15}N). In a protein, single site labelling may be unrestricted (all histidines labelled in this way) or residue selective (only the histidine at a given position is labelled). The labelling may convert a prochiral group into a chiral one, (e.g. X-CH$_2$-Y into X-CH^{2}H-Y).

(f) Block labelling. Two or more atoms within an amino acid are labelled with a particular nucleus (or nuclei). [Multiple-atom, intraresidue, single site labelling is used frequently enough to warrant a special name for this labelling pattern.] Block labelling may convert prochiral sites to chiral ones.

(g) Metabolic labelling. This term can be used, along with specification of the labelled feedstock and growth conditions, to describe unusual protein labelling patterns that result from feeding labelled metabolites to organisms. For example, feeding [^{13}C$^\alpha$]glycine to *E. coli* generally leads to incorporation of the label into the protein at glycines (^{13}C$^\alpha$ position) and serines (^{13}C$^\alpha$ and ^{13}C$^\beta$ positions). Metabolic labelling also may convert prochiral sites to chiral ones.

(h) Exchange labelling. Readily exchangeable hydrogens can be labelled by solvent exchange. If a protein at natural abundance is exchanged in ^2H_2O, signals from the exchanged sites will be missing in the ^1H NMR spectrum; alternatively, if a perdeuterated protein is back exchanged in H_2O, the exchanged sites will appear in the ^1H spectrum.

(i) Differential labelling of protein constituents. If a protein is composed of more than one subunit or has a cofactor, the different constituents can be labelled differentially.

(j) Peptide bond CN labelling. This is a strategy for identifying unique dipeptides (X-Y). The peptide carbonyl of amino acid X is labelled with ^{13}C, and the peptide amide nitrogen of amino acid Y is labelled with ^{15}N. Following their incorporation into a protein, all X-Y dipeptides can be identified by detecting the single bond ^{13}C-^{15}N coupling.

2.4 Labelling methods

Stable isotope labelling of proteins prepared by peptide synthesis comes down to a question of purchasing or synthesizing the appropriate blocked and activated amino acids in labelled form. Many labelled amino acids are available commercially. Numerous procedures have been published for the preparation of isotopically labelled amino acids by chemical synthesis (12) and by biosynthesis (10–14). Algae or cyanobacteria frequently are the organisms of choice for preparing labelled protein that is to be hydrolysed for use in a culture medium (13, 14). The large scale separation of mixtures (10 g) of labelled amino acids prepared from protein hydrolysates can be carried out by displacement chromatography (15). An algal culture apparatus has been described recently that achieves excellent yields of labelled protein (16). Methods for preparing labelled amino acids themselves will not be discussed in further detail here. Instead, we will focus on the introduction of stable isotopes into proteins by biosynthetic methods that utilize labelled amino acids or other commercially available labelled metabolites.

A variety of host organisms can be used for producing stable isotope labelled proteins.

(a) Bacterial cells. *Escherichia coli* is usually the first host to be considered for protein labelling because of the wealth of information about its genetics (17). Several efficient heterologous over-production systems have been developed for *E. coli*, and mutants are available from the ATCC or Yale repositories for an overwhelming majority of metabolic steps. *E. coli* strains have been adapted to grow on a number of precursors that can be ^{13}C-labelled (glucose, glycerol, acetate, amino acids) or ^{15}N-labelled (ammonium chloride, ammonium sulfate, or amino acids). The yield of a recombinant protein usually depends critically on the medium used. If the yield is higher on a rich medium than on a minimal medium, as often is the case, this will increase the cost of labelling, since a labelled rich medium (mixture of amino acids and possibly peptides) is more expensive than a labelled minimal medium (such as [^{13}C] glucose and [^{15}N] ammonia). It frequently pays to experiment with the conditions and possibly with the gene construct and expression system being used so as to improve the yield on a minimal medium. In cases where the gene product is toxic to *E. coli*, it is adantageous to use a very tightly regulated expression system, such as the T7 system (18), so that the cells can be grown up first before specific protein production is induced. If attempts to produce the protein in good yield in *E. coli* are unsuccessful, over-production in other bacteria (for example *Bacillus*) may provide better results.

(b) Yeast cells. Yeast cells often are useful hosts for the production of proteins that contain disulfide bridges or require post-translational processing. Yeast can be grown on a variety of labelled metabolic precursors including glucose, acetate, and methanol.

(c) Fungal cells. High protein yields can be provided by fungal expression systems; they also support post-translational processing such as glycosylation.

(d) Insect cells. Excellent over-production along with post-translational processing can be achieved by the use of baculovirus vectors in insect cell hosts. The cells require a rich medium.

(e) Mammalian cells in tissue culture. Certain proteins are over-produced by transformed cells. This method has been used to label immunoglobulins for NMR studies (19, 20). The cells require a specialized rich medium for growth.

(f) Algae or cyanobacteria. Photosynthetic organisms offer advantages for stochastic labelling of proteins because they can be grown on relatively inexpensive carbon ($^{13}CO_2$) or nitrogen ($^{15}NO_3^-$ or $^{15}NH_3$) sources and can be adapted to growth in pure ^2H_2O. A few over-production systems are available, but the genetics of these organisms are not yet as well understood as those of *E. coli* or yeast.

Protocols are given below for labelling proteins produced in *E. coli*. *Protocol 1* is used for complete labelling with ^{15}N and/or ^{13}C.

Protocol 1. Complete ^{15}N- and/or ^{13}C-labelling of proteins over-produced in *E. coli* grown on minimal medium

1. Grow a 5 ml LB (17) culture, supplemented with appropriate antibiotic(s), of the over-producing strain, transformed with the appropriate plasmid(s), until a cell density of about 10^8 cells/ml is reached.
2. Pellet the cells and resuspend them in complete minimal medium. Trans-

fer the cells to a 3 litre Fernbach flask filled with 1 litre of complete minimal medium.

Complete minimal medium

- weigh out:
  6 g Na_2HPO_4
  3 g KH_2PO_4
  0.5 g NaCl
  1.0 g NH_4Cl^a
- bring volume to 1.0 litre by adding distilled water, mix well, and autoclave
- let solution cool to room temperature and add:
  20 ml 20% D-glucose (autoclaved)^b
  2 ml 1 M $MgSO_4$
  appropriate antibiotics (sterile filtered)
  any needed auxotrophic supplements (sterile filtered)

3. Grow the cells in an incubator shaker (37°C, 200 r.p.m.) until the cell density reaches 10^8 cells/ml ($OD_{600} \geq 1.0$).

4. Induce protein over-expression and continue growing the cells for 3 h.

5. Pellet the cells by centrifugation, resuspend in an appropriate buffer, and store at −20°C.

6. Thaw the cell suspension, disrupt the cell membrane, and hydrolyse the highly viscous genomic DNA using DNase I. Pellet the debris and use the remaining supernatant as a crude lysate in the first step of protein purification.

^{a} 1.0 g $^{15}NH_4Cl$ is substituted to achieve uniform ^{15}N-labelling.
^{b} 10 ml 20% U-[^{13}C]glucose is substituted to achieve uniform ^{13}C-labelling.

Factors that need to be taken into consideration when labelling a protein selectively with an amino acid are dilution of the label from the selectively labelled amino acid supplied and possible conversion of the labelled amino acid to some other amino acid residue. Dilution of the label depends on the size of the pool of unlabelled amino acid present in the organism and the rates of synthesis and degradation of the amino acid. In some cases, adequate labelling can be achieved by adding an excess of the labelled amino acid plus sufficient other unlabelled amino acids to shut down pathways leading to and from the labelled compound. Since moving to an auxotrophic strain represents an added complication, it may be worth a trial to see whether it is necessary. Note that blocks may be needed in steps leading to the synthesis and degradation of the amino acid. [Table I of reference (9) presents a useful compilation of *E. coli* host genotypes.] Aminotransferases (or transaminases) catalyse the transfer of amino groups between α-amino acids and α-keto acids. When selective labelling of α-amino groups or $C^\alpha H$ groups is desired, it may be necessary to use a host strain defective in these: *TrA* (aspartate), *TrD* (aromatic), *TrB* (branched), *TrC* (alanine ⟷ valine) (9). The auxotrophic behaviour of the host should be determined by several trials (with unlabelled amino acids) in which the amino acid to be incorporated is first omitted and then supplied at increasing levels. No (or very low) growth should be observed when the amino acid is omitted, but near normal growth should be observed when the amino acid is supplied at the level to be used for labelling. Typically, 25 mg of the amino acid per litre of culture is required for optimal growth with an auxotrophic phenotype.

Protocol 2. Amino acid selective labelling of proteins over-produced in *E. coli* grown on a minimal medium—(the conditions given are ones shown to be applicable for over-production of a protein with the T7 expression system (18) under control of the *lac* operon)

1. If needed, transform your over-producing plasmid(s) into a strain of *E. coli* which is auxotrophic for the amino acid you wish to label.

2. Prepare medium:
   Per litre of M9 medium (17):
   - 6 g glucose
   - 30 ml sterile filtered amino acids (omit the amino acid to be incorporated)
   - 3 ml filtered vitamins (saturated solution of 'vitamin diet fortification mix')
   - 50 ml filtered cofactors:
     50 mg ampicillin (substitute antibiotic required)
     25 mg nicotinic acid
     60 mg adenine
     25 mg thiamine
     25 mg of the labelled amino acid to be inserted
     dissolved in water to a volume of 50 ml

3. Grow labelled culture.
   (a) Remove culture from freezer stock, plate out on M9, and grow up.
   (b) Use this to inoculate a 5 ml M9 culture.
   (c) Use the 5 ml culture to inoculate 1 litre of the labelled medium described above, grow in incubator shaker at 37°C.
   (d) Induce protein production by addition of 200 mg/litre IPTG when $OD_{600} \geq 1.0$.
   (e) Harvest 3 h later.

Several deuteration strategies have been used with proteins (for a more complete review see (11)).

(a) Perdeuteration (complete deuteration) (*Protocol 3*). Katz and Crespi pioneered the growth of micro-organisms on deuterated substrates (14). The traditional source for perdeuterated amino acids has been hydro-lysates of algae grown in ^2H_2O (13, 14). This method can be used to remove the signal of one component of a protein complex as for example the use of perdeuterated calmodulin in studies of its complex with [^1H]mellitin (21). Following perdeuteration, the backbone hydrogens can be back exchanged. Thus perdeuteration can be used to improve the resolution of signals in hydrogen exchange studies (22) or for improved resolution of H^N-H^N NOE interactions (23).

(b) Residue selective hydrogen labelling. This general approach was first termed selective deuteration (24). A useful distinction, from an experi-mental standpoint, is to reserve the term selective deuteration for deuter-ation in a protonated background and to use selective protonation for protonation in a deuterated background (11). These approaches obviously blend together in intermediate situations.

Selective deuteration can be achieved in several different ways. One method is by exchange. Certain sites can be exchanged in the native protein as in the case of the backbone amide hydrogens or the histidine $H^{\varepsilon 2}$ (25) or in acid denatured protein as in the case of the tryptophan $H^{\zeta 2}$ (26). Alternatively, deuterium can be introduced biosynthetically by pro-viding [^2H]amino acids (11, 27). Some success has been achieved in resolving signals from labelled residues by difference spectroscopy with reference to a protein at natural abundance (19).

Selective protonation (*Protocol 3*) also can be achieved in different ways. One approach is use a mixture of deuterated and protonated amino acids (with no amino acid in both categories) in the growth medium. This requires (partially) isolated deuterated amino acids (24). A simpler alter-native is to use an unfractionated mixture of deuterated amino acids and add an excess of one or more protonated amino acids (13).

(c) Random fractional deuteration (*Protocol 4*) at levels between 50% and 85% (6, 11). This approach provides a remarkable increase in ^1H resolu-tion when used with proteins of moderate size (up to M_r 20000). The gains result from suppression of passive coupling in 2D shift correlation experiments and from slower ^1H relaxation resulting from the lower proton density in the protein. Reduction in spin diffusion results in more reliable NOE data for larger proteins. In conjunction with ^{13}C- and ^{15}N-labelling, random fractional deuteration may prove useful with still larger proteins.

(d) Chiral deuteration of side chains (11). Chiral β-deuteration is of interest for stereospecific assignments and in the determination of side chain angles from spin-spin coupling measurements. Chiral β-deuteration also leads to improved resolution and sensitivity because the predominant contribution to the relaxation of a methylene proton comes from its geminal neighbour. The prochiral methyls of valine and leucine also can be labelled chirally to advantage.

Protocol 3. Perdeuteration and selective protonation

1. Adapt the *E. coli* strain that produces the protein of interest to growth on ^2H_2O by growing the culture repeatedly on a rich medium made up in increasingly higher concentrations of ^2H_2O.

2. Grow an overnight culture (25 ml) on a minimal salt medium (such as M9) (17) in ^2H_2O with 0.2% glucose, 0.1 mM $CaCl_2$, 1 mM $MgSO_4$, 50 μg/ml thiamine, 50 μg/ml biotin, and 2–4 mg/ml deuterated algal hydrolysate.

3. Use this to inoculate a larger culture of the same medium (500 ml or 1 litre) for production of the protein of interest.

4. Induce protein expression (additional algal hydrolysate can be added at this time if it is found to improve the protein yield).

5. Continue growth until production of the protein of interest has peaked.

6. Pellet the cells by centrifugation, resuspend in an appropriate buffer, and store at −20°C until the protein is isolated.

7. Disrupt the cells and isolate the protein.

Note: in cases where one or more [^1H]amino acids are to be incorporated, these should be added in excess (200–500 μg/ml) at steps **2** and **4**.

LeMaster and Richards (28) found that [^2H]succinate and [^2H]alanine can be used as inexpensive deuterated substrates for growth of *E. coli*. Labelled succinic acid is prepared by means of two exchanges of succinic anhydride in deuterated acetic acid (29, 15). Alanine is exchanged by refluxing a four molal solution of the amino acid with pyridoxal in ^2H_2O at pH 5 (30, 31). *Protocol 4* assumes that 50% random fractional deuteration is desired. The growth medium must contain a higher level of ^2H than the target value, since there is about a 5–10% discrimination against the label during biosynthesis (28).

Protocol 4. Random fractional deuteration (50% U)

1. Prepare the deuterated medium: dissolve M63 salts (17) in 8 litres 55% ^2H_2O supplemented with 30 g of sodium [55% ^2H]succinate and 20 g of [55% ^2H]-DL-alanine.

2. Grow the bacterial strain that produces the protein of interest on the deuterated medium; when the cell density is appropriate for induction, induce production of the protein.

3. Harvest the cells at the time protein production has peaked.

4. Continue as in *Protocol 3*.

A convenient strategy for obtaining stereospecific assignments of the prochiral methyl ^1H and ^{13}C signals from leucine and valine makes use of a metabolic labelling scheme (*Protocol 5*). The protein of interest is produced in *E. coli* grown on a mixture of natural abundance glucose and [^{13}C-6]glucose. As a result of the biosynthetic pathways, the only ^{13}C methyl of valine with a directly bonded labelled carbon corresponds to the $^{13}C^{\beta}$-$^{13}C^{\gamma 1}$ unit and the only ^{13}C methyl of leucine with a directly bonded labelled carbon corresponds to the $^{13}C^{\gamma}$-$^{13}C^{\delta 1}$ unit. These units can be identified by reference to coupling patterns observed in standard ^1H-^{13}C correlation data.

Protocol 5. Chiral ^{13}C-labelling of valine and leucine methyl carbons (32)

1. Prepare 1.0 litre minimal medium containing:
   1.2 g $(NH_4)_2SO_4$
   10.6 g K_2HPO_4
   5.1 g KH_2PO_4
   5.9 g glucose monohydrate
   2.4 mg vitamin B1
   1.2 mg biotin
   120 mg $MgSO_4\cdot7H_2O$
   3.5 mg $FeSO_4\cdot7H_2O$
   antibiotics as needed for retention of the plasmid

2. Inoculate the medium with the expression system culture.

3. 150 min after the start of the culture growth, add 470 mg [^{13}C-6]glucose monohydrate.

4. After another 60 min induce the production of the protein of interest.

5. Harvest the cells when protein production has peaked.

6. Continue as in steps **5** and **6** of *Protocol 1*.

---

3. Data acquisition and analysis

3.1 Introduction

A co-evolution of pulse sequences and assignment strategies has been occurring over the past few years. Since the most prevalent spin = 1/2 nuclei in proteins (^1H, ^{13}C, and ^{15}N) have been exploited by now, one may expect stabilization and gradual refinement of the methodology.

Table 2 contains a menu of NMR pulse sequences that are relevant to stable

Table 2. 2D, 3D, and 4D data acquisition protocols for stable isotope-assisted NMR investigations[a]

Experiment class[a] X = C or N	Experiment name[b]	Reference
$(t_1,t_2) = (H,H)$	COSY	(1)
	RELAY	(2)
	HOHAHA = TOCSY	(3)
	NOESY	(4)
	ROESY	(5)
$(t_1,t_2,t_3) = (H,H,H)$	HOHAHA-NOESY	(6)
$(t_1,t_2) = (X,H)$	HMQC	(7)
	HSQC	(8)
	CT-HSQC	(9)
	HSMQC	(10)
	HMBC	(11)
	HETERO-RELAY	(12)
	HETERO-NOE	(13)
	H-X EXCSY	(14)
$(t_1,t_2) = (C,C)$	CC-DQC	(15, 16)
	CC-COSY	(16)
	XX-EXCSY	(14)
$(t_1,t_2) = (N,C)$	CN-MBC	(17)
$(t_1,t_2,t_3) = (H,X,H)$	NOESY-HMQC	(18)
	HOHAHA-HMQC	(19)
	SE-TOCSY-HSQC	(20)
	SE-HSQC-TOCSY	(20)
	SE-NOESY-HMQC	(20)
	SE-NOESY-HSQC	(20)
$(t_1,t_2,t_3) = (H,C,H)$	HCCH-COSY	(21)
	CT-HCCH-COSY	(22)
	HCCH-TOCSY	(21)
$(t_1,t_2,t_3) = (C,C,H)$	HCACO	(23)
	CT-HCACO	(24)
	C TOCSY-REVINEPT	(25)
$(t_1,t_2,t_3) = (N,C,H)$	HNCO	(23)
	CT-HNCO	(26)
	HNCA	(23, 27)
	CT-HNCA	(26)
	HN(CO)CA	(25)
	CT-HN(CO)CA	(26)
	CT-HN(CA)CO	(28)
$(t_1,t_2,t_3) = (C,N,H)$	HCA(CO)N	(23)
	CT-HCA(CO)N	(24)
$(t_1,t_2,t_3) = (H,N,H)$	HA(CA)NNH	(29)
$(t_1,t_2,t_3) = (N,N,H)$	$^{15}N/^{15}N$-edited ^1H-^1H NOESY	(30)
$(t_1,t_2,t_3) = (N,H,H)$	HN(CA)HA	(31)
$(t_1,t_2,t_3,t_4) = (C,H,N,H)$	$^{13}C/^{15}N$-edited ^1H-^1H NOESY	(32)
	HNCAHA	(33)
	HN(CO)CAHA	(33)
$(t_1,t_2,t_3,t_4) = (C,H,C,H)$	$^{13}C/^{13}C$-edited ^1H-^1H NOESY	(34)
$(t_1,t_2,t_3,t_4) = (C,H,C,H)$	HCACON	(35)

Table 2. *Continued*

[a] Pulse sequences are organized by the nuclei that are operative in the various acquisition periods. The number of time-domains (t_i values) establishes the dimensionality of the spectrum. Following multi-dimensional Fourier transformation, each of these time-domains is converted to a frequency axis in the n-dimensional spectrum: for example, (C,H) yields a 2D spectrum with one ^{13}C and one ^1H axis, and (C,H,N,H) yields a 4D spectrum with one ^{13}C axis, two ^1H axes, and one ^{15}N axis.

[b] *Abbreviations:* CC-DQC, ^{13}C-^{13}C double quantum correlation; CC-COSY, ^{13}C-^{13}C COSY; CN-MBC, ^{13}C-^{13}C multiple bond correlation spectroscopy; COSY, correlated spectroscopy; CT, constant time (refers to the way in which the pulse sequence is organized); HOHAHA, homonuclear Hartmann–Hahn spectroscopy; HMBC, heteronuclear multiple bond correlation spectroscopy; HMQC, heteronuclear multiple quantum correlation spectroscopy; HSQC, heteronuclear single bond correlation; H-X EXCSY, ^1H-detected X-X exchange spectroscopy, where X is ^{13}C or ^{15}N; NOE, nuclear Overhauser effect; NOESY, laboratory frame NOE spectroscopy; RELAY, relayed coherence transferred COSY; REVINEPT, reverse INEPT; ROESY, rotating frame NOE spectroscopy; TOCSY, total correlated spectroscopy; XX-EXCSY, X-X exchange spectroscopy, where X is ^1H, ^{13}C, or ^{15}N. The following terms refer to the corresponding moieties in the peptide backbone: HA, H^α; CA, $^{13}C^\alpha$; NH, H^N; N, $^{15}N^\alpha$; CO, $^{13}C'$. Brackets mean that the indicated carbon, while on the coherence transfer pathway, is not observed. HCCH refers to a ^1H-^{13}C-^{13}C-^1H moiety within an amino acid residue.

1. Nagayama, K., Anil Kumar, Wüthrich, K., and Ernst, R. R. (1980). *J. Magn. Reson.*, **40,** 321; Rance, M., Sørensen, O. W., Bodenhausen, G., Wagner, G., Ernst, R. R., and Wüthrich, K. (1983). *Biochem. Biophys. Res. Commun.*, **117,** 479.
2. King, G. and Wright, P. E. (1983). *J. Magn. Reson.*, **54,** 328; Wagner, G. (1983). *J. Magn. Reson.*, **55,** 151.
3. Braunschwieler, L. and Ernst, R. R. (1983). *J. Magn. Reson.*, **53,** 521; Bax, A. and Davis, D. G. (1985). *J. Magn. Reson.*, **65,** 355.
4. Anil Kumar, Ernst, R. R., and Wüthrich, K. (1980). *Biochem. Biophys. Res. Commun.*, **95,** 1; Macura, S., Wüthrich, K., and Ernst, R. R. (1982). *J. Magn. Reson.*, **46,** 269.
5. Bothner-By, A. A., Stephens, R. L., Lee, J., Warren, C. D., and Jeanloz, R. W. (1984). *J. Am. Chem. Soc.*, **106,** 811; Bax, A. and Davis, D. G. (1985). *J. Magn. Reson.*, **63,** 207; Davis, D. G. and Bax, A. (1986). *J. Magn. Reson.*, **64,** 533.
6. Oschkinat, H., Cieslar, C., and Griesinger, C. (1990). *J. Magn. Reson.*, **86,** 453; Griesinger, C., Sørensen, O. W., and Ernst, R. R. (1989). *J. Magn. Reson.*, **84,** 14.
7. Müller, L. (1979). *J. Am. Chem. Soc.*, **101,** 4481; Bax, A., Griffey, R. H., and Hawkins, B. L. (1983). *J. Magn. Reson.*, **55,** 301.
8. Bodenhausen, G. and Ruben, D. J. (1980). *Chem. Phys. Lett.*, **69,** 185; Cavanagh, J., Palmer, A. G. III, Wright, P. E., and Rance, M. (1991). *J. Magn. Reson.*, **91,** 429; Palmer, A. G. III, Cavanagh, J., Wright, P. E., and Rance, M. (1991). *J. Magn. Reson.*, **92,** 151.
9. Santoro, J. and King, G. C. (1992). *J. Magn. Reson.*, **97,** 202; Vuister, G. W. and Bax, A. (1992). *J. Magn. Reson.*, **98,** 428.
10. Zuiderweg, E. R. P. (1990). *J. Magn. Reson.*, **86,** 346.
11. Bax, A. and Summers, M. L. (1986). *J. Am. Chem. Soc.*, **108,** 2093.
12. Brühwiler, D. and Wagner, G. (1986). *J. Magn. Reson.*, **69,** 546; Oh, B.-H., Westler, W. M., and Markley, J. L. (1989). *J. Am. Chem. Soc.*, **111,** 3083.
13. Shon, K. and Opella, S. J. (1989). *J. Magn. Reson.*, **82,** 193; Wang, J., Hinck, A. P., Loh, S. N., and Markley, J. L. (1990). *Biochemistry,* **29,** 102.
14. Alexandrescu, A. T., Loh, S. N., and Markley, J. L. (1990). *J. Magn. Reson.*, **87,** 523.
15. Levitt, M. L. and Ernst, R. R. (1983). *Mol. Phys.,* **50,** 1109; Oh, B.-H., Westler, W. M., Darba, P., and Markley, J. L. (1988). *Science,* **240,** 908.
16. Westler, W. M., Kainosho, M., Nagao, H., Tomonaga, N., and Markley, J. L. (1988). *J. Am. Chem. Soc.*, **110,** 4093.
17. Westler, W. M., Stockman, B. J., Hosoya, Y., Miyake, Y., and Kainosho, M. (1988). *J. Am. Chem. Soc.*, **110,** 6265; Mooberry, E. S., Oh, B.-H., and Markley, J. L. (1989). *J. Magn. Reson.*, **85,** 147.
18. Kay, L. E., Marion, D., and Bax, A. (1989). *J. Magn. Reson.*, **84,** 72.
19. Marion, D., Kay, L. E., Sparks, S. W., Trochia, D. A., and Bax, A. (1989). *J. Am. Chem. Soc.*, **111,** 1515.
20. Palmer, A. G. III, Cavanagh, J., Byrd, R. A., and Rance, M. (1992). *J. Magn. Reson.*, **96,** 416.

113

Table 2. *Continued*

21. Clore, G. M., Bax, A., Driscoll, P. C., Wingfield, P. T., and Gronenborn, A. M. (1990). *Biochemistry*, **29**, 8172.
22. Ikura, M., Kay, L. E., and Bax, A. (1991). *J. Biomolec. NMR*, **1**, 299.
23. Kay, L. E., Ikura, M., Tschudin, R., and Bax, A. (1990). *J. Magn. Reson.*, **89**, 496.
24. Powers, R., Gronenborn, A. M., Clore, G. M., and Bax, A. (1991). *J. Magn. Reson.*, **94**, 209.
25. Fesik, S. W., Eaton, H. L., Olejniczak, E. T., and Zuiderweg, E. R. P. (1990). *J. Am. Chem. Soc.*, **112**, 886.
26. Grzesiek, S. and Bax, A. (1992). *J. Magn. Reson.*, **96**, 432.
27. Farmer, B. T. II, Venters, R. A., Spicer, L. D., Wittekind, M. G., and Müller, L. (1992). *J. Biomolec. NMR*, **2**, 195.
28. Clubb, R. T., Thanabal, V., and Wagner, G. (1992). *J. Magn. Reson.*, **97**, 213.
29. Kay, L. E., Ikura, M., and Bax, A. (1991). *J. Magn. Reson.*, **91**, 84.
30. Ikura, M., Bax, A., Clore, G. M., and Gronenborn, A. M. (1990). *J. Am. Chem. Soc.*, **112**, 9020.
31. Clubb, R. T., Thanabal, V., and Wagner, G. (1992). *J. Biomolec. NMR*, **2**, 203.
32. Kay, L. E., Clore, G. M., Bax, A., and Gronenborn, A. M. (1990). *Science*, **249**, 411.
33. Kay, L. E., Wittekind, M., McCoy, M. H., Friedrich, M. S., and Mueller, L. (1992). *J. Magn. Reson.*, **98**, 443.
34. Clore, G. M., Kay, L. E., Bax, A., and Gronenborn, A. M. (1991). *Biochemistry*, **30**, 12; Zuiderweg, E. R. P., Petros, A. M., Fesik, S. W., and Olejniczak, E. T. (1991). *J. Am. Chem. Soc.*, **113**, 370.
35. Kay, L. E., Ikura, M., Zhu, G., and Bax, A. (1991). *J. Magn. Reson.*, **91**, 422.

isotope-assisted spectroscopy. The following sections summarize protocols that make use of different labelling patterns and pulse sequences.

3.2 Data collection

Complete descriptions of data collection methodology are beyond the scope of this chapter. The procedures are dependent on the pulse sequence to be used and how it can be implemented on the spectrometer at hand. Spectrometer set-up procedures and pulse programs are described in manuals and information sheets from spectrometer manufacturers and in the published literature; some general information is provided in Chapter 3. Data collection methods can be tailored to the particular needs of an experiment. For example phase-cycling can be omitted in a ^{1}H-^{15}N HMQC experiment in order to speed up data collection during a hydrogen exchange experiment (33).

3.3 Data processing and display

Software packages for processing multinuclear, n-dimensional NMR data are available from spectrometer manufacturers (e.g. Bruker, Jeol, Varian) and software houses (e.g. BIOSYM, Molecular Simulations, Tripos) as well as individual research groups (34–36). Ideally, one wants a versatile software package that will carry out a variety of data processing operations (e.g. Fourier transforms, linear prediction, maximum entropy analysis, convolutions, baseline corrections, signal filtering). Graphical displays and plotting software are highly desirable. Automated peak picking and peak sorting are desirable, and some software packages have useful utilities for keeping a data base of intermediate and final results.

4. Isotope-assisted assignment strategies

4.1 Goals and expectations (full or partial assignment)

The full assignment of a protein of moderate size (M_r 10–20000) can be a challenging undertaking. Powerful approaches have been developed recently that make the problem more tractable. These include the optimization of 3D and 4D NMR approaches to heteronuclear chemical shift correlations in proteins labelled completely with ^{13}C and/or ^{15}N and the development of computer software to analyse the results in terms of concerted assignment algorithms (37, 8). Fundamental limitations (primarily rapid proton transverse relaxation rates) appear to limit this general strategy to proteins of M_r less than 30–40000. (This is only an estimate, since the upper M_r limit has not been tested rigorously yet.) Strategies that include perdeuteration along with selective ^1H-, ^{15}N-, and/or ^{13}C-labelling appear attractive for larger proteins (38, 39). At present, selective labelling appears to be the only strategy that will be applicable to still larger proteins ($M_r > 40000$). Although it is not possible to determine full three-dimensional structures of such larger proteins by NMR, it is possible, in favourable cases, to use stable isotope-assisted NMR to map out binding sites and investigate local geometry.

4.2 Assignment procedures

The complete assignment of the NMR spectrum of a protein involves three tasks:

- sorting of NMR peaks by amino acid type
- linking of the peaks (sorted by amino acid type) by sequential NMR connectivities
- alignment of these linked groups of peaks with the protein sequence.

The three tasks can be carried out separately, but they usually are combined in iterative searches as the assignment proceeds. Secondary structural elements (α-helix, β-sheet, and turns) give rise to characteristic patterns of sequence-dependent NOEs; such additional information can be used to verify or extend assignments. Typically, the amino acid sequence is known, but sufficient information can be obtained from spectra of labelled proteins to verify the sequence from NMR data alone.

Amino acid selective stable isotope labelling provides a powerful means of accomplishing the first task (sorting resonances by amino acid type). If the labelling is truly selective (no transfer of the label to other amino acid types), the approach provides unambiguous results. With small proteins at natural abundance or proteins labelled uniformly with ^{13}C and/or ^{15}N, sorting is accomplished by classifying resonances from each residue according to its spin system type. The degree of classification obtainable by this approach is limited by the amount of spectral information available. By ^{1}H NMR alone (no labelling), the sorting process leads to the unambiguous identification of resonances from only 8–15 of the 20 amino acids (depending on the extent of ^{1}H NMR information available) (1). With uniform ^{13}C-labelling, it is possible to sort spin systems of 18 amino acids (40); and with dual ^{13}C/^{15}N-labelling, signals can be sorted cleanly for all 20 amino acids (41, 42, 8).

Without stable isotope labelling, the sorted spin systems are linked to those of their nearest neighbours by means of nuclear Overhauser effect (NOE) data that identify pairs of protons that are closer than about 4.5 Å; if one partner in the proton pair is located in residue *i* and the other in residue *i* + 1, the NOE connectivity establishes a nearest neighbour link. These linked spin systems are then aligned with the protein sequence (1). *Figure 1* illustrates the way in which 2D NOESY data are used to establish sequential links between parts of sorted spin systems; for further details, see Chapter 4.

In larger proteins, this strategy becomes more difficult because of spectral overlap and spin diffusion. As shown in *Figure 2*, which shows the spectrum of recombinant staphylococcal nuclease complexed to a mononucleotide bisphosphate, the problems can be overcome to some extent by random fractional deuteration at levels between 50% and 75%. In the case of an α-helix, where H^N-H^N NOEs provide the sequential assignments, a useful

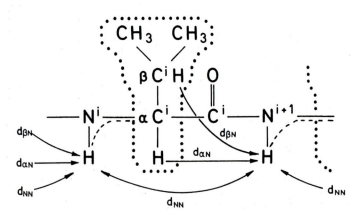

Figure 1. Schematic representation of the assignment strategy used with small proteins at natural abundance. Protons within each residue form a spin system and are identified by through-bond chemical shift correlation experiments such as COSY and HOHAHA. (These are represented in the figure by *dotted lines* along with the *dashed line* to the H^N.) Sequential links between adjacent residues are determined from through-space connectivities provided by the NOESY data. To be observed, an interproton distance (d_{NN}, $d_{\alpha N}$, or $d_{\beta N}$) must be less than 4.5–5.0 Å. (These distances are represented in the figure by *solid lines*.) Reproduced from reference (1) with permission.

116

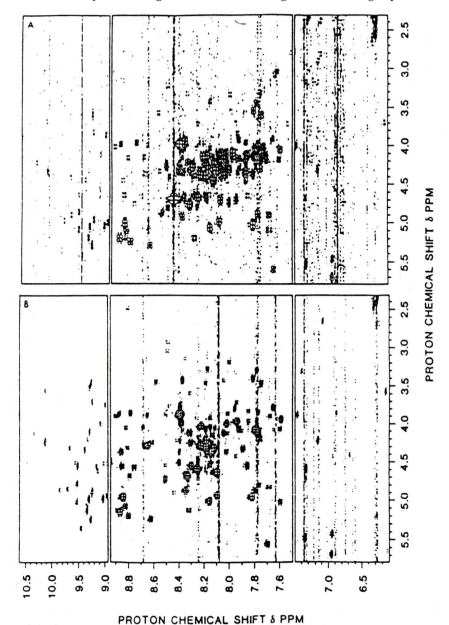

PROTON CHEMICAL SHIFT δ PPM

Figure 2. Illustration of the effect of selective deuteration. COSY fingerprint ($^1H^N$–$^1H^\alpha$ spin-spin connectivity) regions of staphylococcal nuclease (H124L) are compared with (A) natural abundance protein and (B) [50% U-^{2}H]protein (44). The natural abundance sample was 3.5 mM nuclease dissolved in H_2O containing 10.5 mM pdTp, 21 mM $CaCl_2$, and 300 mM KCl, pH 5.1 at 45°C. Deuterated nuclease was 5 mM in 15 mM pdTp, 30 mM $CaCl_2$, and 300 mM KCl, pH 5.1 at 45°C. Reproduced from reference (4) with permission.

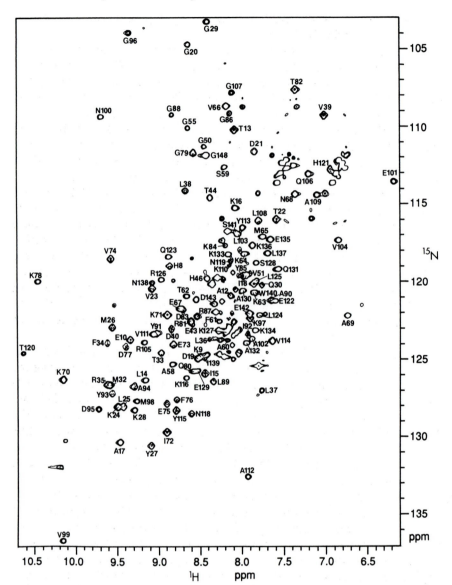

Figure 3. ¹H-¹⁵N single bond correlation spectrum of 5 mM [95% U-¹⁵N]staphylococcal nuclease H124L in 90% H_2O/10% ^2H_2O containing 300 mM KCl, pH 5.1 at 45°C. The HMQC pulse sequence was used (see *Table 3*). Assigned cross-peaks are labelled. Reproduced from reference (45) with permission.

approach is complete deuteration of all C-H groups followed by back exchange of the peptide backbone H^Ns (43).

As noted above, the resolution of amide proton resonances is critical to sequential assignments obtained by means of H^N-H^N NOEs. Resolution of the amide signals can be increased dramatically by means of ^{15}N-labelling and two-dimensional ^1H-^{15}N correlation experiments. *Figure 3* illustrates the 2D ^1H-^{15}N (HMQC) map obtained for the same protein (uncomplexed) whose H^N-H^α (COSY) map is shown in *Figure 2A*. By adding a ^1H-^1H NOE step to the ^1H-^{15}N correlation, the resolution of the NOE can also be improved. This can be implemented as a two-dimensional experiment (46), but maximal resolution is obtained if it is carried out at a three-dimensional experiment (47). In experiments of this type, for example, 3D NOESY-HMQC (48), one axis represents the ^1H chemical shift of the amide NH, one axis represents the ^{15}N chemical shift of the amide NH, and the third represents the chemical shifts of protons that show NOE connectivities to H^Ns (*Figure 4*). Rather than following the assignment pathway in three dimensions, strips along the ^{15}N dimension are sorted according to their sequential connectivities (*Figure 5*). Information about secondary structure also is provided by 2D or 3D NOE data from proteins labelled uniformly with ^{15}N (46, 49, 50) or ^{13}C (51, 52). Labels in adjacent α-carbons can be used in the detection of *cis*-peptide bonds by isotope-filtered NOEs (53).

The multinuclear spin systems of proteins labelled with ^{13}C and ^{15}N extend from one residue to the next by virtue of $^1J_{CC}$ and $^1J_{CN}$ coupling along the peptide backbone (*Figure 6*). If amino acid selective carbonyl carbon ($^{13}C'$) and peptide amide nitrogen ($^{15}N^\alpha$) labelling is used the spectra usually are simple and can be analysed by one-dimensional ^{13}C or ^{15}N spectroscopy (55). This method is robust, carries over from one state of ligation or conformation to another, and is applicable to proteins as large as M_r 150000 (20).

Recently, heteronuclear 2D (41), 3D (56-58), and 4D (59, 60) experiments have been developed that exploit coupling along the peptide backbone to provide sequence-specific assignments in proteins labelled completely with ^{13}C and ^{15}N. *Table 4* lists the various strategies used for sequential assignments and provides references to applications of the pulse sequences to proteins. In principle the 4D triple-resonance approach should provide backbone assignments with as few as two data sets. The 3D approach requires more data sets but may be applicable to larger proteins. (Current experience indicates that the protein size limit with this approach is about M_r 30000.) Finally the 1D and 2D approaches with selective labelling will provide assignments for the largest proteins.

Table 3 lists various NMR protocols used in making backbone assignments. In summary, sequence-specific assignments can be derived, with the help of stable isotope labelling, from:

(a) Unique stable isotope labelling. Unique labelling can be achieved by chemical synthesis, by enzymatic semi-synthesis, by cofactor or inhibitor

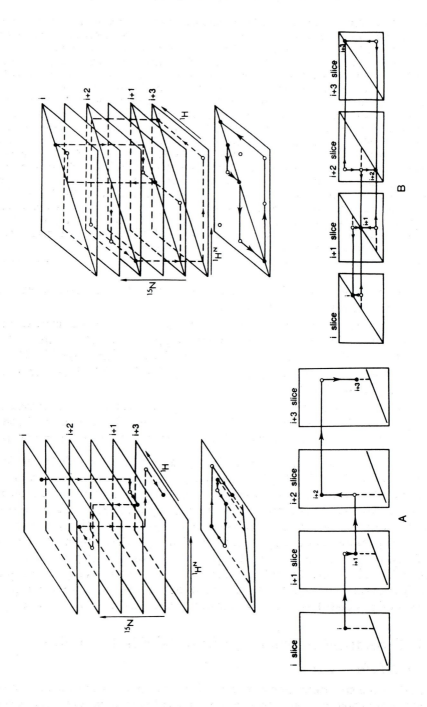

Figure 4. (A) Schematic representation of the $d_{\alpha N}$ sequential connectivities in a 3D ^{1}H-^{15}N NOESY-HMQC spectrum. *Top*: Three-dimensional view. The *vertical arrows* indicate the sequential pathway of ^{15}N resonances. The *arrows* within the ^{1}H-^{1}H^N slices follow sequential connectivities that link H$^\alpha$ and H$^N_{i+1}$. *Middle*: Projection of the sequential pathway in the 3D ^{1}H-^{1}H^N slices on to the ^{1}H-^{1}H^N plane gives a pattern identical to the sequential walk obtained by analysing a conventional 2D ^{1}H-^{1}H NOESY (H$_2$O) spectrum. *Bottom*: Selected 3D ^{1}H-^{1}H^N slices arranged in sequential order. The $d_{\alpha N}$ walk is shown by *solid lines* with *arrows*. Intraresidue NOE cross-peaks are represented by *dots*, and interresidue NOE cross-peaks are represented by *circles*. (B) Schematic view of the d_{NN} connectivities in a 3D ^{1}H-^{15}N NOESY-HMQC spectrum. *Top*: Three-dimensional view. The *arrows* illustrate the sequential pathway of the connected amide protons. The d_{NN} NOE cross-peaks appear as symmetry-related triangles on either side of the sequential ^{1}H-^{1}H^N slices along the diagonal. *Middle*: Projection of the linked 3D peaks on to the ^{1}H-^{1}H^N plane yields a pattern equivalent to the d_{NN} walk obtained by analysing a conventional 2D ^{1}H-^{1}H NOESY (H$_2$O) spectrum. *Bottom*: Selected slices arranged in sequential order. The *solid lines* with *arrows* trace out the NOE connectivity pattern between amide protons in consecutive slices. The diagonal peaks are represented by *dots*. The NOE cross-peaks are represented by *circles*. (Reproduced from reference (54) with permission.)

---

labelling, and (trivially) by residue specific labelling of residues with a single copy number in the protein.

(b) Sequential, stable isotope-assisted, through-space (NOE) connectivities. Common types of sequential, stable isotope-assisted NOEs are: $(^{15}$N-^{1}H$)_i$...$(^1$H-^{15}N$)_{i+1}$, $(^{13}$C$^{\alpha,\beta,\gamma}$-^{1}H$)_i$...$(^1$H-^{15}N$)_{i+1}$, $(^{13}$C$^\alpha$-^{1}H$)_i$-$(^{13}$C$^\gamma$-^{1}H$)_{i+1}$ (for a *trans* X-proline dipeptide), and $(^{13}$C$^\alpha$-^{1}H$)_i$-$(^{13}$C-^{1}H$)_{i+1}$ (for a *cis* peptide bond) (53).

(c) Sequential, stable isotope-assisted, through-bond connectivities. These are provided by a range of 1D–4D experiments, where redundancy of connectivities or higher dimensionality are used to resolve overlap problems.

Additional isotope-assisted 3D and 4D methods support side chain (60–63) (*Table 4*) and NOE (64–66) (*Table 5*) assignments. In so far as possible, it is best to adopt a concerted approach to assignments in which all data concerning spin system identifications, side chain assignments, and sequential connectivities are taken into account. An approach commonly used with larger proteins is to simplify the assignment problem by collecting high dimensional (3D and 4D) NMR data. Such data have fewer redundancies in assignment pathways than 2D data so that the solution can be found without testing all possible assignment pathways against all the data. Of course, once an assignment model has been postulated, all available data can be tested against it for consistency.

5. Structural information from labelled proteins

5.1 NOEs

Most of the structural information available for proteins in solution comes from evaluation of proton–proton cross-relaxation (the nuclear Overhauser

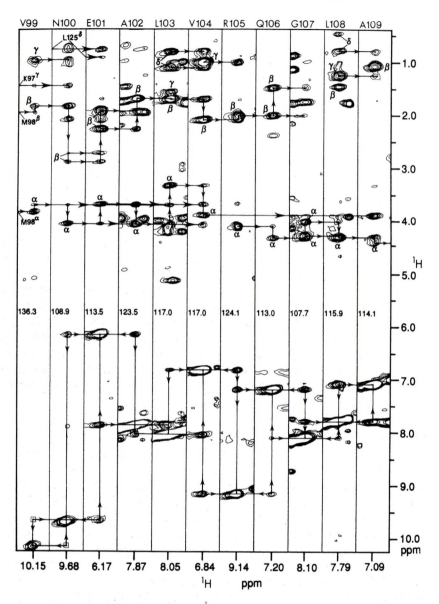

Figure 5. Series of strips selected from ^{1}H-^{1}H^N slices of the 3D ^{1}H-^{15}N NOESY-HMQC spectrum. The strips have been arranged in sequential order. Each strip corresponds to one residue in the V99-A109 α-helix. Because of poor resolution along the ^{15}N dimension, some strips show cross-peaks from other residues with similar ^{15}N chemical shifts in addition to the NOE cross-peaks of the selected residue. The assignment of each strip is given at the *top* by residue number and one-letter amino acid code. Identified intraresidue NOE cross-peaks between side chain protons and amide protons are indicated by *Greek letters. Solid vertical lines* with *arrows* denote NOE connectivities to the ^{1}H^N of the residue

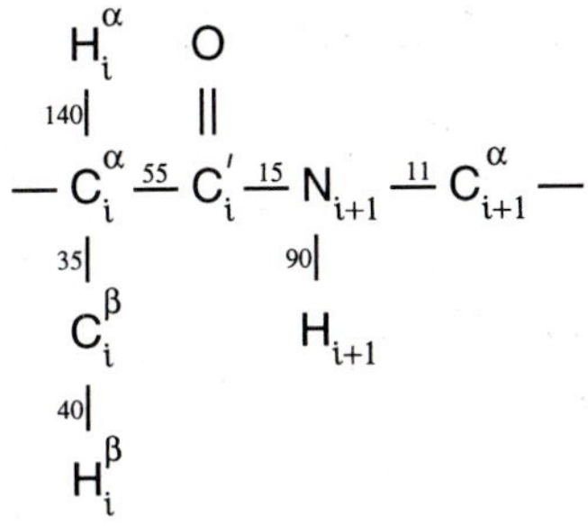

$$
\begin{array}{c}
H_i^\alpha \quad\quad O \\
{\scriptstyle 140}| \quad\quad \| \\
-\,C_i^\alpha \xrightarrow{\scriptstyle 55} C_i' \xrightarrow{\scriptstyle 15} N_{i+1} \xrightarrow{\scriptstyle 11} C_{i+1}^\alpha\,- \\
{\scriptstyle 35}| \quad\quad\quad {\scriptstyle 90}| \\
C_i^\beta \quad\quad H_{i+1} \\
{\scriptstyle 40}| \\
H_i^\beta
\end{array}
$$

Figure 6. Schematic representation of the coupling pathways that can be exploited in making sequential assignments along the peptide backbone from through-bond coupling connectivities with proteins labelled with ^{13}C and ^{15}N.

effect or NOE). Stable isotope labelling can be used in conjunction with pulse sequences that incorporate laboratory-frame cross-relaxation (NOE) or rotating-frame (ROE) cross-relaxation steps (67, 68). The isotopes can be used to select for or filter the cross-relaxation to be detected or to increase the resolution of the cross-relaxation by exploiting the dispersion of a ^{13}C or ^{15}N nucleus that is spin-spin coupled to a cross-relaxing ^{1}H. The stable isotope selection or filtering can be achieved in experiments of different dimensionality (1D–4D) (*Table 5*). In general, higher sensitivity and greater digital resolution are provided by decreasing dimensionality, but fewer overlaps (and hence ambiguities in NOE interpretation) derive from increasingly higher dimensionality (59, 60, 64, 65).

5.2 Coupling constants

Three bond spin-spin constants provide useful structural constraints plus an indication of the presence and extent of conformational averaging. For flexible parts of molecules, the more information one can gain that can be related to a given dihedral angle for a rotatable bond (^{3}Js and ^{1}H-^{1}H NOEs), the better one can specify the probability distribution of rotamers (69). Heteronuclear coupling constants in proteins can sometimes be observed directly from the multiplicity of peaks in 1D spectra. Coupling constants also can be observed and selected for in 2D and 3D experiments. For example, three bond H^N-H^α coupling constants were measured in [99% ^{15}N]staphylococcal nuclease by a variation of the 2D ^{1}H-^{15}N HMQC experiment; dihedral angles

whose amide ^{15}N is selected. *Horizontal lines* with *arrows* indicate NOEs from protons on adjacent residues. The vertical axis gives the ^{1}H chemical shifts in p.p.m. The $^1H^N$ chemical shift (in p.p.m.) of the selected residue is marked on the horizontal axis of each strip. The approximate ^{15}N chemical shift of the slice from which the strip was taken is noted in each strip. The $^1H^N$ chemical shift values used here are the more accurate ones taken from the 2D NOESY data. (Reproduced from reference (54) with permission.)

Table 3. NMR protocols used for backbone assignments

Labelling pattern	NMR data acquisition pulse sequences		Reference
	For connectivites within residues	For connectivities between residues	
Natural abundance	2D COSY	2D NOESY	(1)
	2D RELAY		(2)
	2D HOHAHA		(3)
Natural abundance	3D HOHAHA-NOESY		(4)
U-^{15}N	3D HOHAHA-HMQC	3D NOESY-HMQC	(5)
U-^{15}N	3D SE-TOCSY-HMQC	3D SE-NOESY-HSQC	(6)
U-^{15}N	3D TOCSY-NOESY / 3D NOESY-TOCSY		(7)
(^{15}N$^{\alpha}$)-(^{13}C′)	—	1D ^{13}C observe ($^{1}J_{NC}$ coupling)	(8)
U-^{15}N^{i}/U-^{13}C^{i+1}	3D H̲Ṅ(CA)C̈O	3D H̲ṄCO	(9)
	4D H̲ṄCAḦA	4D H̲Ṅ(CO)C̈AḦA	(10)

Abbreviations: COSY, correlated spectroscopy; HMQC, heteronuclear multiple quantum correlation (aka 'forbidden echo'); HSQC, heteronuclear single quantum correlation (also known as 'Overbodenhausen'); HOHAHA, heteronuclear Hartmann–Hahn experiment (also known as TOCSY, total correlated spectroscopy); NOESY, nuclear Overhauser effect spectroscopy; RELAY, relay correlated spectroscopy; SE, sensitivity enhanced. N$^{\alpha}$ is a peptide amide nitrogen. In the multi-dimensional NMR pulse sequence designations, the underlined nucleus is the one observed in the final evolution period, and nuclei with dots above them are detected in previous evolution periods.

1. Anil Kumar, Ernst, R. R., and Wüthrich, K. (1980). *Biochem. Biophys. Res. Commun.*, **95**, 1; Wüthrich, K. (1986). *NMR of proteins and nucleic acids.* Wiley, New York.
2. King, G. and Wright, P. E. (1983). *J. Magn. Reson.*, **54**, 328; Wagner, G. (1983). *J. Magn. Reson.*, **55**, 151.
3. Braunschweiler, L. and Ernst, R. R. (1983). *J. Magn. Reson.*, **53**, 521; Bax, A. and Davis, D. G. (1985). *J. Magn. Reson.*, **65**, 355.
4. Oschkinat, H., Cieslar, C., Holak, T. A., Clore, G. M., and Gronenborn, A. M. (1989). *J. Magn. Reson.*, **83**, 450; Oschkinat, H., Cieslar, C., and Griesinger, C. (1990). *J. Magn. Reson.*, **86**, 453.
5. Marion, D., Driscoll, P. C., Kay, L. E., Wingfield, P. T., Bax, A., Gronenborn, G. M., and Clore, G. M. (1989). *Biochemistry*, **29**, 6150.
6. Palmer, A. G., III, Cavanagh, J., Byrd, R. A., and Rance, M. (1992). *J. Magn. Reson.*, **96**, 416.
7. Mueller, L., Campbell-Burk, S., and Domaille, P. (1992). *J. Magn. Reson.*, **96**, 408.
8. Kainosho, M. and Tsuji, T. (1982). *Biochemistry*, **24**, 6273.
9. Ikura, M., Kay, L. E., and Bax, A. (1990). *Biochemistry*, **29**, 4659; Clubb, R. T., Thanabal, V., and Wagner, G. (1992). *J. Magn. Reson.*, **97**, 213.
10. Kay, L. E., Wittekind, M., McCoy, M. A., Friedrichs, M. S., and Mueller, L. (1992). *J. Magn. Reson.*, **98**, 443.

calculated from the Karplus equation were compared with corresponding ones from the X-ray structure of the same protein (70). Approaches that lead to heteronuclear E.COSY (71) coupling patterns provide a way of resolving small three bond coupling constants in labelled proteins (which can be related to dihedral angles). Direct, single bond coupling provides the large diagonal offset that allows one to resolve a connectivity to a remote nucleus by NOESY (72, 73) or TOCSY (62). Complete ^{15}N-labelling is the optimal strategy for measurements of $^{3}J_{NH}$ (72–76). Single site or block ^{13}C-labelling or stochastic labelling at about 30% is optimal for measurements of $^{3}J_{CH}$ (62). Dual ^{13}C-labelling (preferably at ~ 99%) is optimal for measurements of

Table 4. NMR protocols used for side chain assignments

Labelling pattern	NMR data acquisition pulse sequence	Reference
Natural abundance	2D HOHAHA	(1)
U-^{15}N	2D NOESY-HMQC	(2)
	3D HOHAHA-HMQC	(3)
U-^{13}C*	2D ^{13}C{^{13}C}DQC	(4)
	3D HCCH TOCSY	(5)
	3D HCCH COSY	(5)

Abbreviations: COSY, correlated spectroscopy; DQC, double quantum correlation; HMQC, hetero-nuclear multiple quantum correlation; HOHAHA, heteronuclear Hartmann–Hahn experiment (also known as TOCSY, total correlated spectroscopy); NOESY, nuclear Overhauser effect spectroscopy.
* Samples can also have U-^{15}N labelling.

1. Braunschwieler, L. and Ernst, R. R. (1983). *J. Magn. Reson.*, **53**, 521; Bax, A. and Davis, D. G. (1985). *J. Magn. Reson.*, **65**, 355.
2. Kay, L. E., Marion, D., and Bax, A. (1989). *J. Magn. Reson.*, **84**, 72.
3. Marion, D., Kay, L. E., Sparks, S. W., Torchia, D. A., and Bax, A. (1989). *J. Am. Chem. Soc.*, **111**, 1515.
4. Oh, B.-H., Westler, W. M., Darba, P., and Markley, J. L. (1988). *Science*, **240**, 908.
5. Clore, G. M., Bax, A., Driscoll, P. C., Wingfield, P. T., and Gronenborn, A. M. (1990). *Biochemistry*, **29**, 8172.

$^3J_{CC}$. A method has been presented recently for accurate measurement of $^3J_{HH}$ in a protein labelled with ^{13}C (77).

5.3 Chemical shifts

Stable isotope labelling can be highly useful in detecting and assigning the chemical shifts of atoms in proteins. 'Random coil' values for ^{1}H, ^{13}C, and ^{15}N chemical shifts have been reevaluated recently on the basis of data from proteins whose structures are known (78). Deviations from 'random coil' chemical shifts provide information about local structure in a protein. Factors involved include paramagnetism (for proteins that have unpaired electron density) (79), ring current effects from aromatic groups (80), the magnetic anisotropy of the peptide group, and backbone electrostatic contributions (81). Interesting correlations have been noted between the chemical shifts of H$^\alpha$ (82, 83) and ^{13}C$^\alpha$ (83) nuclei and secondary structure. There is also a suggestion that H^N chemical shifts may hold useful information about hydrogen bonding (78). Changes in chemical shifts that accompany ligand binding or altered protein conformation are commonly interpreted in kinetic and thermodynamic terms.

5.4 Solvent accessibility

Fesik and co-workers (84) have recently developed a method for probing the solvent accessible regions of a protein or protein complex that has been labelled uniformly with ^{13}C. 2D ^{1}H-^{13}C T_1-HMQC spectra are acquired as a function of the T_1 delay period in the absence and presence of increasing

Table 5. NMR protocols used for isotope-assisted H · · · H distance measurements

Filtered NOE experiment (nuclei observed)	Multiplicity	Reference
Single X-filter		
X-H · · · NOE · · · H′	1D broadband X (H)	(1)
	1D selective X (H)	(2)
	2D (H,H′)	(3)
	2D (X,H)	(4)
	3D (H,X,H′)	(5)
Double X-filter		
X-H ... NOE ... H′-X′	2D (H,H′)	(6)
	4D (H,X,X′,H′)	(7)
Double X,Y-filter		
X-H ... NOE ... H′-Y	2D (H,H′)	(8)
	3D (H,X,H′)	(9)
	4D (H,X,Y,H′)	(10)

Abbreviations: X and Y are ^{15}N or ^{13}C. Nuclei in parentheses are the ones observed on the axes of the multi-dimensional spectrum. Note that selective labelling can often be used to advantage in isotope filtered experiments: for example, one can selectively label one or both partners in a protein–ligand or protein–protein complex (only one partner can be labelled with ^{13}C or ^{15}N, or, optimally, one partner can be labelled with ^{13}C and the other with ^{15}N).

1. Freeman, R., Mareci, T. H., and Morris, G. A. (1981). *J. Magn. Reson.*, **42**, 341; Bendall, M. R., Pegg, D. T., Doddrell, D. M., and Field, J. (1981). *J. Am. Chem. Soc.*, **103**, 934; Fesik, S. W. (1988). *Nature*, **332**, 865.
2. Teller, C., Williams, H., Gao, Y., Ortiz, C., Stolowich, N. J., and Scott, A. I. (1990). *J. Magn. Reson.*, **90**, 600.
3. Otting, G., Senn, H., Wagner, G., and Wüthrich, K. (1986). *J. Magn. Reson.*, **70**, 500; Bax, A. and Weiss, M. A. (1987). *J. Magn. Reson.*, **71**, 571.
4. Shon, K. and Opella, S. J. (1989). *J. Magn. Reson.*, **82**, 193; Wang, J., Hinck, A. P., Loh, S. N., and Markley, J. L. (1990). *Biochemistry*, **29**, 102.
5. Fesik, S. W. and Zuiderweg, E. R. P. (1988). *J. Magn. Reson.*, **78**, 588.
6. Otting, G. and Wüthrich, K. (1989). *J. Magn. Reson.*, **85**, 586.
7. Zuiderweg, E. R. P., Petros, A. M., Fesik, S. W., and Olejniczak, E. T. (1991). *J. Am. Chem. Soc.*, **113**, 370; Clore, G. M., Kay, L. E., Bax, A., and Gronenborn, A. M. (1991). *Biochemistry*, **30**, 12.
8. Fesik, S. W., Eaton, H. L., Olejniczak, E. T., and Gampe, R. T. Jr. (1990). *J. Am. Chem. Soc.*, **112**, 112.
9. Kay, L. E., Marion, D., and Bax, A. (1989). *J. Magn. Reson.*, **84**, 72.
10. Kay, L. E., Clore, G. M., Bax, A., and Gronenborn, A. M. (1990). *Science*, **249**, 411.

amounts of a paramagnetic probe molecule (4 mM to 10 mM HyTEMPO). Solvent accessible sites show a large effect of HyTEMPO on cross-peak volumes; those that are solvent inaccessible do not. The method can be used to probe contact regions of protein–protein (85) or protein–drug (84) interactions.

6. Strategies for very large proteins

6.1 Properties of carbonyl ^{13}C resonances

Many of the isotope-assisted strategies described above have been developed in order to extend NMR analysis to larger proteins. Complete ^{15}N- and/or ^{13}C-labelling and 2D–4D NMR appear to be suitable for proteins as large as M_r 30 000. Many proteins of interest to structural biology are still larger. Their slow tumbling rates give rise to short T_2 relaxation times which lead to reduced sensitivity in multi-dimensional experiments. For the largest proteins, one wishes to observe a signal that is relatively slowly relaxing and can be resolved in a 1D spectrum. An ideal signal for this purpose comes from the carbonyl carbons of a selectively labelled protein (55). Such a spectrum can be achieved by incorporating one or more [^{13}C′] amino acids. An important advantage of the ^{13}C′ approach is that sequential assignments can be made to unique Xaa-Zaa dipeptides by means of dual ^{13}C/^{15}N-labelling in which [^{13}C′]-Xaa and [^{15}N$^\alpha$]-Zaa are incorporated biosynthetically. Although signals from carbonyls of all Xaa residues will appear in the ^{13}C NMR spectrum, only those from Xaa-Zaa dipeptides will show ^{13}C-^{15}N coupling. In favourable cases this can be observed with proteins as large as M_r 150 000 (20). At lower magnetic fields, the signal from the carbonyl carbon is sharper than those from protonated carbons since the ^{1}H-^{13}C dipolar interaction is the predominant relaxation mechanism. At higher magnetic fields, however, the chemical shift anisotropy mechanism comes into play which is stronger for the carbonyl carbon than for other carbons. For this reason, there is an optimal field strength for ^{13}C NMR studies of carbonyl carbons in proteins: for proteins around M_r 50–100 000, this corresponds to about 75 MHz for ^{13}C (300 MHz for ^{1}H).

Much of the carbonyl carbon strategy of use with larger proteins was developed in studies of Streptomyces subtilisin inhibitor (SSI) (86). This proteinase inhibitor is a dimer of two identical subunits (M_r 23 000 for the dimer). The active site of SSI resembles those of other substrate-like protein proteinase inhibitors; it includes a scissile peptide bond between the P_1 and $P_{1'}$ residues (Met73 and Val74 in the case of SSI). The NMR signal from the carbonyl of Met73 provides critical information on the state of the scissile bond. The signal from the Met73 carbonyl can be observed in 1D NMR spectra of the free inhibitor and in the inhibitor–proteinase 2:2 complex. The chemical shift of Met73 carbonyl in the complex indicates that the peptide bond is intact and trigonal.

Since selectively labelled ^{13}C′ backbone atoms yield resolved and readily assignable signals, they can serve as probes for many processes:

- individual pK_a values of titrating groups and the dependence of the protein conformation on its ionization state
- conformational changes that accompany amino acid replacements by mutagenesis
- structural changes that occur on ligand binding
- protein folding

6.2 The CN labelling strategy

Figure 7 shows an application of the CN labelling approach (55). The bottom spectrum is that of SSI produced from bacteria fed [^{13}C′]Phe. The three

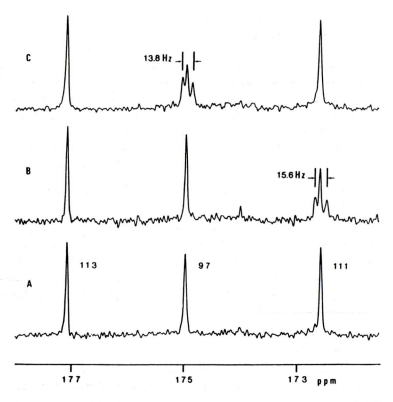

Figure 7. Illustration of the CN assignment strategy. Proton-decoupled 75.4 MHz ^{13}C NMR spectra of isotopically labelled SSI molecules: (A) [F]SSI, (B) [F,A]SSI, (C) [F,GS]SSI. Each labelled SSI (18–20 mg) was dissolved in 0.5 ml of 0.05 M deuterophosphate buffer, pH 7.3. The sample temperature was 50 °C. A total of 40 000 transients were accumulated and Fourier transformed after being multiplied by a line-broadening factor of 0.5 Hz. The 'trios' observed in (B) and (C) arise from the superimposed signals from ^{13}C-^{15}N and ^{13}C-^{14}N species. These data led to the assignment of the phenylalanine carbonyl resonances of SSI as described in the text. (Reproduced from reference (87) with permission.)

strong carbonyl signals come from the three phenylalanine residues of SSI located at positions 97, 111, and 113 in the protein sequence. Since all three phenylalanine residues are located in unique dipeptides (Phe97-Ser98, Phe111-Ala112, and Phe113-carboxyl terminus), their signals can be assigned unambiguously by the CN labelling method (87). The middle spectrum is that of SSI produced from bacteria fed [^{13}C']Phe and [^{15}N]Ala. By comparison to the bottom trace, one sees that the signal at 172.5 p.p.m. is partially split by characteristic single bond ^{15}N-^{13}C (15.6 Hz) coupling. This identifies the coupled ^{13}C signal as arising from Phe111. Dual labelling with [^{13}C]Phe and [^{15}N]Gly, leads to ^{15}N-labelling of Gly and Ser (because of scrambling of the label), leads to splitting of the middle Phe $^{13}C'$ peak at 175 p.p.m., and identifies it as the signal from Phe97.

The carbonyl and nitrogen assignments, which are conveniently made by this method, can be extended by reference to additional connectivities. Overall, the backbone carbonyl resonance may be the best point to begin assignments of proteins larger than about M_r 30000. Selective ^{13}C block labelling methods can be used to extend carbonyl carbon assignments to the rest of the residue. Numerous variations are possible, including specifically (site-, regio-, or stereospecifically) labelled, or uniformly labelled amino acids with ^2H, ^{15}N, ^{13}C, or even ^{12}C, at various isotope enrichment levels.

6.3 Carbonyl carbon signals as probes for chemical modification

The assigned and enhanced carbonyl carbon signals of a protein selectively labelled with a single type of 1-[^{13}C]amino acid have been found to be useful probes for investigating structural changes induced by chemical modification (88). The probes can be used to pinpoint the sites of modification and to monitor changes (some quite subtle) that accompany chemical modification. Four case studies are presented:

- selective oxidation of surface methionine residues
- oxidation of internal methionine residues
- selective reduction and oxidation of disulfide bonds (89)
- renaturation of a protein following complete reduction of disulfides and GdnCl denaturation (88)

These are illustrated by experimental protocols and results for Streptomyces subtilisin inhibitor (SSI).

6.3.1 Oxidation of methionine residues

Proteins generally contain only a few methionine residues. Even with larger proteins, it is relatively easy to assign all the carbonyl carbon signals observed for a protein selectively labelled with 1-[^{13}C]Met. For this reason it is interesting to explore methods that utilize this structural probe. The oxidation rates of individual methionyl residues, depend on the relative accessibilities of the methionine S^δ to an oxidant such as molecular oxygen or hydrogen peroxide. Thus determination of the relative rates of methionine oxidation (*Protocol 6*) can provide information on the microenvironment and tertiary structure of a protein.

Protocol 6. Oxidation of methionine residues as followed by ^{13}C NMR— the procedure is one that has been used for studies of SSI (H. Sato and M. Kainosho, unpublished results)

1. Prepare a stock solution of 0.03% peroxide by 100-fold dilution of 30% aqueous hydrogen peroxide with ^2H_2O.

Protocol 6. *Continued*

2. Dissolve 10 mg of [M]SSI (all three methionyl residues labelled selectively with [98% 1-^{13}C]Met) in 0.5 ml of 50 mM deuterophosphate buffer, pH 7.3. Transfer the solution to a 5 mm NMR tube.

3. Add two equivalents (for oxidation of the two surface residues per subunit to the sulfoxide) of the 0.03% deuterium peroxide solution (20 µl in the present case) to the protein solution and incubate it at 50°C for 6 h.

4. Obtain and process 1D ^{13}C NMR spectra.

5. Carbonyl ^{13}C NMR signals from the modified surface methionines (Met70 and Met73) were shifted upfield by about 0.8 p.p.m.; that from the un-modified methionine (Met103) showed no change (*Figure 8*).

6. Addition of a second 20 µl aliquot of the 0.03% deuterium peroxide solution led, after incubation for 20 days at 60°C, to oxidation of the third (buried) methionine (*Figure 9*).

Each SSI subunit contains three methionine residues: Met70, Met73 (the P$_1$ site residue in the scissle bond), and Met103. Met70 and Met73 are surface residues and are expected to be oxidized easily to methionine sulfoxide. Met103, on the other hand, is buried in the interior and should be resistant to oxidation. This was exactly what was observed experimentally. Addition of limited amounts of hydrogen peroxide (or deuterium peroxide) to SSI in solution led to selective oxidization of the surface methionines (Met70 and Met73) (*Figure 8*). The interior methionine (Met103), however, was not oxi-dized at all under the conditions used to oxidize the surface methionines; it could be oxidized by adding an additional excess amount of the reagent (*Figure 9*).

As in all such probe experiments, proper interpretation of the results requires that one distinguish between chemical shift changes resulting from chemical modification and those from conformational changes. The lack of a chemical shift change for the unmodified methionine (Met103) signal is con-sistent with (but not adequate proof for) no induced conformational changes around this residue. In the case of SSI, kinetic analysis of the results showed that the two surface methionines are oxidized at exactly the same rate. Another feature of the results is that the carbonyl carbon resonances of Met70 sulfoxide and Met73 sulfoxide are split into closely spaced doublets with approximately equal intensities (*Figure 8*). This result suggests that equal amounts of the R and S sulfoxide enantiomers were formed. The splitting is caused by interaction of the oxidized side chain with the surface of the protein. Since the intensities of the two forms are equal, the environment of each of the surface methionines shows no stereoselectivity for the oxidation reaction.

Oxidation of the internal methionine appears to take place when the

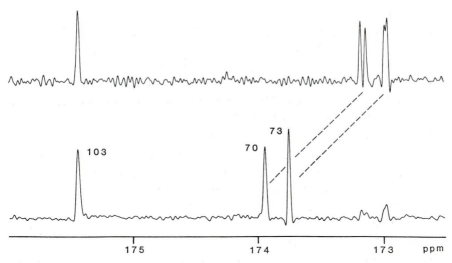

Figure 8. Use of ^{13}C NMR to follow selective methionine oxidation. The example shown is streptomyces subtilisin inhibitor labelled with ^{13}C′ Met, [M]SSI. The sample conditions were: 10 mg protein in 0.5 ml of 50 mM deuterophosphate buffer, pH 7.3. The 75.4 MHz ^{13}C NMR spectrum of the native (unmodified) protein (*bottom spectrum*) shows separate signals from the three methionines in the protein (Met70, Met73, and Met103). Oxidation was carried out *in situ* at 60°C for 8 h by the addition of 10 μl of 0.03% deuterium peroxide in deuterium oxide. Upon oxidation, the carbonyl carbon signals from Met70 and Met73 shifted upfield by about 0.8 p.p.m. and split into double peaks of approximately equal intensity; the signal from Met103 remains unsplit and unshifted (*top spectrum*). The results indicate that two methionine residues (Met70 and Met73 which are known to be at the protein surface) are oxidized preferentially. Peak splitting arises from conversion of the prochiral sulfur to a chiral sulfur; since the peaks have nearly equal intensity, sulf-oxidation must take place with nearly random stereoselectivity, as might be expected for surface residues. Since the chemical shift of Met103 (buried) remains unchanged, it appears to be unaffected by the modification. (H. Sato and M. Kainosho, unpublished results.)

protein is in a native (or native-like) structure because the reaction shows partial stereoselectivity: the carbonyl carbon resonance of Met103 sulfoxide appeared as a 2:1 doublet (*Figure 9*). Slight changes (consistent with loss of chiral chemical shift differences) were observed for the carbonyl resonances of Met70 sulfoxide and Met73 sulfoxide upon oxidation of Met103; these indicate that the environment of these residues is altered by a conformational change that accompanies oxidation of the buried methionine.

The reaction course of methionine oxidations also can be followed by monitoring the ^{1}H NMR signals from the methionine methyls, which typically are shifted downfield by about 0.6 p.p.m. upon modification. It usually is more difficult to assign the methionine methyl resonances (particularly with a larger protein) than the methionine ^{13}C′ resonances (which can be assigned

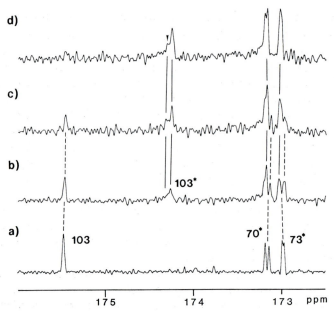

Figure 9. Study of complete methionine oxidation. The sample whose spectrum is shown in *Figure 8* was oxidized further by an addition 20 µl (two-fold excess) of 0.03% deuterium peroxide in deuterium oxide. The reaction was allowed to proceed for 30 days at 60 °C, and its progress was followed by ^{13}C NMR. Spectra were obtained (a) before and (b) 6, (c) 14, and (d) 28 days after addition of peroxide. The carbonyl signal from oxidized Met103 is an unsymmetrical doublet (about 3:1; the *arrow* indicates the position of the minor modified form) centred about 1.2 p.p.m. upfield of the signal from the unoxidized residue. The results indicate that the modification reaction of the buried methionine shows partial stereoselectivity. Modification of Met103 alters the environment of the two surface methionines as shown by reduction of the chemical shift difference of the label from molecules with different chirality at the site of modification. (H. Sato and M. Kainosho, unpublished results.)

by selective double labelling). The ^{13}C assignments, along with ^{1}H and ^{13}C NMR results from oxidation experiments, were used in the case of SSI to distinguish the ^{1}H resonances of the surface methionines (Met70 and Met73) from that of the internal methionine (Met103).

^{13}C′ signals can be used as probes to monitor structural changes at different sites along the peptide backbone. Several kinds of 1-[^{13}C] amino acids were incorporated into SSI, and changes in their chemical shifts were used to map out the extent of the conformational change (discussed above) induced by oxidation of the interior methionine (Met103). The same approach can be used for evaluating the effects of other chemical modification reactions including site-specific mutagenesis. This method is especially useful for larger proteins whose native structures are known from X-ray crystallography so that the spectral changes can be interpreted more readily.

6.3.2 Detection of disulfide bridge oxidation and reduction

It has been found that the $^{13}C'$ chemical shift of a cysteine residue is perturbed upon oxidation or reduction of a disulfide bridge in which it participates (89). This leads to a strategy for assigning disulfide bond pairing in proteins by stable isotope-assisted NMR. The procedure is to incorporate 1-$[^{13}C]$Cys into a protein. The chemical shifts of the Cys $^{13}C'$ resonances are then monitored as the disulfides are cleaved progressively by reduction. Provided that individual disulfides are reduced differentially, pairwise changes will be observed in the Cys $^{13}C'$ chemical shifts which permit them to be assigned to disulfide pairs. The Cys $^{13}C'$ signals are finally assigned by the $^{13}C'_i$-$^{15}N_{i+1}$ double labelling method (55) or other sequential assignment method.

The mechanism by which correct pairs of disulfide bonds form from a fully reduced protein is a lively and controversial research topic. The processes are usually followed from the fully reduced unfolded state. Some of the intermediates in the processes are considered to be transient species with short lifetimes and relatively small populations. Conventional studies of disulfide bond formation utilize chromatographic separation of chemically trapped intermediates. It should be of interest to follow disulfide formation by ^{13}C NMR of cysteinyl carbonyl carbon resonances, since this approach does not require any chemical trapping and separation steps and, thus, should be free of suspected artefacts.

Protocol 7 contains a procedure for studying differential reduction of disulfide bridges by ^{13}C NMR (89). As with methionine, cysteine is present in proteins in low abundance, and thus the ^{13}C NMR spectrum of a protein labelled with $[^{13}C']$Cys is usually simple and easily assigned. SSI has four cysteines per subunit which form two disulfide bridges in the native protein: Cys35–Cys50 and Cys71–Cys101. The four enhanced carbonyl resonances of [C]SSI (SSI with cysteinyl residues labelled selectively with 1-$[^{13}C]$ Cys) were assigned by double labelling experiments and site-directed mutagenesis (89). The reduction and reoxidation of [C]SSI was followed by monitoring the $^{13}C'_i$ NMR signals of the labelled cysteines. (The $^{13}C'$ resonances of other residues are affected by the oxidation/reduction reactions, but the interpretation of these changes is more difficult.)

Protocol 7. Use of ^{13}C NMR to follow selective oxidation of disulfide bridges—the procedure describes the method used to study SSI (89)

1. Prepare [C]SSI by culturing *Streptomyces albogreseolus* S-3253 in a medium containing a complete mixture of amino acids, with L,L-1,1'-$[^{13}C_2]$cystine as the only labelled amino acid.

2. Prepare a 354 mM solution of dithiothreitol (DTT) in 0.1 M borate buffer,

Protocol 7. *Continued*

pH 8.3. A 10 μl aliquot of this solution is sufficient to reduce one disulfide bond per subunit in a 20 mg sample of SSI. (The quantity of reductant needed for partial reduction was determined in advance by using the Ellman method to assay free sulfhydryl.)

3. Dissolve 20 mg of [C]SSI (98% ^{13}C) in 0.5 ml of 0.1 M borate buffer, pH 8.3 containing 10% ^{2}H$_2$O for the lock signal. Transfer the sample to a 5 mm o.d. NMR sample tube.

4. Add 10 μl of a freshly prepared DTT solution. Fill the NMR tube with nitrogen gas and seal it. Mix the solution by shaking, and let it stand 1.5 h at room temperature.

5. Acquire a 1D ^{13}C NMR spectrum at 75 MHz by accumulating 2000 transients with a recycling time of 2 sec (1 h data collection period).

6. Repeat with another aliquot of DTT. Wait for the reaction, and acquire another spectrum.

7. Repeat step **6** until full reduction is achieved.

On gradual addition of the DTT solution, it was observed by ^{13}C NMR that the bond between Cys71 and Cys101 is reduced preferentially. Increasing amounts of DTT reduced the Cys71–Cys50 bond, but at a much slower rate (*Figure 10*).

The ^{13}C NMR spectra of [C]SSI at 23°C, *Figure 10a*, gave three lines, since the signals from Cys50 and Cys101 are coincident at this temperature. By adding a two-fold molar excess of DTT to the NMR sample, the intensities of Cys71 and the lower field overlapped peak decreased significantly with the concomitant appearance of new small peaks with asterisks at the lower field side of those signals (*Figure 10b*). These new signals are due to reduced Cys71 and reduced Cys101. These signals become larger at the expense of signals from the native protein. Finally, after addition of a ten-fold molar excess of DTT, the Cys71–Cys101 bond was completely reduced (*Figure 10d*). Note that the intensity of Cys35 at the final stage decreased significantly, showing that the Cys35–Cys50 bond was partially reduced. The differential reduction of the two disulfide bridges in SSI can be explained on the basis of the solvent accessiblities of their sulfur atoms as seen in the crystal structure: Cys35 (1%), Cys50 (0%), Cys71 (7%), and Cys101 (89%).

It frequently is difficult to establish disulfide pairing patterns in proteins in the absence of an X-ray structure. Provided that disulfides can be reduced selectively, the ^{13}C method described above can be used to determine the pairings. It is necessary, of course, to have assigned the Cys ^{13}C′ signals in advance.

Protocol 8 presents the analogous procedure used for following folding and oxidation of cysteines from the fully reduced, unfolded state.

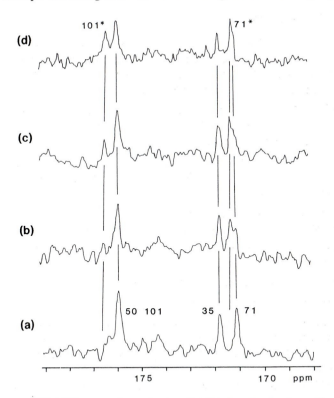

Figure 10. Use of ^{13}C NMR to monitor selective disulfide bond reductions. The example shown is Streptomyces subtilisin inhibitor labelled with ^{13}C' Cys, [C]SSI. The samples contained 20 mg [C]SSI dissolved in 0.5 ml 0.1 M borate buffer (10% ^{2}H$_2$O) containing 0.1 M KCl, pH 8.3. The samples contained varying amounts of the reductant dithiothreitol (DTT) added to the NMR tube: (a) no DTT, (b) two molar equivalents (theoretical amount needed to reduce both disulfide bonds in both subunits of the SSI dimer), (c) four equivalents, (d) ten equivalents. 75.4 MHz ^{13}C NMR spectra were acquired at room temperature (23 °C) immediately after each addition of DTT. The data acquisition time was about one hour (2000 transients with a recycle time of 2 sec). (Reproduced from reference (89) with permission.)

Protocol 8. Use of ^{13}C NMR to follow renaturation of a protein that has been fully reduced and denatured—the procedure is one used with SSI (89)

1. Dissolve [C]SSI (98% ^{13}C) in 1 ml of 0.1 M borate buffer, pH 8.3, containing 6 M guanidium hydrochloride (GdnHCl). Add a 1 ml aliquot of a 100 mM DTT solution in 0.1 M borate buffer under nitrogen, and allow this to react for 1.5 h at room temperature. (The concentration of GdnHCl in this solution was 3 M. SSI was completely unfolded and both disulfide bridges were reduced as determined from NMR spectra.)

Protocol 8. *Continued*

2. Remove the denaturant by the following procedure carried out in a nitrogen atmosphere. Apply the unfolded reduced protein solution to a 1.2 × 20 cm Bio-Gel P-4 (Bio-Rad) column equilibrated with 0.1 M borate buffer. Monitor fractions by a flow cell UV detector at 280 nm and collect fractions containing fully reduced SSI. Concentrate the reduced SSI by ultrafiltration with a YM-5 membrane filter (Amicon, cut-off molecular weight of M_r 5000). Substitute this buffer by ultrafiltration with a 0.1 M borate buffer containing 10% ^2H_2O, pH 8.3, and adjust the final volume to 0.5 ml. Transfer this solution to a 5 mm o.d. NMR sample tube.

3. Follow the time-course of the reoxidation process by ^{13}C NMR at 75.4 MHz (*Figure 11*). In order to facilitate air oxidation, open the NMR tube and keep it unsealed during the reaction.

Because of the small surface area, the oxidation rate was very slow; it took more than 200 hours to regenerate the two disulfide bonds completely. The first spectrum (*Figure 11a*) was recorded three hours after removing GdnHCl and DTT from the reaction mixture. Since all the sample preparation procedures were carried out under nitrogen, the carbonyl carbon signals arise from reduced SSI (all cysteines are Cys-SH). Apparently, the secondary structure as well as the overall conformation around these four cysteines are almost identical to those of native SSI. The ^{13}C NMR spectrum changes after the solution is exposed to oxygen in the air (*Figure 11*).

Within several hours, signals due to *Cys71 and *Cys101 (where * indicates reduced) diminished, and new signals due to Cys71 and Cys101 (oxidized) grew in. The chemical shifts of the latter were almost identical to those of native SSI (*Figure 11b*). At this time, *Cys35 still has almost full intensity, showing that the Cys35–Cys50 bond is not formed yet. Therefore, the predominant species in this period is Cys35/Cys50 reduced SSI. This half-reduced SSI is different from the one observed by selective reduction of native SSI. After a day, approximately half of the protein molecules are in the fully oxidized state (*Figure 11c*); and finally, after 216 hours, all molecules are in the native form.

The same time-course may be studied by monitoring other carbonyl resonances whose chemical shifts change with the redox states of the disulfide bonds. Such studies bring us information on the extent of conformational changes that accompany disulfide cleavage and regeneration of disulfide bonds.

7. Thermodynamic information from labelled proteins

7.1 Titration curves

Histidine titration curves often can be determined by ^1H NMR at natural abundance by observation of the His $H^{\epsilon 2}$ and/or $H^{\delta 1}$ signals (90). In some

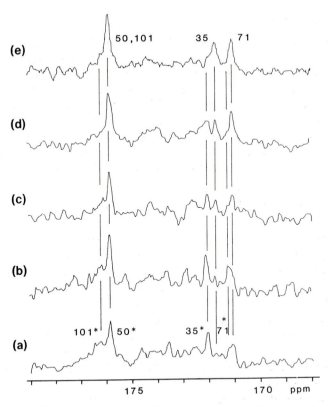

Figure 11. Use of ^{13}C NMR to follow disulfide bond formation. The example shown is the regeneration of disulfide bonds in [C]SSI that had been fully reduced with DTT. The protein sample (20 mg in 0.5 ml of 0.1 M borate buffer containing 10% ^2H_2O and 0.1 M KCl, pH 8.3) was transferred to a 5 mm (o.d.) NMR tube without a cap to facilitate air oxidation, and incubated for the times specified at room temperature (23°C). The data acquisition time was about one hour (2000 transients with a recycle time of 2 sec). Incubation times were: (a) 0 h, (b) 4 h, (c) 22 h, (d) 69 h, (e) 213 h. Assignments are shown in the figure: numbers without asterisks (*top*) refer to reformed disulfides; numbers with asterisks (*bottom*) refer to reduced cysteines. (Reproduced from reference (89) with permission.)

cases, however, the proton signals are poorly resolved so that ^{13}C- or ^{15}N-labelling is advantageous. Complete ^{15}N-labelling, which generally is easier than specific introduction of [^{15}N]His, is particularly useful for histidine titration curves. The His ring nitrogens appear in a unique spectral region and do not overlap with those of other residues. A particular benefit of uniform ^{15}N-labelling is the ability to observe strong one bond and two bond ^1H-^{15}N connectivities. Thus in cases where the exchange lifetime of a His ring N-H is long compared to $1/^1J_{15N-^1H}$ (~ 10 msec), one can determine whether one or both ring nitrogens are protonated, and, in the case of single protonation,

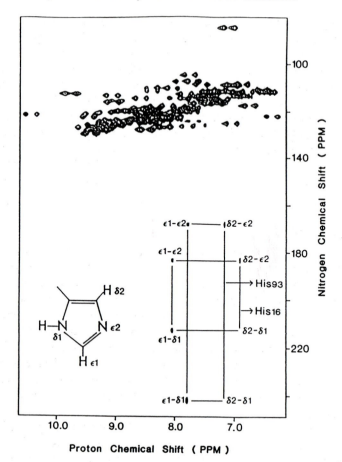

Figure 12. Example of concerted isotope-assisted assignments in the side chain rings of histidine. The figure shows a portion of the single bond ^{1}H-^{15}N correlation spectrum (HMQC pulse sequence, see *Table 3*) of [98% U-^{15}N]ferredoxin (from *Anabaena* 7120). The pattern of histidine peaks (*boxed in*) provides simultaneous assignments for both ring protons and ring nitrogens. The analysis relies on the fact that multiple bond cross-peaks are attenuated in this experiment. Thus the strongest peaks come from two bond coupling (^{1}H$^{\epsilon 1}$-^{15}N$^{\delta 1}$, ^{1}H$^{\epsilon 1}$-^{15}N$^{\epsilon 2}$, ^{1}H$^{\delta 2}$-^{15}N$^{\epsilon 2}$) and the weakest peak comes from three-bond coupling (^{1}H$^{\delta 2}$-^{15}N$^{\delta 1}$). Since the missing corner in each rectangle identifies the chemical shifts of H$^{\delta 2}$ and ^{15}N$^{\delta 1}$, the chemical shifts of the other protons and nitrogens can be immediately determined. The spectrum was obtained at 500 MHz ^{1}H (50.68 MHz ^{15}N). The spectral width in the ^{15}N dimension was 259.1 p.p.m., which included all downfield ^{15}N resonances. The spectrum was recorded without ^{15}N decoupling during acquisition; as a result, cross-peaks appear as doublets. For clarity only one tautomeric form of the neutral imidazole ring of histidine is shown. A total of 480 FID blocks were collected as 4096 data points; each represented the average of 80 transients. The experiment time was 18 hours. The sample was 0.5 ml of 6.5 mM protein in 90% H_2O/10% ^2H_2O containing 50 mM phosphate buffer, pH 7.1. (Reproduced from reference (42) with permission.)

which nitrogen has the bound proton (37). The imidazole ring protons and ring nitrogens can be assigned concertedly: one C-H proton (H^ϵ) will show strong (two bond) connectivities to two nitrogens; the other (H^δ) will show one strong (two bond) connectivity to one of these nitrogens ($^{15}N^\epsilon$) and one weak (three bond) connectivity to the other ($^{15}N^\delta$). This is illustrated by results with oxidized *Anabaena* ferredoxin (42) in *Figure 12*.

For studies of other titratable groups, side chain ^{13}C-labelling (Asp, Glu, Tyr, Cys, Lys, Arg) and ^{15}N-labelling (Lys, Arg) are indispensable. This approach has been used, for example, in studies of the pH dependence of the enzyme staphylococcal nuclease (91). The chemical shifts of backbone ^{15}N and $^{13}C'$ are sometimes responsive to the protonation state of a side chain and can be used to monitor its titration.

7.2 Ligand binding

Numerous ligand binding strategies have been developed that involve NMR and stable isotope labelling. In one approach, the ligand and protein are labelled differentially, and NMR methods are used to distinguish between signals from the labelled and unlabelled components of the complex. Binding effects can be monitored by measuring chemical shifts or relaxation parameters, and these can be used to evaluate the thermodynamics and kinetics of binding. The mechanism of binding can be investigated by stable isotope directed NOE measurements. In some cases the small molecule is easier to label than the protein; with current methodology, however, it may in fact be more economical to enrich the protein biosynthetically than to label a drug by chemical synthesis.

Figure 13 shows a diagrammatic representation of various approaches used in ligand binding studies (92). The key to interpretation of the NOE results is the proper assignment of the signals; stable isotope labelling can assist in this as well. Recent applications of this strategy have included studies of the conformation of cyclosporin A (a cyclic peptide) bound to cyclophilin (a M_r 17 700 protein with peptidyl-prolyl isomerase activity) (93–95), investigations of the FMN binding site of flavodoxin (96), and localization of bound water in the solution structure of reduced human thioredoxin (97). The water binding study made use of a 3D ^1H-^{15}N ROESY-HMQC experiment with protein uniformly labelled with ^{15}N. Chapter 7 gives a detailed discussion of NMR methods for studying protein–ligand interactions and further examples.

7.3 Conformational equilibria

NMR spectroscopy is probably the best available technique for detailed investigations of protein conformational changes. Isotope-assisted NMR has been used in thermodynamic and kinetic studies of protein folding (discussed above), changes in conformation that accompany ligand binding (45, 54), and changes in *cis* and *trans* configuration about Xaa-Pro peptide bonds (53). In

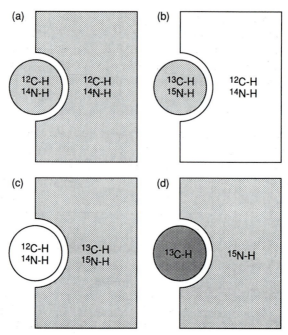

Figure 13. Principle of spectral simplification by isotopic enrichment. (a) When both the receptor and ligand (represented by the *circle*) are at natural isotopic abundance, they are both observed in ^{1}H NMR spectra (indicated by shading), and neither can be observed exclusively. Enrichment of ligand (b) or receptor (c) with ^{13}C or ^{15}N allows selection for signals from only the ligand or receptor, respectively (as indicated by differential shading). (d) Complementary enrichment of the ligand with ^{13}C and the receptor with ^{15}N allows selection for either species independently (as indicated by two shades of grey). (Reproduced from reference (92) with permission.)

favourable cases, one can determine the rates of these processes in addition to equilibrium values.

8. Protein dynamics and hydrogen exchange rates from labelled proteins

8.1 Relaxation measurements

NMR relaxation parameters provide information about overall molecular tumbling and internal dynamics in proteins. ^{1}H relaxation is difficult to analyse because the multiple mechanisms come into play. ^{13}C relaxation of carbons with directly bonded hydrogens are easier to interpret because the dominant mechanism is ^{1}H–^{13}C dipolar coupling. ^{15}N relaxation is of interest because the NOE changes sign depending on the level of internal mobility. Although methods have been proposed for measuring ^{13}C relaxation at natural abundance in proteins (most recently through proton detection schemes), ^{13}C-labelling is mandatory for achieving the level of signal-to-noise required for accurate results. Promising new methods achieve sensitivity and resolution by means of 2D or 3D experiments that provide ^{1}H detection of ^{13}C and ^{15}N relaxation rates (98–102).

8.2 Chemical exchange

For macromolecules, ^{1}H-^{1}H cross-relaxation and chemical exchange appear identical in NOESY spectra. Thus the two processes may be difficult to distinguish, unless comparison is made to rotating-frame (ROESY) data (103). Because NOEs involving ^{13}C and ^{15}N are negligible, chemical exchange studies of stable isotope enriched sites provide unambiguous results (104). In addition, the ^{13}C or ^{15}N labels can extend the time scale for chemical exchange measurements because their T_1 values often are longer than those for ^{1}H. Prototype experiments have shown results on folding and unfolding of ^{13}C-(104) and ^{15}N-labelled (105) proteins.

8.2.1 ^{15}N assisted H exchange studies

Hydrogen exchange in proteins has long been studied by ^{1}H NMR. Sites that can be studied include backbone amide, side chain amides of Asn and Gln, Trp H$^{\epsilon 2}$, His H$^{\epsilon 2}$, Tyr H$^{\eta}$, and Arg H$^{\eta 1}$, H$^{\eta 2}$, 1D methods can be used for the few signals that are well resolved, but 2D methods (COSY and NOESY) are generally needed. Stable isotope-assisted methods are required, however, with larger proteins. The ^{1}H-^{15}N single bond correlation map is often nearly completely resolved for proteins up to M_r 20000. Thus with complete ^{15}N-labelling, one can make detailed hydrogen exchange measurements provided that the ^{1}H-^{15}N single bond correlation map has been assigned. It takes no longer to collect a ^{1}H-^{15}N single bond correlation spectrum than a 2D homonuclear proton spectrum, but the N–H region of the former is frequently fully resolved whereas the latter has severe overlaps.

Protocol 9. Determination of hydrogen exchange protection factors from HSQC data (106) with a protein uniformly labelled with ^{15}N

1. Prepare uniformly ^{15}N-labelled protein (see *Protocol 1*).

2. Dissolve the protein in H_2O under the solvent conditions desired for the exchange experiment (500 µl sample volume, 1–5 mM protein concentration, desired ionic strength, and pH).

3. Transfer the sample to ^2H_2O. This is most easily done by lyophilizing the protein and then dissolving it in ^2H_2O, but if the protein of interest can not be lyophilized, it can be done by passing the sample through a column that has been pre-equilibrated with the ^2H_2O solution desired. In this case,

Protocol 9. *Continued*

allowance must be made for dilution of the protein and one may wish to carry out the exchange at a pH near 3.5 where N–H exchange is minimal and then titrate the protein solution to the pH at which the measurements are to be made. Transfer the sample to a 5 mm NMR tube.

4. The spectrometer should have been set-up in advance of dissolving the protein in ^2H_2O. The sample temperature should be controlled with a calibrated variable temperature unit. Collect HSQC spectra (106) at various times (at increasingly spaced intervals that span the half-time for exchange of hydrogens of interest). Collect 512 t_1 increments of 2048 complex t_2 points. Suppress solvent by means of a 1.1 sec DANTE pulse sequence, and decouple ^{15}N during acquisition by GARP phase modulation.

5. Process the data by zero-filling to yield a final digital resolution of 2 Hz/pt on the ^1H axis and 1 Hz/pt on the ^{15}N axis. Use linear prediction fitting on both ends of the free induction decays to achieve a better base plane following Fourier transformation.

6. Measure peak volumes to obtain the fractional proton occupancy. These can be referenced to the standard volume provided by a capillary insert containing [^{15}N]acetyl glycine in 50% $^1H_2O/^2H_2O$ (107).

7. Fit the peak intensities as a function of the exchange time by non-linear least squares to first-order kinetics to obtain the half-time for exchange.

8. Divide this by the half-time for exchange from model data (108) to obtain the protection factor.

The ^{15}N assisted approach is being used to probe the generation of hydrogen exchange protection during protein folding and hence to obtain structural information on intermediates. See references (109 and 110) for discussions of these methods. *Protocol 9* is easily modified for measurements (at equilibrium) of $^1H/^2H$ fractionation factors at individual sites in proteins (107).

8.2.2 ^{13}C assisted H exchange studies

As noted above, the ^{15}N assisted experiment can not be used to follow exchange rates whose half-times are shorter than the data collection period. By monitoring effects on carbonyl carbon chemical shifts and lineshapes, however, a much wider range of amide hydrogen exchange rates can be measured.

The ^{13}C chemical shift of an amide carbonyl carbon in a polypeptide, δ ($^{13}C_i'$), is influenced by the isotope on both nearest backbone nitrogens (111), by the hydrogen isotope on the nitrogen of the same residue (N_i), and by the hydrogen isotope bound to the preceding residue (N_{i+1}) (*Figure 14*). Since the isotope effect on the chemical shift is attenuated by intervening chemical

i-th   (i+1)th residue

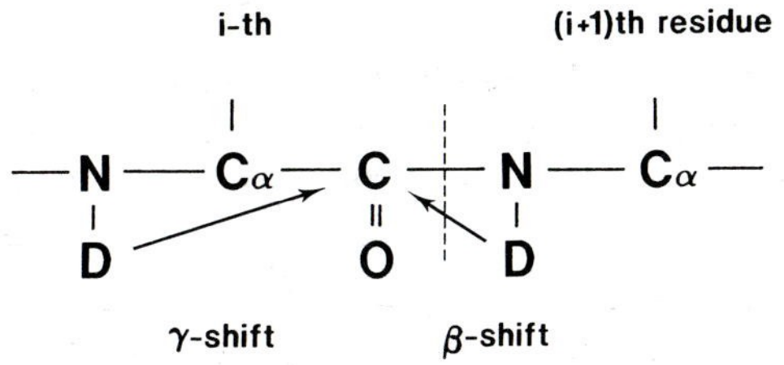

$$-N-C\alpha-C+N-C\alpha-$$

γ-shift   β-shift

Figure 14. Definition of the β- and γ-shift of the peptide carbonyl ^{13}C NMR signals of the ith residue induced by deuterium substitution of adjacent amide hydrogens. Typical values at 75 MHz ^{13}C (300 MHz for ^{1}H) are 4.7 Hz for β-shifts and 3 Hz or less for γ-shifts.

bonds, the interresidue shift (β-shift) is generally larger than the intraresidue shift (γ-shift). Typically, the β-shift is 0.05–0.09 p.p.m. (4–7 Hz at 75 MHz ^{13}C frequency), and the γ-shift is > 0.03 p.p.m. (> 2 Hz at 75 MHz) (55, 87).

By analogy with a vicinal coupling constant, the γ-shift depends on the conformation (dihedral angles ϕ_i and ψ_i). For unfavourable conformations, the isotope shift can not be resolved. On the other hand, the larger β-shift has only a slight dependence on secondary structure (the peptide bond is usually nearly planar) and can be exploited usefully for studies of hydrogen exchange.

The sum of the β- and γ-shift can be obtained by comparing the chemical shift of $^{13}C'_i$ in H_2O and ^2H_2O, assuming isotopic exchange equilibrium. In 50% $H_2O-^2H_2O$ the isotope shift would be half this value, again assuming fast exchange and an $^{1}H/^{2}H$ fractionation factor of 1.0. In general, the ^{13}C NMR lineshapes of the amide carbonyl carbons of a protein dissolved in 50% $^2H_2O-H_2O$ depend on:

- the magnitudes of the β- and γ-isotope shifts
- on the $^{1}H/^{2}H$ composition of the solution
- the lifetimes of the two neighbouring N-H protons
- on the exchange time if exchange is not at (dynamic) equilibrium
- on the $^{1}H/^{2}H$ fractionation factor.

[The fractionation factor is generally assumed to be unity, but values in proteins have recently been found to vary between 0.4 and 1.6 (107).]

Procedures have been developed for the two limiting cases of fast and slow N-H exchange. The steady state DEALS method (deuterium exchange effect on amide line shape), in which the steady state lineshapes of backbone carbonyl carbons of a protein dissolved in a 50% $^2H_2O-H_2O$ mixture are monitored, is suitable for measurements of fast amide hydrogen exchange; the time-dependent DEALS method, in which the relative areas of isotope shifted backbone carbonyl carbon signals are followed as a function of exchange time, is suitable for slow exchange processes. Both methods utilize 1D ^{13}C NMR signals from protein backbone carbonyls. The protein can be labelled uniformly with ^{13}C, but for reasons of resolution the protein generally is labelled selectively by incorporating one or a few 1-[^{13}C']-labelled amino acids. Although DEALS experiments require more (labelled) protein than techniques that employ ^{1}H detection, the approach works with large proteins. Since the N-H protons do not need to be resolved or assigned, DEALS can be applied to regions of a protein that can not be analysed by other methods.

i. Steady state DEALS experiment

This method is illustrated by results on [L]SSI (SSI in which all nine leucine residues per subunit have been selectively labelled with [98% 1-^{13}C]Leu). Since the dimer is symmetrical, the ^{13}C spectrum shows only nine enhanced carbonyl ^{13}C NMR signals in 100% ^{2}H$_2$O or 100% H$_2$O. These signals have been unambiguously assigned to individual residues by double labelling and mutagenesis.

Protocol 10. Steady state DEALS method

1. Assign the enhanced carbonyl carbon resonances by double labelling methods, site-directed mutagenesis, or other methods (112).

2. Dissolve 20 mg of [L]SSI in 0.5 ml solvent (the resulting protein concentration is about 5.6 mM as monomeric SSI). In order to determine the magnitudes of the isotope shifts, spectra have to be obtained in H$_2$O and ^{2}H$_2$O. For exchange rate determination, the solvent composition should be 50% ^{2}H$_2$O–H$_2$O. It may be useful to collect exchange data at a variety of pH values since N-H exchange rates are pH dependent. Adjustment of the pH of samples in 50% ^{2}H$_2$O–H$_2$O is best achieved by adding small amounts of 1 N 50% ^{2}HCl–HCl or 1 N 50% NaO^{2}H–NaOH as needed so as not to disturb the isotope ratio of the solvent. (In the case illustrated here, the pH was adjusted —in successive experiments—to 6.3, 7.3, and 10.0). Transfer the sample to a 5 mm NMR tube.

3. Collect 75 MHz 1D ^{13}C NMR spectra for the labelled protein sample in each solvent at each pH value. Here it may be of interest to acquire data at a variety of temperatures. (In the case shown, the temperatures were 40, 50, 60, and 70 °C.) Decouple protons by continuous irradiation. In order to provide adequate signal-to-noise and digital resolution, each spectrum consisted of 40 000 transients at an acquisition time of 2 sec (1 kHz or 13.3 p.p.m. spectral width). The total time for each NMR measurement of each spectrum was about 22 h.

4. Use resolution enhancement, (e.g. multiplication by a shifted sine-bell) to permit observation of the small γ-isotope shift; the β-shift normally can be resolved without resolution enhancement.

5. Examine the spectra for information about N-H exchange rates. (See the discussion below.)

Figure 15 shows the resolution enhanced spectra obtained by the above experiments for the solution at pH 7.3. The carbonyl carbon patterns from Leu33 and Leu14 each consist of four resolved lines. This is the pattern expected for slow exchange where the peaks (from low to high field) correspond to: (N-H)$_i$-γ-^{13}C$'_i$-β-(N-H)$_{i+1}$, the limiting shift in H$_2$O; (N-^{2}H)$_i$-γ-^{13}C$'_i$-β-(N-H)$_{i+1}$);

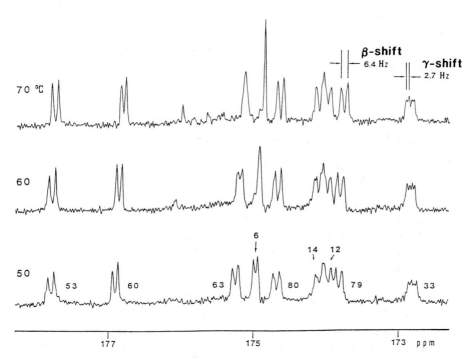

Figure 15. The steady state DEALS experiment. The example shown is *Streptomyces subtilisin* inhibitor labelled with ^{13}C$'$ Leu, [L]SSI. Spectra were acquired with samples at pH 7.3 at a ^{13}C NMR frequency of 75.4 MHz. The temperatures of the samples are given in the figure (see the text for additional details). Note that the doubled peaks observed for the carbonyls of Leu6 and Leu63 at 50°C are lost at 60 and 70°C; this happens because the backbone H^N exchange rates at the adjacent residues (Tyr7 and Thr34) become faster than the inverse of the β-shift values. The Leu60 carbonyl signal broadens slightly at 70°C and becomes a time-averaged single line at higher pH (see text). The four peak pattern of Leu33 arises from a large β-shift and a smaller γ-shift; this indicates that the exchange rates of the amide hydrogens of Thr34 and Thr32 are slow under these conditions. (M. Kainosho and H. Nagao, unpublished results.)

$(N-H)_i$-γ-$^{13}C'_i$-β-$(N-^2H)_{i+1}$; and $(N-^2H)_i$-γ-$^{13}C'_i$-β-$(N-^2H)_{i+1}$, the limiting shift in ^2H_2O. The magnitudes of the β- and γ-shifts can be readily determined for these residues from the single spectrum. The results indicate slow exchange (at pH 7.3 and between 50°C and 7°C) for the N-Hs of residues Leu14 and Leu33 and their $i + 1$ neighbours Thr15 and Thr34. By contrast, all the other leucines of SSI exhibit only at pH 7.3 and 50°C. Lack of resolved γ-shifts can indicate either that the isotope shift is small (as the result of the backbone conformation) or that exchange is fast at the leucine N-Hs. The presence of resolved β-shifts indicates that exchange is slow at all $(N-H)_{i+1}$. At 70°C, the β-shifts of the Leu63 and Leu6 carbonyls are averaged by fast exchange at the respective $i + 1$ N-Hs (Tyr7 and Thr64). At pH 10 (spectrum not shown), the β-shift for Leu60 is averaged out indicating rapid exchange of the Asn61 N-H. The other β-shifts do not collapse even at pH 10 and 70°C (spectrum not shown). For the β-shifts, slow exchange corresponds to lifetimes in excess of $1/\beta$-isotope shift (Hz). Thus, N-Hs that exhibit resolved β-shifts have lifetimes greater than about 300 msec, and N-Hs that exhibit collapsed β-shifts have lifetimes less than about 100 msec. In the X-ray crystal structure of SSI, Leu6, Leu60, and Leu63 are on the protein surface, and all other Leu residues are located in the internal hydrophobic core. The steady state DEALS results are consistent with this structure. The experiment can provide interesting structural information for proteins whose structure is not known.

The only limit on the length of the data acquisition in the steady state DEALS experiment is the stability of the protein. Hence the method can be used for proteins that are not highly soluble. The time resolution is limited only by resolution of the isotope shifts. With small proteins at high magnetic fields, it should be possible to study exchange rates on the order of a few milliseconds. The experiment shows the existence of slow N-H exchange but does not provide more than a lower limit for the exchange rate.

ii. Time-dependent DEALS experiment

This method can be used to study N-H exchange rates with half-times greater than or similar to the NMR data collection time. It involves measuring the intensities of isotope shifted carbonyl carbon peaks of a protein as a function of time following transfer between H_2O and ^2H_2O (K. Uchida and M. Kainosho, to be published). This experiment also requires selectively labelled protein, for example [L]SSI. Note that the experimental procedure is similar to more conventional amide hydrogen exchange experiments (which utilize ^1H or $^1H\{^{15}N\}$ detection) except that detection is by the isotope effect on the directly observed ^{13}C NMR spectrum.

Protocol 11. Time-dependent DEALS experiment

1. To 60 mg of [L]SSI, add 2.0 ml of 0.05 M phosphate buffer, pH 7.1. Adjust the pH to 7.1. Lyophilize to dryness.

2. Dissolve the lyophilized sample by adding 2.0 ml of 100% ^2H_2O at 0°C. (The final volume will be approximately 2.6 ml.) Quickly transfer the sample to a 10 mm NMR tube and insert into the NMR probe.

3. The ^{13}C NMR probe should have been pre-equilibrated to the desired temperature (50°C in the case shown here) and shimmed with another protein solution of the same volume and salt concentration as the sample to be studied ('dummy solution').

4. Immediately acquire the first 1D 75 MHz ^{13}C NMR spectrum. Acquisition parameters: single pulse experiment; spectral width 1 kHz (13.3 p.p.m.); 1024 transients; acquisition time 2 sec; total of 34 min for each spectrum.

5. After the first spectrum has been accumulated, a few minutes can be taken to optimize the field. Additional spectra are then acquired at increasingly spaced intervals.

6. The areas of the $^{13}C'$ peaks are measured from spectra normalized to the same gain. The $(N-H)_i$-γ-$^{13}C'_i$-β-$(N-H)_{i+1}$ peak will decrease in area, and one or more isotope shifted peaks will grow in intensity. The peak areas are fitted to first-order kinetics by non-linear least squares analysis to provide the exchange rates. Time points for each spectrum are taken as the time elapsed between dissolving the protein in ^2H_2O and the midpoint of each spectral accumulation.

Figure 16 is a typical time-dependent DEALS result for [L]SSI. Note that the lineshapes of the carbonyl signals assigned to Leu6, Leu60, and Leu63 remained constant with time. This indicates that each $(N-H)_{i+1}$ reached equilibrium before the first spectrum was collected (as would be predicted from the steady state DEALS results presented above). The relatively rapid development of a β-shift satellite to the Leu53 peak provides the exchange rate for the N-H of its $i + 1$ neighbour Ala54, which is located in an α-helix. The $i + 1$ neighbours of Leu12, Leu14, Leu33, Leu79, and Leu80, which are all located in β-sheet, exchange more slowly. Leu33 is the only residue that develops a γ-shift satellite (2.7 Hz upfield); this indicates that the N-H of Leu33 exchanges more rapidly than that of Thr34 or that the N-H of Thr34 has already reached equilibrium by the time that the first spectrum is obtained. (These two possibilities can be distinguished on the basis of the chemical shift of the $^{13}C'$ peak in pure ^2H_2O.)

Acknowledgements

The authors thank their co-workers for supplying many of the practical procedures reported here. Supported by NIH grants RR02301 and GM35976 (to JLM), by a grant-in-aid from the Ministry of Education, the Science and Technology Agency (to MK), and a NEDO International Grant (to JLM and MK).

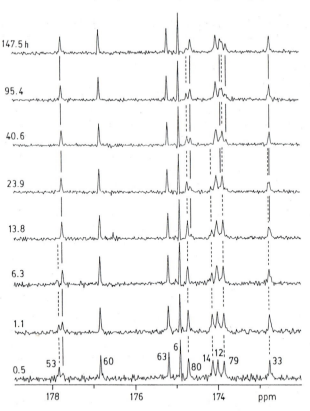

Figure 16. The time-dependent DEALS experiment with (H → ²H) exchange. The example shown is of [L]SSI (see the caption to *Figure 15*) at pH 7.1, 50°C, and a ¹³C NMR frequency of 75.4 MHz (see text for details). The *vertical broken lines* indicate the positions of signals that decrease in intensity, and the *vertical solid lines* indicates signals that grow in intensity during the exchange experiment. The carbonyl signals assigned to Leu6, Leu60, and Leu63 show no changes over the time scale of the experiment. As shown by the steady state DEALS experiment (*Figure 15*), the amide hydrogens of the N-terminal neighbours of these residues show rapid hydrogen exchange. (K. Uchida and M. Kainosho, unpublished results.)

References

1. Wüthrich, K. (1986). *NMR of proteins and nucleic acids*. Wiley, New York.
2. Wagner, G., Hyberts, S. G., and Havel, T. F. (1992). *Annu. Rev. Biophys. Biomol. Struct.*, **21**, 167.
3. Markley, J. L. (1989). In *Methods in enzymology*, Vol. 176 (ed. N. J. Oppenheimer and T. L. James), pp. 12–14. Academic Press.
4. Stockman, B. J. and Markley, J. L. (1990). In *Advances in biophysical chemistry*, Vol. 1, (ed. C. A. Bush), pp. 1–46. JAI Press Inc., Greenwich, Connecticut.
5. McIntosh, L. P. and Dahlquist, F. W. (1990). *Q. Rev. Biophys.*, **23**, 1.
6. LeMaster, D. R. (1989). *Annu. Rev. Biophys. Biophys. Chem.*, **19**, 243.

7. Fesik, S. W. and Zuiderweg, E. R. P. (1990). *Q. Rev. Biophys., 23,* 97.
8. Clore, G. M. and Gronenborn, A. M. (1991). *Annu. Rev. Biophys. Biophys. Chem., 20,* 29.
9. Muchmore, D. C., McIntosh, L. P., Russell, C. B., Anderson, D. E., and Dahlquist, F. W. (1989). In *Methods in enzymology*, Vol. 177 (ed. N. J. Oppenheimer and T. L. James), pp. 44–73. Academic Press.
10. Hibler, D. W., Harpold, L., Dell'Acqua, M., Pourmotabbed, T., Gerlt, J. A., Wilde, J. A., and Bolton, P. H. (1989). In *Methods in enzymology*, Vol. 177 (ed. N. J. Oppenheimer and T. L. James), pp. 74–85. Academic Press.
11. LeMaster, D. M. (1989). In *Methods in enzymology*, Vol. 177 (ed. N. J. Oppenheimer and T. L. James), pp. 23–43. Academic Press.
12. Ott, G. G. (1981). In *Syntheses with stable isotopes,* pp. 224, Wiley-Interscience, New York.
13. Markley, J. L. (1972). In *Methods in enzymology*, Vol. 26 (ed. C. H. W. Hirs and S. N Timasheff), pp. 605–627. Academic Press.
14. Crespi, H. L. and Katz, J. J. (1972). In *Methods in enzymology*, Vol. 26 (ed. C. H. W. Hirs and S. N. Timasheff), pp. 627–637. Academic Press.
15. LeMaster, D. M. and Richards, F. M. (1982). *Anal. Biochem., 122,* 238.
16. Sørensen, P. and Poulsen, F. (1992). *J. Biomolec. NMR, 2,* 99.
17. Sambrook, J., Fritsch, E. F., and Maniatis, T. (ed.) (1989). *Molecular Cloning, A Laboratory Manual*, 2nd edn, Cold Spring Harbor Laboratory Press.
18. Studier, F. W. and Moffatt, B. A. (1986). *J. Mol. Biol., 189,* 113.
19. Anglister, J., Bond, M. W., Frey, T., Leahy, D., Levitt, M., McConnell, H. M., Rule, G. S., Tomasello, J., and Whittaker, M. (1987). *Biochemistry, 26,* 6058.
20. Kato, K., Matsunaga, C., Nishimura, Y., Waelchi, M., Kainosho, M., and Arata, Y. (1989). *J. Biochem.* (Tokyo), *105,* 867.
21. Seeholzer, S. H., Cohn, M., Putkey, R. A., Means, A. R., and Crespi, H. L. (1986). *Proc. Natl Acad. Sci. USA., 83,* 3634.
22. Crespi, H. L., Kostka, A. G., and Smith, U. H. (1974). *Biochem. Biophys. Res. Commun., 61,* 1407.
23. Torchia, D. A., Sparks, S. W., and Bax, A. (1988). *J. Am. Chem. Soc., 110,* 2320.
24. Markley, J. L., Putter, I., and Jardetzky, O. (1968). *Science, 161,* 1249.
25. Markley, J. L. and Cheung, S. M. (1973). (*Proc. Conf. Stable Isotopes in Biol. Chem. and Med.*, Argonne, IL, Mary 1973), U.S. Atomic Energy Commission CONF-730525, 103.
26. Bradbury, J. H. and Norton, R. S. (1976). *Mol. Cell. Biochem., 13,* 113.
27. Feeney, J., Roberts, G. C. K., Birdsall, B., Griffiths, D. V., King, R. W., Scudder, P., and Burgen, A. S. V. (1977). *Proc. R. Soc. London, Ser. B, 196,* 267.
28. LeMaster, D. M. and Richards, F. M. (1988). *Biochemistry, 27,* 142.
29. Stella, V. J. (1973). *J. Pharm. Sci., 62,* 634.
30. Tenenbaum, S. W., Witherup, T. H., and Abbott, E. H. (1974). *Biochim. Biophys. Acta., 362,* 308.
31. LeMaster, D. M. and Richards, F. M. (1982). *J. Labelled Compd. Radiopharm., 19,* 639.
32. Neri, D., Szyperski, T., Otting, G., Senn, H., and Wüthrich, K. (1989). *Biochemistry, 28,* 7510.

33. Marion, D., Ikura, M., Tschudin, R., and Bax, A. (1989). *J. Magn. Reson.*, **85**, 393.

34. Appendix: Computer Programs Related to Nuclear Magnetic Resonance. (1989). In *Methods in enzymology*, Vol. 177 (ed. N. J. Oppenheimer and T. L. James) pp. 455–467. Academic Press.

35. Hoch, J., Poulsen, F., and Redfield, C. (ed.) (1991). *Computational, aspects of the study of biological macromolecules by nuclear magnetic resonance spectroscopy.* Plenum Press, New York.

36. Eccles, C., Güntert, P., Billeter, M., and Wüthrich, K. (1991). *J. Biomolec. NMR,* **1**, 111.

37. Stockman, B. J., Reily, M. D., Westler, W. M., Ulrich, E. L., and Markley, J. L. (1989). *Biochemistry,* **28**, 230.

38. Oda, Y., Nakamura, H., Yamazaki, T., Nagayama, K., Yoshida, M., Kanaya, S., and Ikehara, M. (1992). *J. Biomolec. NMR,* **2**, 137.

39. Pachter, R., Arrowsmith, C. H., and Jardetzky, O. (1992). *J. Biomolec. NMR,* **2**, 183.

40. Oh, B.-H., Westler, W. M., Darba, P., and Markley, J. L. (1988). *Science,* **240**, 908.

41. Westler, W. M., Stockman, B. J., Hosoya, Y., Miyake, Y., and Kainosho, M. (1988). *J. Am. Chem. Soc.,* **110**, 6265.

42. Oh, B.-H., Mooberry, E. S., and Markley, J. L. (1990). *Biochemistry,* **29**, 4004.

43. Torchia, D. A., Sparks, S. W., and Bax, A. (1988). *Biochemistry,* **27**, 5135.

44. Wang, J., LeMaster, D. M., and Markley, J. L. (1990). *Biochemistry,* **29**, 88.

45. Wang, J., Hinck, A. P., Loh, S. N., LeMaster, D. M., and Markley, J. L. (1992). *Biochemistry,* **31**, 921.

46. Shon, K. and Opella, S. J. (1989). *J. Magn. Reson.,* **82**, 193.

47. Fesik, S. and Zuiderweg, E. R. P. (1988). *J. Magn. Reson.,* **78**, 588.

48. Marion, D., Kay, L. E., Sparks, S. W., Torchia, D. A., and Bax, A. (1989). *J. Am. Chem. Soc.,* **111**, 1515.

49. Zuiderweg, E. R. P. and Fesik, S. W. (1989). *Biochemistry,* **28**, 2387.

50. Oh, B.-H. and Markley, J. L. (1990). *Biochemistry,* **29**, 3993.

51. Wang, J., Hinck, A. P., Loh, S. M., and Markley, J. L. (1990). *Biochemistry,* **29**, 102.

52. Zuiderweg, E. R. P., McIntosh, L. P., Dahlquist, F. W., and Fesik, S. W. (1990). *J. Magn. Reson.,* **86**, 210.

53. Torchia, D. A., Sparks, S. W., Young, P. E., and Bax, A. (1989). *J. Am. Chem. Soc.,* **111**, 8315.

54. Wang, J., Mooberry, E. S., Walkenhorst, W. F., and Markley, J. L. (1992). *Biochemistry,* **31**, 911.

55. Kainosho, M. and Tsuji, T. (1982). *Biochemistry,* **24**, 6273.

56. Kay, L. E., Ikua, M., Tschudin, R., and Bax, A. (1990). *J. Magn. Reson.,* **89**, 496.

57. Powers, R., Gronenborn, A. M., Clore, G. M., and Bax, A. (1991). *J. Magn. Reson.,* **94**, 209.

58. Bax, A. and Ikura, M. (1991). *J. Biomolec. NMR,* **1**, 99.

59. Kay, L. E., Clore, G. M., Bax, A., and Gronenborn, A. M. (1990). *Science,* **249**, 411.

60. Kay, L. E., Ikura, M., Zhu, G., and Bax, A. (1991). *J. Magn. Reson.,* **91**, 422.

61. Fesik, S. W., Eaton, H. L., Olejniczak, E. T., Zuiderweg, E. R. P., McIntosh, L. P., and Dahlquist, F. W. (1990). *J. Am. Chem. Soc.*, **112**, 886.
62. Edison, A. S., Westler, W. M., and Markley, J. L. (1991). *J. Magn. Reson.*, **9**, 434.
63. Baldisseri, D. M., Pelton, J. G., Sparks, S. W., and Torchia, D. A. (1991). *FEBS Lett.*, **281**, 33.
64. Ikura, M., Bax, A., Clore, G. M., and Gronenborn, A. M. (1990). *J. Am. Chem. Soc.*, **112**, 9020.
65. Clore, G. M., Kay, L. E., Bax, A., and Gronenborn, A. M. (1991). *Biochemistry*, **30**, 12.
66. Zuiderweg, E. R. P., Petros, A. M., Fesik, S. W., and Olejniczak, E. T. (1991). *J. Am. Chem. Soc.*, **113**, 370.
67. Griffey, R. H. and Redfield, A. G. (1987). *Q. Rev. Biophys.*, **19**, 51.
68. Otting, G. and Wüthrich, K. (1990). *Q. Rev. Biophys.*, **23**, 39.
69. Dzakula, Z., Westler, W. M., Edison, A. S., and Markley, J. L. (1992). *J. Am. Chem. Soc.*, **114**, 6207.
70. Kay, L. E., Brooks, B., Sparks, S. W., Torchia, D. A., and Bax, A. (1989). *J. Am. Chem. Soc.*, **111**, 5488.
71. Griesinger, C., Sørensen, O. W., and Ernst, R. R. (1985). *J. Am. Chem. Soc.*, **107**, 6494.
72. Montelione, G. T. and Wagner, G. (1989). *J. Am. Chem. Soc.*, **111**, 5474.
73. Montelione, G. T., Winkler, M. E., Rauenbeuhler, P., and Wagner, G. (1989). *J. Magn. Reson.*, **82**, 198.
74. Wider, G., Neri, D., Otting, G., and Wüthrich, K. (1989). *J. Magn. Reson.*, **85**, 426.
75. Wagner, G., Schmieder, P., and Thanabal, V. (1991). *J. Magn. Reson.*, **93**, 436.
76. Schmieder, P., Thanabal, V., McIntosh, L. P., Dahlquist, F. W., and Wagner, G. (1991). *J. Am. Chem. Soc.*, **113**, 6323.
77. Gemmecker, G. and Fesik, S. W. (1991). *J. Magn. Reson.*, **95**, 208.
78. Wisehart, D. S., Sykes, B. D., and Richards, F. M. (1992). *J. Mol. Biol.*, **222**, 311.
79. Bertini, I. and Luchinat, C. (1986). *NMR of paramagnetic molecules in biological systems.* Benjamin/Cummings, Menlo Park, CA.
80. Perkins, S. J. (1982). *Biol. Magn. Reson.*, (ed. L. J. Berliner and J. Reuben), **4**, 193.
81. Ösapay, K. and Case, D. A. (1991). *J. Am. Chem. Soc.*, **113**, 9436.
82. Wisehart, D. S., Sykes, B. D., and Richards, F. M. (1992). *Biochemistry*, **31**, 1647.
83. Spera, S. and Bax, A. (1991). *J. Am. Chem. Soc.*, **113**, 5490.
84. Fesik, S. W., Gemmecker, G., Olejniczak, E. T., and Petros, A. M. (1991). *J. Am. Chem. Soc.*, **113**, 7080.
85. Petros, A. M., Neri, P., and Fesik, S. W. (1992). *J. Biomolec. NMR*, **2**, 11.
86. Hiromi, K., Akasaka, K., Mitsui, Y., Tonomura, B., and Murao, S. (ed.) (1985). *Protein protease inhibitor—the case of Streptomyces subtilisin inhibitor.* Elsevier, Amsterdam.
87. Kainosho, M., Nagao, H., and Tsuji, T. (1987). *Biochemistry*, **26**, 1068.
88. Kainosho, M., Nagao, H., Imamura, Y., Uchida, K., Tomonaga, N., Nakamura, Y., and Tsuji, T. (1985). *J. Mol. Struct.*, **126**, 549.

89. Uchida, K., Miyake, Y., and Kainosho, M. (1991). *J. Biomolec. NMR,* **1,** 49.
90. Markley, J. L. (1975). *Accts. Chem. Res.,* **8,** 70.
91. Grissom, C. B. and Markley, J. L. (1989). *Biochemistry,* **28,** 2116.
92. Stockman, B. J. and Markley, J. L. (1992). *Curr. Opinion Struct. Biol.,* **2,** 52.
93. Fesik, W. W., Gampe, R. T. Jr., Holzman, T. F., Egan, D. A., Edalji, R., Luly, J. R., Simmer, R., Helfrich, R., Kishore, V., and Rich, D. H. (1990). *Science,* **250,** 1406.
94. Weber, C., Wider, G., von Freyberg, B., Traber, R., Braun, W., Widmer, H., and Wüthrich, K. (1991). *Biochemistry,* **30,** 6563.
95. Fesik, S. W., Gampe, R. T. Jr., Eaton, H. L., Gemmecker, G., Olejniczak, E. T., Neri, P., Holzman, T. F., Egan, D. A., Edalji, R., Simmer, R., Helfrich, R., Hochlowski, J., and Jackson, M. (1991). *Biochemistry,* **31,** 6574.
96. Stockman, B. J., Krezel, A. M., Markley, J. L., Leonhardt, K. G., and Straus, N. A. (1990). *Biochemistry,* **30,** 9600.
97. Forman-Kay, J. D., Gronenborn, A. M., Wingfield, P. T., and Clore, G. M. (1991). *J. Mol. Biol.,* **220,** 209.
98. Wagner, G., Nirmala, N. R., Montelione, G. T., and Hyberts, S. (1990). In *Frontiers of NMR in molecular biology* (ed. D. Live, I. M. Armitage, and D. Patel) pp. 129–143. Alan R. Liss, Inc.
99. Clore, G. M., Driscoll, P. C., Wingfield, P. T., and Gronenborn, A. M. (1990). *Biochemistry,* **29,** 7387.
100. Kay, L. E., Torchia, D. A., and Bax, A. (1989). *Biochemistry,* **28,** 8972.
101. Peng, J. W., Thanabal, V., and Wagner, G. (1991). *J. Magn. Reson.,* **94,** 82.
102. Peng, J. W., Thanabal, V., and Wagner, G. (1991). *J. Magn. Reson.,* **95,** 421.
103. Fejzo, J., Westler, W. M., Macura, S., and Markley, J. L. (1990). *J. Am. Chem. Soc.,* **112,** 2574.
104. Alexandrescu, A. T., Loh, S. N., and Markley, J. L. (1990). *J. Magn. Reson.,* **87,** 523.
105. Wider, G., Neri, D., and Wüthrich, K. (1991). *J. Biomolec. NMR,* **1,** 93.
106. Zuiderweg, E. R. P. (1990). *J. Magn. Reson.,* **86,** 386.
107. Loh, S. N., and Markley, J. L. (1993). In *Techniques in protein chemistry III,* (ed. R. Angeletti). Academic Press, New York; Loh, S. N. (1993) Ph.D. thesis, University of Wisconsin, Madison.
108. Molday, R. S., Englander, S. W., and Kallen, R. G. (1972). *Biochemistry,* **11,** 150.
109. Baldwin, R. L. and Roder, H. (1991). *Curr. Biol.,* **1,** 218.
110. Briggs, M. S. and Roder, H. (1992). *Proc. Natl Acad. Sci. USA,* **89,** 2017.
111. Feeney, J., Partington, P., and Roberts, G. C. K. (1974). *J. Magn. Reson.,* **13,** 268.
112. Westler, W. M., Kainosho, M., Nagao, H., Tomonaga, N., and Markley, J. L. (1988). *J. Am. Chem. Soc.,* **110,** 4093.

6

Effects of chemical exchange on NMR spectra

LU-YUN LIAN and GORDON C. K. ROBERTS

1. Introduction—general considerations

In the context of NMR, 'chemical exchange' refers to any process in which a nucleus exchanges between two or more environments in which its NMR parameters, (e.g. chemical shift, scalar coupling, or relaxation) differ. These may be intramolecular or intermolecular processes.

Intramolecular exchange processes include:

- motions of side chains in proteins (1)
- helix-coil transitions of nucleic acids (2)
- unfolding of proteins (3)
- conformational equilibria (4)

Intermolecular exchange processes include:

- binding of small molecules to macromolecules (5)
- protonation/deprotonation equilibria of ionizable groups (6)
- isotope exchange processes (notably the exchange of labile protons of a macromolecule with solvent) (7)
- enzyme catalysed reactions (8)

(The references are to single illustrative examples of each kind of exchange.) In all these cases, the effect of the exchange process on the NMR spectrum depends upon its rate, *relative to the magnitude of the accompanying change in NMR parameters*.

Thus if an ionization process, for example, was accompanied by a change in chemical shift of 250 Hz, an exchange rate of $>1000 \text{ sec}^{-1}$ would be regarded as *fast*, one of $\approx 250 \text{ sec}^{-1}$ as *intermediate*, and one of $<100 \text{ sec}^{-1}$ as *slow* on the 'NMR time scale'. Each of these regions of exchange leads to characteristic effects on the NMR spectrum (*Figure 1*). In slow exchange, separate resonances are seen for the nucleus of interest in each of the two states. By

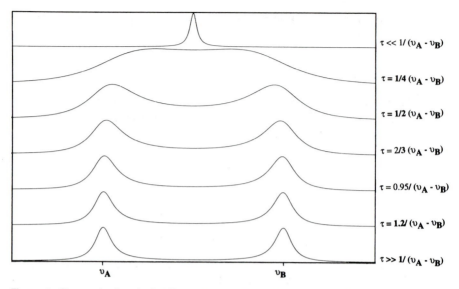

Figure 1. Change in chemical shifts and linewidths in the presence of chemical exchange between two equally populated environments. *Bottom* to *top*: increasing exchange rates as denoted in the figure.

contrast, in fast exchange a single averaged resonance is observed. In the intermediate exchange region, there is a complex series of changes in appearance of the spectrum as the two separate signals coalesce into one. Bearing in mind the typical values of NMR parameters, the range of rates which can be studied in approximately 0.05–5000 sec^{-1}, which can be extended to faster rates by the use of nuclei with larger chemical shift ranges than ^{1}H, notably ^{19}F and ^{13}C; within this range, analysis of the spectrum can provide estimates of the rate constants of exchange. A major advantage of NMR in this context is its ability to provide values for rate constants for a reaction *at equilibrium*.

Detailed descriptions of these exchange effects in NMR spectra can be found in a number of textbooks and review articles (9–15). In this chapter we shall describe the basic techniques commonly used for the qualitative and quantitative analysis of exchange effects in NMR spectra of biological macro-molecules, drawing our examples from proteins, although the same approach can of course be applied to other systems. For simplicity, and because it represents by far the most commonly encountered case, we shall describe the methods in terms of the situation in which the chemical shift of the nucleus is the main parameter differing between the states. Because of the typical magnitudes of scalar couplings (for ^{1}H-^{1}H couplings, 0–20 sec^{-1}) and relaxation rates (1–300 sec^{-1}), these are commonly seen to be in the fast exchange regime—that is, as averages. Some of the consequences of this are discussed briefly in Section 4.4. We shall also restrict discussion largely to exchange

154

between two states, while indicating in general terms how the methods can be extended to more complex processes.

2. Definitions

In discussing the effects of exchange on NMR spectra, an important parameter is the lifetime of a particular stae, defined as follows:
For a two-site first-order exchange

$$A \underset{k_{-1}}{\overset{k_{+1}}{\rightleftharpoons}} B \tag{1}$$

$$\text{Lifetime of state A, } \tau_A = 1/k_{+1} \tag{2}$$

$$\text{Lifetime of state B, } \tau_B = 1/k_{-1} \tag{3}$$

For a two-site second-order exchange, for example the binding of a ligand, L, to an enzyme, E (see also reference 15)

$$E + L \underset{k_{-1}}{\overset{k_{+1}}{\rightleftharpoons}} EL \tag{4}$$

For a nucleus in the *ligand* molecule:

$$\text{Lifetime in state EL, } \tau_{EL} = 1/k_{-1} \tag{5}$$

$$\text{Lifetime in state L, } \tau_L = 1/k_{+1}[E], \tag{6}$$

where [E] is the concentration of free enzyme; since [E] will vary with the concentration of ligand, so will τ_L. A more convenient form of eqn 6 can be obtained by using the definition of the equilibrium constant for the reaction

$$K_d = \frac{k_{-1}}{k_{+1}} = \frac{[E][L]}{[EL]} \tag{7}$$

and substituting for [E] to give

$$\tau_L = \frac{[L]}{k_{-1}[EL]} = \frac{p_L}{k_{-1}p_{EL}} \tag{8}$$

where

$$p_L = \frac{[L]}{[L] + [EL]} \tag{9}$$

and

$$p_{EL} = \frac{[EL]}{[L] + [EL]} \tag{10}$$

are, respectively, the fraction of the ligand in the free and bound states. (A similar nomenclature will be used for the fractional populations of other species; we also define E_T and L_T as the total concentrations of enzyme and ligand, respectively.)

Similarly for a nucleus in the *enzyme*:

$$\text{Lifetime in state EL, } \tau_{EL} = 1/k_{-1}$$

$$\text{Lifetime in state E, } \tau_E = \frac{1}{k_{+1}[L]} = \frac{p_E}{k_{-1}p_{EL}} \qquad [11]$$

In many discussions of exchange effects, a single lifetime is used to characterize the process,

$$1/\tau = 1/\tau_A + 1/\tau_B = k_{+1} + k_{-1} \qquad [12]$$

and similarly for a second-order exchange

$$1/\tau = 1/\tau_{EL} + 1/\tau_L = k_{-1} (1 + p_{EL}/p_L). \qquad [13]$$

The NMR parameters characteristic of state A are chemical shift, δ_A, or resonance frequency, ν_A, scalar coupling constant, J_A, relaxation rates, $1/T_{1A} (=R_A)$, $1/T_{2A}$, with a similar nomenclature for the other states (the linewidth, $LW = 1/\pi T_2$). Note that the resonance frequency (chemical shift) is expressed in hertz rather than the dimensionless p.p.m. units; it is often also expressed as $\omega_A = 2\pi\nu_A$ radians sec^{-1}. *Table 1* shows the definition of the slow, intermediate, and fast exchange in terms of these parameters. It is clear that the three parameters define different, although partially overlapping, time scales. A system in fast exchange at a low frequency can sometimes be found to be in intermediate or even slow exchange on higher field spectrometers. Similarly, it is entirely possible for the same exchange process to lead to fast exchange behaviour of one signal in the spectrum and slow exchange behaviour of another, dependin simply

Table 1. The NMR time scale

	Exchange rate		
Time scale	Slow	Intermediate	Fast
Chemical shifts, δ [a]	$k \ll \delta_A - \delta_B$	$k = \delta_A - \delta_B$	$k \gg \delta_A - \delta_B$
Coupling constant, J [b]	$k \ll J_A - J_B$	$k = J_A - J_B$	$k \gg J_A - J_B$
T_2 relaxation [c]	$k \ll \dfrac{1}{T_{2,A}} - \dfrac{1}{T_{2,B}}$	$k = \dfrac{1}{T_{2,A}} - \dfrac{1}{T_{2,B}}$	$k \gg \dfrac{1}{T_{2,A}} - \dfrac{1}{T_{2,B}}$

[a] $(\delta_A - \delta_B)$ is typically in the order of hundreds of hertz.
[b] J is typically in the order of 1–10 Hz.
[c] $\dfrac{1}{T_{2,A}} - \dfrac{1}{T_{2,B}}$ is typically in the order of 1–20 Hz for protons; the linewidth at half-height, $\Delta = 1/\pi T_2$.

ing simply on the extent to which their NMR parameters are affected by the process in question.

3. Initial qualitative analysis

The first-stage in the analysis of exchange effects in an NMR spectrum is to establish to which region of exchange the spectrum corresponds. This should, for example, be the first-step in any study of protein–ligand interactions by NMR.

The general experimental methods and precautions regarding sample preparation (Chapter 2) and the acquisition of the NMR data (Chapter 3) apply also when exchange effects are to be analysed. Some points are, however, worth reiterating. The sensitivity of many rate processes to pH and temperature makes it important that these parameters are accurately controlled. In studying intermolecular exchange processes, the accuracy of the rate constants obtained will depend directly on the accuracy with which the concentrations of the ligand and the protein are measured. Finally, many of the analyses depend upon measurements of lineshape or intensity, and it is important to avoid saturation effects—that is, to use a sufficient pulse interval—and to obtain spectra with adequate digital resolution.

To distinguish between fast, intermediate, and slow exchange one must alter the position of the equilibrium and/or the rate of exchange, most easily by varying the temperature or, for second-order exchange, the concentration. As wide a range as possible of temperature or concentration should be used; it is usually easy to identify the exchange region for a well-resolved signal, but in the complex spectra of macromolecules it may only be possible to see the signal of interest over a limited range of temperature or concentration. The experimental procedures are described in *Protocols 1* and *2* for first- and second-order exchange, respectively.

Protocol 1. To determine the exchange regime for a first-order
exchange process

1. Obtain spectra at intervals of 10°C, starting at 5–10°C and going up to as high a temperature as possible, consistent with the stability of the protein.

2. Look for changes in the spectrum; the nature of these progressive changes as the temperature is increased reveals the region of exchange.

   (a) Many resonances in a protein NMR spectrum show some temperature dependence of their chemical shift. A particularly marked temperature dependence for one or a few signals may indicate an exchange process in *fast* exchange, but this is difficult to identify unambiguously.

   (b) Resonance broadens markedly, often becoming unobservable—*intermediate* exchange.

Protocol 1. *Continued*

    (c) Resonance broadens as the temperature is increased, or first sharpens and then broadens, remaining as a discrete resonance; may also change in intensity—*slow* exchange.

    (d) Resonance changes in intensity, with little detectable change in linewidth—*very slow* exchange.

Protocol 2. To determine the exchange regime for a second-order exchange process

(A protein–ligand binding interaction is used as an example. The same protocol can be used in a pH titration experiment.)

1. Obtain spectra of the protein without ligand present and of the free ligand.

2. Add the ligand to the protein sample in small increments, obtaining a spectrum after each addition. It is important that the first addition should give a ligand concentration much less than that of the protein (approximately 0.1–0.2 molar equivalents).

3. Look for changes in the spectrum; the nature of these progressive changes as the ligand concentration is increased reveals the region of exchange.

    (a) New resonance appears, increases in intensity, and shifts progressively as the ligand concentration is increased, approaching the chemical shift of a ligand resonance at high ligand concentration—*ligand resonance in fast exchange*.

    (b) Resonance shifts progressively without change in intensity—*protein resonance in fast exchange*.

    (c) In either of the above cases, the resonance may broaden at intermediate ligand concentrations if the exchange rate is only '*moderately fast*'.

    (d) Existing resonance (protein) or new resonance (ligand), sharp at very low ligand concentrations, broadens markedly as ligand concentration increases, often becoming undetectable at around [ligand]/[protein] ~ 2; a signal reappears at high ligand concentration, and may shift somewhat as the concentration is further increased—*intermediate* exchange.

    (e) New resonance appears and increases in intensity without changing chemical shift—resonance from either *ligand* or *protein* in *slow* exchange (assignment requires magnetization transfer experiments—see below).

4. Repeat steps **2** and **3** until no further changes are observed in the NMR spectrum; a spectrum should be obtained at a [ligand]/[protein] ratio of at least five, higher for weakly binding ligands.

5. If the results are not clear-cut, examine the temperature effects on the spectrum, as outlined in *Protocol 1*, at an intermediate ligand concentration, typically [ligand]/[protein] $\approx$ 2.

Qualitative examination of changes in chemical shift or linewidth (relaxation rate) thus allows one to identify the exchange region; more detailed analysis of these same changes leads to quantitative estimates of the rates of exchange. The following sections describe methods for obtaining such quantitative information.

The methods used for the analysis of exchange effects fall into two main classes:

- lineshape analysis
- magnetization transfer (slow exchange only)

The details of the analysis depend on the exchange region and on whether the exchange is a first- or second-order process. Each of these situations will be considered in turn.

4. Lineshape analysis

4.1 General

The equations describing the shape of resonance lines in the presence of exchange are complicated, and the only general approach to lineshape analysis is the iterative fitting of computer simulated lineshapes to experimental spectra. Considerable simplifications are possible for the fast and slow exchange cases, and these are described below in Sections 4.2 and 4.3 (see also 16, 17).

In carrying out a lineshape simulation, it is important to distinguish between cases which involve changes in scalar coupling from those which do not. The spectra shown in *Figure 2* (18) correspond to simple resonances without homonuclear scalar coupling. Exchange effects on lines of this kind can be simulated by using McConnell's modification of the Bloch equations (19). For a two-site exchange, the lineshape (amplitude as a function of frequency) is given by the imaginary part of $G(\nu)$,

$$G(\nu) = \frac{iC[2p_Ap_B\tau - \tau^2(p_A\alpha_B + p_B\alpha_A)]}{p_Ap_B - \tau^2\alpha_A\alpha_B} \qquad [14]$$

where C is a scaling factor, p_A and p_B are the fractional population of species A and B respectively, τ is defined by eqn 12, and

$$\alpha_A = 2\pi i(\nu_A - \nu) + (1/T_{2A}) + p_B/\tau \qquad [15]$$

$$\alpha_B = 2\pi i(\nu_B - \nu) + (1/T_{2B}) + p_A/\tau \qquad [16]$$

Lineshape simulation for the case where there is no scalar coupling thus requires knowledge of the resonance frequencies, populations, relaxation

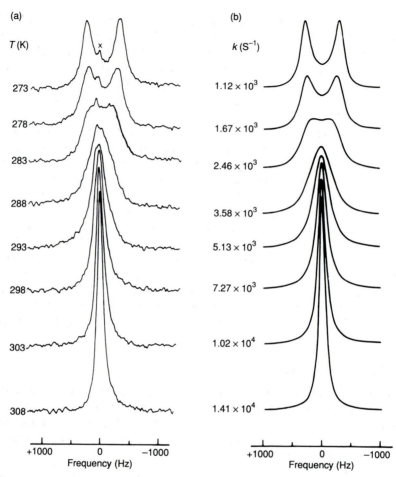

Figure 2. Observed (a) and calculated (b) 188.3 MHz ^{19}F NMR spectra of 3',5'-difluoro-methotrexate bound to *L. casei* dihydrofolate reductase. The sample contained less than one molar equivalent of difluoromethotrexate so that all the ligand is bound to the enzyme. The calculated spectra were derived from best-fit parameters. The 'best-fit' rate and experimental temperature for each pair of experimental and calculated spectra are indicated. (Reproduced with permission from (18).)

rates, and lifetimes of the two states. In principle all these parameters can be estimated by least squares fitting of eqn 14 to the experimental data, using any one of a wide variety of general non-linear regression programs. However, it is important to obtain as good initial estimates of the parameters as possible to ensure successful convergence of the fitting process. Estimates of resonance frequencies and relaxation rates may be obtained at low temperature, where exchange is slow, and initial estimates of population and lifetime may then be obtained from the appearance of the spectrum. It is also desirable, where possible, to fit data at a number of different exchange rates and/or populations—that is, at several temperatures (18) and, for example, ligand concentrations. It is most important to obtain spectra with a long pulse interval, good digital resolution, and good signal-to-noise ratio.

Eqn 14 can still be used when there is (weak) scalar coupling, provided that this does not change as a result of the exchange process; for this situation, eqn 14 must be applied to each transition (multiplet component) of the signals. The more general approach to lineshape simulation, based on density matrix theory, is able to treat cases with or without scalar coupling. A good introduction to the theory, with some biological examples, is given by Nageswara Rao (12). A number of programs which carry out iterative lineshape fitting for the general case are now readily available (see Appendix of reference 12), and these probably represent the best general approach to lineshape analysis, provided that spectra of sufficient quality can be obtained. When a full lineshape analysis is not possible because of some overlap of the signals of interest with others—not an infrequent situation in protein spectra—estimates of the parameters may be made by comparing, for example, the dependence of the frequency of maximum intensity and the linewidth as a function of ligand concentration with those derived from simulated spectra (see *Protocol 4A* below).

4.2 Slow exchange

When exchange between the two states is slow (*Table 1*), separate resonance lines are observed for each state. The exchange rate can be readily measured from the linewidths of the resonances; these yield the apparent spin-spin relaxation rates, $1/T_{2i,obs}$, which are given by the following equations: For first-order exchange

$$1/T_{2A,obs} = 1/T_{2A} + 1/\tau_A = 1/T_{2A} + k_{+1} \qquad [17]$$

and

$$1/T_{2B,obs} = 1/T_{2B} + 1/\tau_B = 1/T_{2B} + k_{-1} \qquad [18]$$

For a second-order exchange, such as ligand binding to an enzyme (eqn 4)

$$1/T_{2L,obs} = 1/T_{2L} + 1/\tau_L = 1/T_{2L} + k_{-1}p_{EL}/p_L \qquad [19]$$

$$1/T_{2E,obs} = 1/T_{2E} + 1/\tau_E = 1/T_{2E} + k_{-1}p_{EL}/p_E \qquad [20]$$

and

$$1/T_{2EL,obs} = 1/T_{2EL} + 1/\tau_{eL} = 1/T_{2EL} + k_{-1} \qquad [21]$$

The procedure for determining the exchange rate under these conditions is described in *Protocol 3*

Protocol 3. Determination of exchange rate under conditions of slow exchange

1. Obtain as good a spectrum as possible using the guide-lines in Section 3; in the case of a second-order exchange, obtain spectra at a series of ligand concentration ratios, over a range of [ligand]/[protein] ratios from ≈ 1 to ≈ 10.

2. If at all possible, process the data without using a window function. Measure the linewidth at half-height (expressed in hertz). (If the signal-to-noise ratio is such that an exponential line-broadening function must be used, this must be subtracted from the measured linewidth.)

3a. For first-order exchange (eqn 1) make use of the temperature dependence of the linewidth. The two terms on the right-hand side of eqn 17 and 18 have opposite temperature dependences; $1/T_2$ ($= \pi.LW$) decreases with increasing temperature, while the exchange rate, of course, increases. For fairly similar values of $1/T_2$ and k, one can observe a minimum in the linewidth as the temperature is increased. An example is shown in *Figure 3* (20); at low temperatures, the linewidth is determined by $1/T_2$, and thus initially decreases, reaching a minimum at about 290 K and then increasing as it becomes determined by k. Values for $1/T_2$ and k and their activation energies, E_w and E_k, can be obtained by fitting the temperature-dependence to eqn 15 or 16 as appropriate, modified so as to contain the temperature-dependence explicitly:

$$1/T_{2,obs} = (1/T_{2,t})\ exp(E_w/RT) + k_t\ exp(-E_k/RT) \qquad [22]$$

where $1/T_{2,t}$ and k_t are the values of $1/T_2$ and k at some reference temperature and T is the absolute temperature. (Note that *Figure 3* shows the logarithmic version of this equation—the familiar Arrhenius plot of $\ln(\pi.LW)$ vs. $1/T$. A fuller version of the equation, in terms of activation enthalpy and activation entropy, could of course be used, but the data are rarely precise enough for this to be worthwhile.)

3b. For second-order exchange (eqn 4), make use of the concentration-dependence of the linewidths indicated by eqn 19 and 20. For a resonance of the free ligand, from eqn 19, a plot of $1/T_{2L,obs}$ ($=\pi.LW$) versus p_{EL}/p_L (calculated from the ligand and protein concentrations and K_d) will have a slope of k_{-1}. Note that, in many cases, $E_T > 10\ K_d$, (i.e. stoichiometric binding), and then for $L_T > E_T$, $p_{EL}/p_L \approx E_T/(L_T - E_T)$. Similarly, for a resonance of the free protein a plot of $1/T_{2E,obs}$ versus p_{EL}/p_E will have a slope of k_{-1}, although this will usually be less useful, since for stoichiometric binding the linewidth will only be concentration-dependent for $L_T < E_T$. $1/T_{2E}$ can, however, be measured from the spectrum in the absence of ligand and k_{-1} calculated from data at a single concentration using eqn 20. It is more difficult to estimate the exchange rate from resonances of either ligand or protein in the complex, since direct measurements of $1/T_{2EL}$ cannot generally be made. For a protein resonance one can assume $1/T_{2EL} \approx 1/T_{2E}$ and use eqn 21 directly. Alternatively, for either protein or ligand resonances, use the method described in step **3a**.

The range of rate constants which can be measured by these methods, roughly $1-100 \text{ sec}^{-1}$, is limited on the one hand by the smallest exchange contribution to the linewidth which can be measured, and on the other by the broadest observable line. In the case of multi-site exchange, the lifetime of each state giving a resolvable resonance can be estimated from the linewidth, but not necessarily the individual rate constants. For example, in the exchange process $A \longleftrightarrow B \longleftrightarrow C$, the lifetime of state B could be estimated, but not the individual rate constants for $B \rightarrow A$ and $B \rightarrow C$. These rate constants can, however, be measured by magnetization transfer (see Section 5).

In principle the equilibrium constant of the reaction can be determined from the intensity ratio of the signals from the two states. For a first-order process, this is usually straightforward for $0.1 < K < 10$. For a second-order

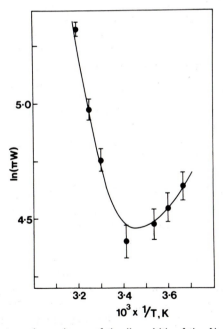

Figure 3. The temperature dependence of the linewidth of the N_1 proton resonance of trimethoprim in its complex with dihydrofolate reductase, presented as an Arrhenius plot [ln ($\pi \times$ linewidth) versus 1/T]. The *points* are experimental data, the *error bars* indicating the estimated maximum uncertainty in linewidth ($\pm$ 2 Hz). (Reproduced with permission from (20).)

process, the need to measure the intensity ratio at a series of concentrations leads to considerable error, and this procedure can not be recommended; in addition, the protein concentration is very often such that $E_T > 10\,K_d$, (i.e. stoichiometric binding), and no information on K_d can be obtained.

4.3 Fast exchange

When exchange is very fast with respect to the difference in chemical shift between the two states (*Table 1*), a single resonance is observed, whose chemical shift is the weighted average of the chemical shifts of the two individual states,

$$\delta_{obs} = \delta_A p_A + \delta_B p_B \qquad [23]$$

where p_A and p_B are the fractional populations of species A and B, respectively, with $p_A + p_B = 1$. Similarly for a ligand binding reaction (eqn 4),

$$\delta_{obs} = \delta_E p_E + \delta_{EL} p_{EL} \qquad [24]$$

for an enzyme resonance, or

$$\delta_{obs} = \delta_L p_L + \delta_{EL} p_{EL} \qquad [25]$$

for a ligand resonance. For very fast exchange, the relaxation rate (linewidth) is described by a precisely similar equation (given here for first-order exchange)

$$1/T_{2,obs} = \pi LW = p_A/T_{2,A} + p_B/T_{2,B} \qquad [26]$$

If, however, the exchange rate is somewhat slower ('moderately fast' exchange), there is an additional exchange contribution to relaxation (21), and eqn 26 is modified to:

$$\frac{1}{T_{2,obs}} = \frac{p_A}{T_{2,A}} + \frac{p_B}{T_{2,B}} + \frac{p_A p_B^2\,4\pi^2(\delta_A - \delta_B)^2}{k_{-1}} \qquad [27]$$

In the case of a first-order exchange process in the fast exchange region, the populations p_A and p_B can not be determined from the spectrum, the exchange term in eqn 27 can not be separated out, and the exchange rate can not be measured. (The temperature dependence is of no help here, since the exchange term is proportional to $1/k$, and this, like $1/T_2$, decreases with increasing temperature.) For second-order exchange, on the other hand, the populations can be varied by varying the concentrations. In terms of the protein–ligand interaction (eqn 4), eqn 27 becomes, for a ligand resonance,

$$\frac{1}{T_{2,obs}} = \frac{p_{EL}}{T_{2,EL}} + \frac{p_L}{T_{2,L}} + \frac{p_{EL} p_L^2\,4\pi^2(\delta_{EL} - \delta_L)^2}{k_{-1}} \qquad [27a]$$

This clearly predicts a different dependence of $1/T_2$ on p_L and p_{EL}—i.e. on ligand concentration—from that predicted by eqn 26. The exchange term passes through a maximum at $p_{EL} = 1/3$, whereas eqn 26 (or the first two terms of eqn 27a) predicts a monotonic decrease in linewidth with increasing ligand concentration (5). Provided that the linewidth in the bound state is not too large, this allows the presence of a significant exchange term to be recognized. This is particularly important, since the simple concentration dependence of the chemical shift given by eqn 24 and 25, which is often used to determine K_d, applies only for very fast exchange, and not for moderately fast exchange. Inappropriate use of eqn 24 and 25 can lead to large systematic errors in the values of K_d derived from the data (22). It is thus important to distinguish very fast and moderately fast exchange even if only measurements of K_d are envisaged. The procedure for analysis of fast exchange spectra is given in *Protocol 4*, with *Protocols 4A* and *4B* applying, respectively, to very fast and moderately fast exchange.

Protocol 4. Analysis of fast exchange spectra for the case of a protein–ligand interaction

1. Obtain a series of spectra, using the guide-lines in Section 3, at increasing ligand concentrations. It is most important to cover an adequate range of concentrations. For protein resonances this should start with $L_T \approx 0.1\ E_T$ and go up to $L_T \approx 10\ E_T$. When ligand resonances are being studied, the lowest concentration which can be examined will be limited by the detectability of the ligand signals, but every effort should be made to obtain data at $L_T < E_T$.

2. Measure the chemical shift and linewidth of the resonance(s) of interest as a function of ligand concentration.

3. *For a protein resonance.*
   (a) If the linewidth remains constant, the resonance is in the *very fast* exchange region; the equilibrium constant can be determined, as described in *Protocol 4A*, but the exchange rate can not (although an approximate lower limit is given by $k_{-1} > 40\ |\delta_E - \delta_{EL}|$).
   (b) If the linewidth passes through a maximum, the resonance is in the *moderately fast* exchange region; the equilibrium constant and exchange rate can be determined as described in *Protocol 4B*. Note that the maximum linewidth will be observed at $p_{EL} \leqslant 1/3$, and the maximum will only be detectable if sufficiently low ligand concentrations are used.

   For a ligand resonance.
   (a) If the linewidth decreases monotonically with increasing ligand concentration, the resonance is likely to be in the *very fast* exchange region; the equilibrium constant can be determined, as described in *Protocol 4A*, but the exchange rate can not.

Protocol 4. *Continued*

(b) If the linewidth passes through a maximum, the resonance is in the *moderately fast* exchange region; the equilibrium constant and exchange rate can be determined as described in *Protocol 4B*.

For a large protein a large value of $1/T_{2EL}$ may partly 'mask' the effects of the exchange term, so that a distortion of the shape of the plot of $1/T_{2,obs}$ (or linewidth) versus L_T is observed, rather than an actual maximum in the linewidth. It is thus worth carrying out the additional test of a plot of linewidth versus p_{EL} (calculating p_{EL} from K_d estimated as described in *Protocol 4A*, although this will not be an accurate estimate for moderately fast exchange). For a very fast exchange, this plot will be linear, with an intercept at $p_{EL} = 0$ equal to the linewidth of the free ligand resonance; in moderately fast exchange it will be non-linear, although the value of the intercept may be a better criterion for the presence or absence of an exchange term (23).

Protocol 4A. Determination of K_d in very fast exchange

Under these conditions, the chemical shift is given by eqn 24 for an enzyme resonance or eqn 25 for a ligand resonance. The analysis will be illustrated for a ligand resonance; the procedures for analysis of the effects of ligand binding or pH titration on an enzyme resonance are exactly analogous.

1. Measure the chemical shift of the ligand resonance at a series of ligand concentrations as described under steps **1** and **2** of *Protocol 4*.

2. From eqn 7, 10, and 25

$$\delta_{obs} - \delta_L = \frac{(\delta_{EL} - \delta_L)\{(E_T + L_T + K_d) - \sqrt{(E_T + L_T + K_d)^2 + 4E_T L_T}\}}{2L_T} \quad [28]$$

Fit eqn 28 to the data as a plot of δ_{obs} versus L_T, using an appropriate non-linear regression program. Values of δ_L and E_T will usually be known accurately, leaving δ_{EL} and K_d to be determined by the regression analysis.

3. Note that values of K_d can be estimated even when a much narrower range of ligand concentrations is used than that indicated in step **1** of *Protocol 4*. Particularly for larger proteins, it may only be possible to obtain data for $L_T \gg E_T$ due to overlap of ligand and protein signals. Under these conditions, a rigorous distinction between very fast and moderately fast exchange can not be made, and the value of K_d obtained should be treated with caution. The values of δ_{EL} and K_d obtained from the regression analysis will be significantly correlated, particularly if data at low L_T are not available—providing another reason for making measurements at as low ligand concentrations as practicable.

Protocol 4B. Determination of K_d and exchange rate in moderately fast exchange

Although a variety of approximations such as eqn 27a can be applied in the moderately fast exchange region, it is never easy to establish whether or not they are strictly applicable to a given set of data. It is generally much more satisfactory to use the full lineshape equation (eqn 14); if detailed lineshape fitting (Section 4.1) is not practical, analysis can be carried out in terms of chemical shift and linewidth measurements alone, as described here (see also 22).

1. Measure the chemical shift and linewidth of the ligand or protein resonance at a series of ligand concentrations as described under steps **1** and **2** of *Protocol 4A*.

2. Make initial estimates of the relevant parameters. Accurate values will be known for E_T and δ_L or δ_E. An estimate for δ_{EL} can be obtained from the chemical shift of the enzyme resonance at the highest ligand concentration or from the chemical shift of the ligand resonance at the lowest ligand concentration, as appropriate. A rough estimate for k_{-1} can be obtained from the maximum value of the linewidth by assuming that the exchange contribution dominates, and thence a rough estimate for K_d is $10^{-9} k_{-1}$ M.

3. Using these initial estimates, calculate the spectra for each value of L_T.

4. Measure the linewidth at half-height and the frequency of maximum intensity from the calculated spectra, and compare these to the corresponding values measured from the experimental spectra, perhaps most usefully by plotting the measurements as a function of L_T. The parameters δ_{EL}, k_{-1}, and K_d can then be adjusted until the two sets of data match. It should be recognized that the estimates of these parameters will be considerably less precise than if a full lineshape analysis is carried out, and elaborate statistical fitting of the data is probably not justified. However, in circumstances when a full lineshape analysis can not be carried out, the estimate of K_d obtained in this way will at least not be subject to the systematic error introduced by inappropriate use of the very fast exchange equations.

4.4 Problems of fast exchange

A significant limitation to the interpretation of NMR spectra in the fast exchange regime should always be borne in mind: the spectra represent an average across all the states which the nuclei experience, but we do not *a priori* know how many states there are, nor what their NMR parameters may be. Interpretation of fast exchange spectra must always, therefore, be carried out in terms of a model, and the conclusions will only be meaningful if the model is correct. In many cases, it is straightforward to choose a reasonable model for interpretation. Changes as a function of pH in the chemical shift of nuclei close to an ionizable group will naturally be analysed in terms of a two-site exchange between the protonated and unprotonated states of the ioniz-able group. Similarly, the three bond scalar coupling constants across a single C–C bond will be analysed in terms of the three classical 'staggered' rotamers about the central bond. It is clearly sensible to interpret the data in terms of the simplest possible model, but it is important to recognize the consequences of using the wrong model. Thus, the simplest model for a ligand binding to a protein is that of eqn 4; however, if the complex exists in two conformations, EL_1 and EL_2, the parameters deduced for EL from analysis in terms of eqn 4 will in fact be the weighted average parameters for EL_1 and EL_2, and not those of any real species. It is always important to obtain data for as many resonances as possible in such a situation, to maximize the chance of picking up any apparent inconsistencies which might indicate that the model being used is incorrect. Further discussion of these issues can be found in reference 11.

5. Magnetization transfer experiments

When exchange is slow, and separate resonances are seen for the nucleus in two (or more) states, a powerful group of methods known as magnetization transfer experiments can be applied to trace the pathways of exchange, and determine the exchange rate constants. In general, the experiment consists of an initial selective perturbation of one of the signals by an external radio-frequency field; that is, the longitudinal magnetization corresponding to this signal is selectively perturbed. Due to the exchange process, this perturbation will be transferred to the other signal. Assuming that the longitudinal relaxa-tion rates in the two sites are not considerably faster than the exchange rate, the transfer can be monitored by observing the change in intensity of one or both of the two signals or, in the case of the two-dimensional experiment, the intensity of a cross-peak between them. These changes are observed either after the magnetization has reached steady state, or as a function of time after the initial perturbation. The sections below describe how this is done with three commonly used magnetization transfer experiments: saturation trans-fer, inversion transfer, and two-dimensional exchange (for reviews, see 13, 14, 24). The methods will be described for a simple first-order two-site exchange process (eqn 1). The same methods can be used for second-order exchange, provided that the different definitions of the lifetimes in the two states (eqn 2 and 3 versus eqn 5 and 6) are taken into account by replacing k_{+1} by $1/\tau$ from eqn 6 or 11 where appropriate; extensions to multi-site exchange are noted for the different methods in the sections below.

5.1 General theory

From the Bloch equations, modified (19) to include the effects of a first-order two-site exchange (eqn 1), the return of M_A and M_B toward equilibrium after a perturbation is given by

$$dM_A/dt = -(M_A - M_A^e)R_A - k_{+1}M_A + k_{-1}M_B \qquad [29]$$

$$dM_B/dt = -(M_B - M_B^e)R_B + k_{+1}M_A - k_{-1}M_B \qquad [30]$$

where M_A^e and M_B^e are the equilibrium magnetizations of A and B respectively, and $R_A (=1/T_{1A})$ and $R_B (=1/T_{1B})$ are the spin-lattice relaxation rates for A and B, measured in the absence of chemical exchange. The general solution to these equations has the form of a double exponential:

$$M_A(t) = C_1\exp(\lambda_+t) + C_2\exp(\lambda_-t) + M_A^e \qquad [31]$$

$$M_B(t) = C_3\exp(\lambda_+t) + C_4\exp(\lambda_-t) + M_B^e \qquad [32]$$

where the apparent rate constants λ_+ and λ_- are each combinations of k_{+1}, k_{-1}, R_A, and R_B. Using the shorthand notation $k_A = k_{+1} + R_A$ and $k_B = k_{-1} + R_B$, and with M_A°, M_B° representing the initial intensities immediately after the perturbation,

$$\lambda_+ = \tfrac{1}{2}\{-(k_A + k_B) + [(k_A - k_B)^2 + 4k_{+1}k_{-1}]^{\frac{1}{2}}\} \qquad [33]$$

$$\lambda_- = \tfrac{1}{2}\{-(k_A + k_B) - [(k_A - k_B)^2 + 4k_{+1}k_{-1}]^{\frac{1}{2}}\} \qquad [34]$$

$$C_1 = [(\lambda_- + k_A)(M_A^e - M_A^\circ) - k_{-1}(M_B^e - M_B^\circ)]/(\lambda_+ - \lambda_-) \qquad [35]$$

$$C_2 = [-(\lambda_+ + k_A)(M_A^e - M_A^\circ) + k_{-1}(M_B^e - M_B^\circ)]/(\lambda_+ - \lambda_-) \qquad [36]$$

$$C_3 = [-k_{+1}(M_A^e - M_A^\circ) - (\lambda_+ + k_B)(M_B^e - M_B^\circ)]/(\lambda_+ - \lambda_-) \qquad [37]$$

$$C_4 = [k_{+1}(M_A^e - M_A^\circ) + (\lambda_- + k_B)(M_B^e - M_B^\circ)]/(\lambda_+ - \lambda_-) \qquad [38]$$

5.2 One-dimensional magnetization transfer

In the one-dimensional experiment, the resonance from one state is selectively saturated or inverted by an appropriate rf pulse or pulses, and changes in intensity of the signals from one or both states measured.

The following guide-lines are applicable to 1D selective saturation and inversion experiments.

(a) The selectivity of a pulse depends on its length, and as a measure of its selectivity the distance (in frequency units) to the first null in its excitation pattern is commonly taken. For a square pulse of duration t, the first null appears approximately $1/t$ Hz away from the carrier frequency. Therefore, in order to achieve a selectivity of 1 Hz, a pulse length of about 1 sec must be used.

(b) The pulse power to be used should be carefully calibrated and adjusted to avoid spill-over effects.

(c) The choice of frequency selective excitation methods includes: the DANTE pulse (25; *Protocol 5*); gated low power decoupling (for saturation only); the 1-1 pulse (90°-*t*-90°, with delay $t = 1/2\Delta$, Δ being the separation between the signal being excited and its exchanging partner; 26); and the shaped pulse (Gaussian pulse; 27).

Protocol 5. The DANTE pulse for selective excitation

A DANTE pulse (25) comprises a train of *m* very short pulses of width *a* separated by delays *t*. This pulse sequence produces selective excitation at the irradiation frequency (with a net flip-angle *ma*) and excitation sidebands separated by $1/t$. The bandwidth of irradiation at each sideband is determined by the total duration of the sequence, that is *mt*. The delay *t* must be such that, with the centreband irradiation frequency set on the region of interest, sidebands are away from peaks which should not be excited.

1. Select the resonance to be saturated or inverted; since complete selectivity is not possible, this should be reasonably well separated from other resonances (say, 50 Hz).

2. Choose a power setting for either the transmitter or decoupler to give a 90° pulse length of 20–40 msec to avoid inconveniently short pulses (attenuation will be needed in both cases).

3. Determine the delay *t* from the desired frequency of the first sideband; for example, if this is to be displaced from the centreband by 1000 Hz, then $t = 1/1000 = 1$ msec.

4. Determine number of pulses, *m*, from the bandwidth required to avoid unwanted excitation of other resonances; for example, for a bandwidth of 100 Hz, $2/mt = 100$, and for *t* as above, $m = 20$.

5. For a 90° pulse at the centreband, $ma = 20$ to 40 msec (calibrated at step **2** above), and hence, in this example, $a = 1$–2 msec. Thus, a train of 20 pulses, each 1–2 msec in length and separated by 1 msec delays will give a selective 90° pulse of 100 Hz bandwidth. Generally, it is desirable to make *a* a very small fraction of the desired total pulse width, subject to *mt* remaining small in comparison with relaxation times. However, if *a* is too short, insufficient overall power is delivered to bring about complete inversion or saturation.

6. Check the selectivity by carrying out a series of irradiations at varying separations from the chosen peak, so that the minimal frequency separation at which spill-over effects can be neglected is determined.

7. (a) For selective saturation, repeat the pulse train for a sufficient number of times (chosen by trial and error).
   (b) For selective inversion, choose *ma* to obtain a 180° pulse.

5.2.1 The saturation transfer experiment (28, 29)

The advantage of the saturation transfer experiment is that the experiment is performed under conditions such that data analysis is simplified. For the two-site exchange case being considered here, the magnetization of the nucleus in state A is saturated, normally with either a DANTE pulse or a lower power gated decoupling pulse, and the intensity of the resonance from state B is monitored. Since, after saturation, $M_A = 0$, the double exponential solution to the Bloch equations (eqn 32) is simplified to a single exponential:

$$M_B = M_B^e[(k_{-1}/\lambda_B)\exp(-\lambda_B t) + R_B/\lambda_B] \qquad [39]$$

where

$$\lambda_B = R_B + k_{-1} \qquad [40]$$

In two variants of the saturation transfer experiment, the intensity of the resonance from state A is monitored either as a function of time as saturation is being transferred by the exchange process (*Protocol 6A*), or simply after steady state has been established (*Protocol 6B*).

Protocol 6A. Saturation transfer time-course experiment

1. Set up the DANTE pulse (as described in *Protocol 5*) or low power decoupling pulse to saturate the signal from the nucleus in state A. Sufficient pulse power must be used to saturate the resonance in a time short compared to $1/(k_{-1} + R_B)$.

2. Saturate resonance A for a time t (typically a few tenths of a second). Collect sufficient scans to give a good signal-to-noise ratio.

3. For the control experiment set the irradiation frequency in a region of the spectrum where there are no signals. This is normally at one end of the spectrum outside the usual spectral width; it is good practice to set this off-resonance frequency at the end nearer to the resonance irradiation frequency. Collect the same number of scans as in step **2**, irradiating for the same time t. *Note*: to obtain a good difference for quantitative analysis, particularly for long-term experiments, it is desirable to interleave the acquisition of the irradiation and control experiments so that each data set samples any long-term changes in temperature or field drift in the same way. Typically, a cycle would consist of 8 or 16 scans at one irradiation frequency, followed by the same number at the other frequency.

4. Repeat steps **2** and **3** for a series of irradiation times t, typically in the range 0.1–10 sec.

5. Measure the saturation transfer to resonance B by integration of the paired control and experimental (off- and on-resonance) spectra. Do not

Protocol 6A. *Continued*

use severe window functions to process the data as this can lead to distortion of the baseline and lineshape.

6. Fit the magnetization ratio M_B/M_B^e as a function of the irradiation time t to eqn 39 by non-linear regression. The time-constant of the exponential decay is $1/(R_B + k_{-1})$, and the intensity ratio at $t \to \infty$ is $R_B/(R_B + k_{-1})$, so that R_B and the exchange rate k_{-1} can be determined (*Figure 4*).

7. The inverse experiment, saturating B and observing A, will provide a value for k_{+1}.

Protocol 6B. Steady state saturation transfer experiment

1. Set up the DANTE pulse (as described in *Protocol 5*) or low power decoupling pulse to saturate the signal from the nucleus in state A. Sufficient pulse power must be used to saturate the resonance in a time short compared to $1/(k_{-1} + R_B)$.

2. Saturate resonance A for a time t (typically a second). Collect sufficient scans to give a good signal-to-noise ratio. To ensure that a steady state has been reached, repeat the experiment with saturation for a time $2t$; if the intensity of resonance B is no less with the longer saturation time, steady state has been reached.

3. Carry out the control experiment as described in step **3** of *Protocol 6A*.

4. Measure the saturation transfer to resonance B by integration of the paired control and experimental (off- and on-resonance) spectra. Do not use severe window functions to process the data as this can lead to distortion of the baseline and lineshape.

5. Measure the $R_{B,obs}$ (spin-lattice relaxation rate) of resonance B using the inversion-recovery method *while resonance A is saturated*.

6. From eqn 32, the steady state magnetization of B when A is saturated for a time $t \to \infty$ is given by

$$M_B^\infty/M_B^e = R_B/(R_B + k_{-1}) \qquad [41]$$

and in addition $R_{B,obs} = R_B + k_{-1}$ and thus k_{-1} and R_B can be calculated separately.

7. The inverse experiment, saturating B and observing A, and with a value for $R_{A,obs}$, will provide a value for k_{+1}.

Saturation transfer can be extended to exchange processes involving more than two sites, but for *n*-site exchange, the resonances of $n - 1$ sites must be saturated simultaneously and the behaviour of the magnetization in the re-

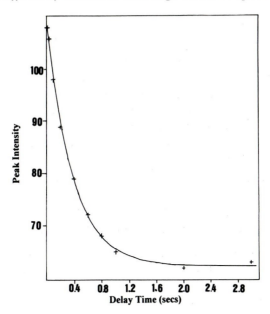

Figure 4. Plot of the time-dependent signal intensities of a set of saturation transfer spectra of Trp repressor in the presence of five molar equivalents of L-tryptophan, pH 7.5, 295 K. The $H^{\delta 1}$ resonance of the bound tryptophan was saturated; the intensity of the $H^{\delta 1}$ resonance of the free ligand was monitored. The curve represents the best fit obtained by a least squares analysis. A dissociation rate constant of 3.45 ($\pm$ 0.5) sec^{-1} was obtained.

maining, *n*th, site followed in a manner analogous to that described above. This yields the lifetime in the *n*th site, but individual rate constants for exchange can only be obtained from a complete set of such experiments, saturating each group of $n - 1$ sites in turn.

5.2.2 Inversion transfer (13, 30)

Equations 29–38 describe the changes in magnetization of a nucleus at each of a pair of exchanging sites after perturbation of one or both of the sites. In the saturation transfer experiments described in the preceding section, the time evolution was simplified by ensuring, through saturation, that the net magnetization of one of the sites remained at zero. This does, however, lead to a loss of information. The maximum information about the exchange process is obtained by transient selective perturbation, such as selective inversion, of the signal from one site, and analysis of the time evolution of the magnetization at both sites. This approach affords more experimental data and covers a larger range of rate constants than saturation transfer.

The experiment is performed in essentially the same manner as the saturation transfer experiment described in Section 5.2.1 and *Protocol 6A*, but with the selective saturation pulse replaced by a selective inversion (180°) pulse; the pulse sequence is thus

$$180°_{\text{selective}} - t - 90°_{\text{non-selective}} - \text{acquire},$$

t being a variable delay (see *Figure 5*). A series of spectra is acquired with different values of the delay *t* (compare with *Protocol 6A*). If resonance B is inverted, then, as *t* is increased, the intensity of resonance A first decreases in intensity, and then returns to its equilibrium intensity, while the intensity of B returns to equilibrium monotonically (*Figure 5*; 31). As predicted by eqn 31 and 32, both recovery curves are double exponentials (*Figure 5*).

The experimental data thus consist of the intensity (peak integral) for resonances A and B as a function of the delay *t*. These are analysed by fitting eqn 31 and 32 to the data by non-linear least squares regression, thus obtaining estimates for the unknown rates k_{+1}, k_{-1}, R_A, and R_B, and for the limiting intensities M_A°, M_A^{e}, M_B°, and M_B^{e}, subject to the constraint that $k_{+1}M_A^{e} = k_{-1}M_B^{e}$ at equilibrium. A word of caution is required in order not to give an over-simplified view of the problem. The two exponents λ_+ and λ_- can be determined independently only if they are sufficiently different; in practice λ_+ and λ_- must differ by a factor of three or four. For $\lambda_+ = \lambda_-$, only

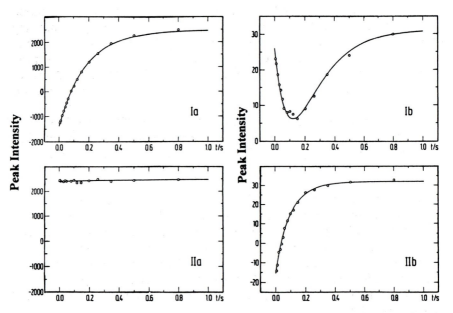

Figure 5. Plots of the peak heights as a function of time of a set of complementary inversion transfer ^{13}C NMR spectra of (90% ^{13}C)-H^{13}CO$_3^-$ (100 mM) and ^{13}CO$_2$ (0.38 mM) in the presence of Mn(II)-human carbonic anhydrase I. The H^{13}CO$_3^-$ and ^{13}CO$_2$ signal heights are shown in a and b, respectively; in experiment I the H^{13}CO$_3^-$ and in experiment II the ^{13}CO$_2$ resonance was selectively inverted. The curves represent the best fit obtained by a least squares analysis. (Reproduced with permission from (31).)

the combination $C_1\lambda_+ + C_2\lambda_-$ can be obtained from the data for **A**, and only the combination $C_3\lambda_+ + C_4\lambda_-$ from the data for **B**. Thus a single experiment may not suffice to determine the rate constants of interest. However, carrying out a complementary experiment in which the other signal is selectively inverted provides another pair of independent combinations of rate constants, and the data from the two experiments together suffice to determine all four rates (exchange and relaxation). Thus for a two-site exchange process, it is in general necessary to perform two inversion transfer experiments. In general, for an *n*-site exchange process, it is necessary to perform *n* experiments, in which the resonance of each of the sites is inverted in turn, and the time evolution of all *n* signals followed. In its ability to deal with multi-site exchange, inversion transfer is clearly superior to saturation transfer (13; see also Section 5.4.)

5.3 Two-dimensional exchange spectroscopy (24)

The effects of exchange processes on the lineshape in one-dimensional spectra described in Section 4 above are also, of course, seen in two-dimensional spectra and can be analysed in much the same way. Here we shall restrict discussion to the slow exchange regime and to magnetization transfer processes in particular. It is well known that experiments to study magnetization transfer due to the nuclear Overhauser effect (NOE) can be carried out very effectively by two-dimensional spectroscopy (Chapter 3), and the same is true for magnetization transfer due to chemical exchange—indeed the two experiments are both formally and practically identical (24). The procedures for carrying out a 2D exchange experiment are identical to those described for NOESY in Chapter 3.

For a quantitative analysis, it is possible to view the 2D NOESY spectrum of an exchanging system in the absence of scalar coupling as the solution to the Bloch–McConnell equations (for a two-site exchange, eqn 29 and 30):

$$\mathbf{M} = \mathbf{M}^\circ \exp(\mathbf{R}t_m) \qquad [42]$$

where **M** is the matrix comprising the peak volumes of the 2D spectrum, **R** is a matrix, of dimensions equal to the number of sites between which exchange is occurring, containing all the exchange rate constants and spin-lattice relaxation rates, and t_m is the mixing time. When the exchange rates dominate over the relaxation rates, the 2D spectrum is in effect a graphic display of the exchange pathways. The very appearance of an exchange cross-peak at ω_i, ω_j is sufficient proof that exchange is taking place between the sites *i* and *j* (*Figure 6(a)*; 32).

The integrated intensity I_{ij} of the 2D absorption peak at ω_i, ω_j is given by (25):

$$I_{ij}(t_m) = \mathbf{M}_j^\circ \exp(-\mathbf{R}t_m)_{ij} \qquad [43]$$

For the simplest case of first-order two-site exchange with equal populations

$(M_A = M_B = M, k_{+1} = k_{-1} = k)$ and equal relaxation rates $(R_A = R_B = R)$, this has the solution:
Diagonal peaks:

$$I_{AA} = I_{BB} = \tfrac{1}{2}\{[1 + \exp(-2kt_m)]\exp(-t_m R)\} \qquad [44]$$

Cross-peaks

$$I_{AB} = I_{BA} = \tfrac{1}{2}\{[1 - \exp(-2kt_m)]\exp(-t_m R)\} \qquad [45]$$

In this simple two-site exchange system, the exchange rate can be determined from the initial rate of build-up of cross-peak intensity at short mixing times (compare with *Figure 6(b)*)

$$(\delta I_{AB}/\delta t_m)_i = M^\circ k \qquad [46]$$

or directly from the ratio of the cross and diagonal peak intensities at short mixing times

$$I_{AA}/I_{AB} = (1 - kt_m)/kt_m \qquad [47]$$

However, the cross-peak intensity may be weak at mixing times short enough to satisfy the initial rate assumption. More generally, it is necessary to record a series of 2D exchange spectra with different t_m values and to fit the dependence of diagonal peak and cross-peak intensity on mixing time to eqn 44 and 45. Care must be taken to ensure that the intensities can be measured accurately, by integrating the volume of a cross-peak. Inaccuracies can arise from peak overlap, baseline distortions, and the arbitrariness of peak boundaries; the latter problem can be alleviated by applying a mild Gaussian window function to the data. When the exchange process is more complex than the very simple two-site case, the equivalent of eqn 44 and 45 can be derived from eqn 42 (with the matrices of appropriate dimensions); the general solution is given in reference 24. This can then be fitted to the peak intensities as a function of t_m in a similar way; the number of parameters is likely to be such that the intensity data for several cross-peaks will have to be analysed together.

5.3.1 Variants of the 2D exchange experiment

The conventional 2D exchange experiment described above suffers from three disadvantages.

(a) The need to obtain data at very short mixing times (initial rate region) means that the cross-peaks may be obscured by dominant diagonal peaks.

(b) The need to acquire data at several mixing times means that the technique is very time consuming. In addition, the comparison of peak amplitudes among a series of 2D spectra may contain intensity inaccuracies due to instrumental instabilities.

(c) The difficulty in distinguishing between cross-peaks arising from exchange and those representing NOE effects.

The accordion experiment is a variant of the basic exchange experiment designed to save time (24, 33). Because a set of 2D exchange experiments with different mixing times t_m is in fact a 3D experiment with three time variables t_1, t_m, and t_2, the proposal is that two of the variables be incremented simultaneously by setting $t_m = Ct_1$, where C is a constant, usually between 10 and 100 and chosen in such a way that maximum t_m is roughly three times the spin-lattice relaxation time T_1. This linkage of t_m and t_1 reduces the 3D experiment to a special form of a 2D experiment with the usual independent variables t_1 and t_2. Because t_1 and t_m are varied together, a Fourier transformation with respect to t_1 is also a transformation with respect to t_m. The accordion experiment, however, has not been widely used since it was first proposed, the main reasons being the inherent difficulty in analysing the data (24).

Other variations of the conventional 2D EXCSY experiment are those designed to distinguish exchange from NOE cross-peaks. Since the same 2D experiment is used for detecting both cross-relaxation (NOE) and exchange effects, it is obvious that both effects will, in principle, appear in the same spectrum. In addition, there may be transferred NOEs, which result from a two-step magnetization transfer, cross-relaxation between two spins in state A being followed by chemical exchange to state B. Care must therefore be taken to distinguish between NOE and exchange cross-peaks. If the resonances of interest are well resolved in the spectrum, or if $(k_{+1} + k_{-1}) \gg R_{A,B}$, so that the exchange cross-peaks are significantly more intense, this may be straightforward. In the case of a protein–ligand interaction, the two kinds of cross-peak can often be distinguished by comparing a NOESY spectrum recorded with $[L_T] < [E_T]$, so that there is a negligible concentration of free ligand, with one recorded at $[L_T] \approx 5 [E_T]$; exchange cross-peaks will only be visible in the latter case. (In principle, the dependence of the lifetime on the ligand concentration also affords a way of making this distinction, but it is usually prohibitively time-consuming to record 2D spectra at a series of ligand concentrations as well as at a series of t_m values.)

A useful method for making the distinction between NOE and exchange cross-peaks is the rotating-frame NOE, or ROESY, experiment (34, 35; see also Chapter 3). In this experiment the magnetization transfer involves not longitudinal magnetization, as in the conventional NOESY experiment, but transverse magnetization during a period of spin-locking. Whereas both NOE and exchange cross-peaks have the same (positive) sign in a NOESY spectrum, they have opposite signs in a ROESY spectrum, (transverse) NOEs being negative and exchange cross-peaks being positive. There are a number of other potential contributions to the cross-peak intensities in ROESY spectra (including coherence transfer and HOHAHA transfer) which make quantitative analysis difficult (35, 36), but this experiment does provide, in many cases, a valuable distinction between NOE and exchange cross-peaks.

Recently, a method for suppressing both cross-relaxation (NOE) and

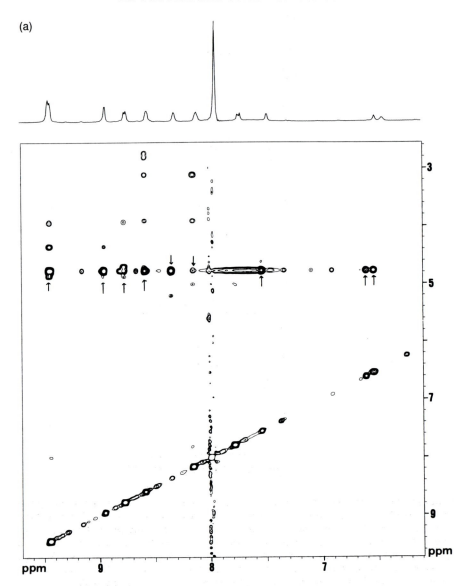

(a)

coherence transfer effects, resulting in 'pure exchange' spectra, has been proposed (37). The elimination of cross-relaxation is based on the principle of forcing transverse and longitudinal magnetization to compensate each other.

5.4 Comparison of magnetization transfer experiments

Each of the three kinds of magnetization transfer experiment described above has advantages and disadvantages, and these are briefly summarized here.

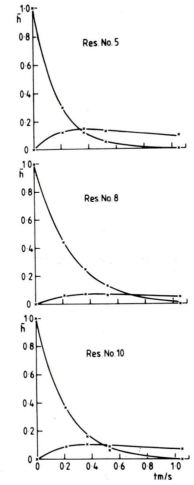

Figure 6. (a) A phase-sensitive contour plot of a 2D exchange experiment showing the exchange of the amide protons of the peptide viomycin with water at 310 K, mixing time 210 msec. The H^N-H_2O exchange cross-peaks are indicated by *arrows*. (b) Representative plots of the variation of peak intensity (h) with mixing time (t_m) for resonance of amide protons at 9.45 p.p.m. (Res. No. 5), 9.49 p.p.m. (Res. No. 8), and 8.60 p.p.m. (Res. No. 10). The crosses correspond to experimental data and curves are simulated using calculated values of (R_{1A} + k), R_{1B}, and $k_{average}$. (Reproduced with permission from (32).)

5.4.1 Saturation transfer

(a) *Advantages*. For two-site exchange, this experiment has the real advantage of simplicity, both experimentally and in terms of data analysis, since the magnetization transfer can be described by a single exponential. It is also somewhat easier to apply in the situation where one of the two sites has only a low population; a resonance from this site can be saturated and the observations made on the resonance of the abundant site.

(b) *Disadvantages*. The selectivity of saturation is less than that of inversion, and markedly less than that of the two-dimensional experiment. It is also the most restricted of the three methods in terms of the range of exchange rates which can be evaluated. The upper limit is set by the need for slow exchange on the chemical shift time-scale. The lower limit is set by the smallest intensity change that can be measured by difference spectroscopy, or, in the case of 2D spectroscopy, by the smallest cross-peak that can be detected; this is in turn governed by the relaxation rates of the species involved. Saturation transfer is limited to $k_{+1} \geqslant R_B$ (for the case where A is saturated), while the other methods can give useful information on somewhat slower exchange processes. (Note that these rates are much slower than those which can be measured by exchange broadening.) Finally, saturation transfer is difficult to apply to exchange processes involving more than two sites, since for n-site exchange, the resonances of $n - 1$ sites must be saturated simultaneously, which is technically difficult.

5.4.2 Inversion transfer

(a) *Advantages*. The inversion transfer experiment has good selectivity, and covers a fairly wide range of exchange rates. It makes maximum use of the information available, and if the complementary experiments (inverting the signal from each site in turn) are carried out, it probably provides the best estimate of the exchange rates. It is also readily applicable to multi-site exchange.

(b) *Disadvantages*. The principal disadvantage of this method is the need to fit the data to multiple exponentials, with the attendant difficulties if the time constants of the exponentials are similar. To overcome this, the full series of complementary experiments must be done (see Section 5.2.2), making this a somewhat time-consuming experiment.

5.4.3 2D exchange

(a) *Advantages*. A major advantage of the 2D experiment is its ability to provide a 'map' of the exchange process(es). All the sites are 'labelled' in a single experiment without previous knowledge of the spectrum, thus allowing all exchange pathways to be observed simultaneously. By using the 'accordion' experiment, all the exchange rate constants can, in principle, be obtained from a single experiment. Furthermore, problems associated with selective excitation in crowded spectra are completely avoided.

(b) *Disadvantages*. The 2D experiment is a time-consuming one if quantitative analysis is to be carried out; some kind of three-dimensional experiment will, in general, have to be performed. This may be achieved either by a long series of individual 2D experiments or by using the 'accordion' technique with its rather more complex data analysis. Even allowing for carrying out the full set of inversion transfer experiments, the precise comparison of the time required depends on the individual exchange process. The 2D experiment is, however, generally more time-consuming than the full set of inversion transfer experiments, so the latter is perhaps to be preferred, provided that the necessary selectivity of inversion can be obtained. The 2D exchange experiment also potentially suffers from the contribution of zero-quantum coherence to cross-peak intensity. In the normal 2D exchange (NOESY) experiment, these can be suppressed by introducing a small random variation in t_m (see Chapter 3). In the accordion experiment, where t_m is linked to t_1, this is not possible, and here small 'satellite' signals, arising from zero-quantum coherence, may be seen for each cross- and diagonal peak (24); for the broad lines seen for proteins these may overlap with, and contribute to the intensity of, the peaks of interest.

Acknowledgement

We thank Dr E. I. Hyde for making *Figure 4* available to us.

References

1. Campbell, I. D., Dobson, C. M., Moore, G. R., Perkins, S. J., and Williams, R. J. P. (1976). *FEBS Lett.,* **70,** 96.
2. Hilbers, C. W. and Patel, D. J. (1975). *Biochemistry,* **14,** 2656.
3. Evans, P. E., Kautz, R. A., Fox, R. O., and Dobson, C. M. (1989). *Biochemistry,* **28,** 362.
4. Gronenborn, A., Birdsall, B., Hyde, E. I., Roberts, G. C. K., Feeney, J., and Burgen, A. (1981). *Mol. Pharm.,* **20,** 145.
5. Hyde, E. I., Birdsall, B., Roberts, G. C. K., Feeney, J., and Burgen, A. S. V. (1980). *Biochemistry,* **19,** 3746.
6. Lenstra, J. A., Bolscher, B. G. J., Stob, S., Beintema, J. J., and Kaptein, R. (1979). *Eur. J. Biochem.,* **98,** 385.
7. Leroy, J. L., Charretier, E., Kochoyan, M., and Gueron, M. (1988). *Biochemistry,* **27,** 8894.
8. Vasadava, K. V., Kaplan, J. I., and Nageswara Rao, B. D. (1984). *Biochemistry,* **23,** 961.
9. Kaplan, J. I. and Fraenkel, G. (1980). *NMR of chemically exchanging systems.* Academic Press, New York.
10. Sandstrom, J. (1982). *Dynamic NMR spectroscopy.* Academic Press, London.
11. Jardetzky, O. and Roberts, G. C. K. (1981). *NMR in molecular biology.* Academic Press, New York.
12. Nageswara Rao, B. D. (1989). In *Methods in enzymology* (ed. N. J. Oppenheimer and T. L. James), Vol. 176, pp. 279–311. Academic Press, London.
13. Led, J. J., Gesmar, H., and Abildgaard, F. (1989). In *Methods in enzymology* (ed. N. J. Oppenheimer and T. L. James), Vol. 176, pp. 311–329. Academic Press, London.

14. Berkowitz, B. and Balaban, R. S. (1989). In *Methods in enzymology* (ed. N. J. Oppenheimer and T. L. James), Vol. 176, pp. 330–341. Academic Press, London.
15. Roberts, G. C. K. (1986). In *NMR in the life sciences* (ed. E. M. Bradbury and C. Nicolini), pp. 73–86. Plenum Press, New York.
16. Leigh, J. S. Jr. (1971). *J. Magn. Reson.*, **4**, 308.
17. McLaughlin, A. C. and Leigh, J. S. Jr. (1973). *J. Magn. Reson.*, **9**, 296.
18. Clore, G. M., Gronenborn, A. M., Birdsall, B., Feeney, J., and Roberts, G. C. K. (1984). *Biochem. J.*, **217**, 659.
19. McConnell, H. M. (1958). *J. Chem. Phys.*, **28**, 430.
20. Bevan, A. W., Roberts, G. C. K., Feeney, J., and Kuyper, L. F. (1985). *Eur. Biophys. J.*, **11**, 211.
21. Reuben, J. and Fiat, D. (1969). *J. Chem. Phys.*, **51**, 4918.
22. Feeney, J., Batchelor, J. G., Albrand, J. P., and Roberts, G. C. K. (1979). *J. Magn. Reson.*, **33**, 519.
23. Smallcombe, S. H., Ault, B., and Richards, J. H. (1972). *J. Am. Chem. Soc.*, **94**, 458.
24. Ernst, R. R., Bodenhausen, G., and Wokaun, A. (1987). In *Principles of nuclear magnetic resonance in one and two dimensions*, Chapter 9. Clarendon Press, Oxford.
25. Morris, G. A. and Freeman, R. (1978). *J. Magn. Reson.*, **29**, 433.
26. Plateau, P. and Gueron, M. (1982). *J. Am. Chem. Soc.*, **104**, 7310.
27. Bauer, C., Freeman, R., Krenkiel, T., Keeler, J., and Shaka, A. J. (1984). *J. Magn. Reson.*, **58**, 442.
28. Forsen, S. H. and Hoffman, R. A. (1963). *J. Chem. Phys.*, **39**, 2892.
29. Forsen, S. H. and Hoffman, R. A. (1964). *J. Chem. Phys.*, **40**, 1189.
30. Vold, R. L., Waugh, J. S., Klein, M. P., and Phelps, D. E. (1968). *J. Chem. Phys.*, **48**, 3831.
31. Led, J. J., Neesgard, E., and Johansen, J. T. (1982). *FEBS Lett.*, **147**, 74.
32. Dobson, C. M., Lian, L. Y., Redfield, C., and Topping, K. D. (1986). *J. Magn. Reson.*, **69**, 201.
33. Bodenhausen, G. and Ernst, R. R. (1982). *J. Am. Chem. Soc.*, **104**, 1304.
34. Bothner-By, A. A., Stephens, R. L., Lee, J., Warren, C. D., and Jeanloz, R. W. (1984). *J. Am. Chem. Soc.*, **106**, 811.
35. Bax, A. and Davis, D. G. (1985). *J. Magn. Reson.*, **63**, 207.
36. Brown, L. R. and Farmer, B. T. II. (1989). In *Methods in enzymology* (ed. N. J. Oppenheimer and T. L. James), Vol. 176, pp. 199–215. Academic Press, London.
37. Fejzo, J., Westler, W. M., Macura, S., and Markley, J. L. (1991). *J. Magn. Reson.*, **92**, 20.

NMR studies of protein–ligand interactions

J. FEENEY and B. BIRDSALL

1. Introduction

Over the last few years there has been an increasing interest in studying interactions between proteins and other molecules in solution. Such non-covalent interactions are of crucial importance in many molecular recognition processes in biology, typical examples being found in complexes of enzymes with substrates and inhibitors, of drugs with their receptors, and of transcription factors with DNA duplexes. In each case, the formation of the non-covalent complex is the initial event triggering off some important biological process. High resolution NMR spectroscopy has proved to be a very useful technique for studying such complexes in solution and the aim of this article is to indicate how the appropriate NMR data can be acquired and analysed to provide detailed information concerning interactions, conformations, and dynamic processes for such protein–ligand complexes. There are numerous examples of such studies in the literature but in the interests of presentational simplicity we will restrict the discussion to only a few systems which illustrate the various applications. Many of the examples will be taken from work in our own laboratory dealing with the binding of antifolate drugs to dihydrofolate reductase (DHFR). This enzyme reduces dihydrofolate **1** (and folate **2** with lower efficiency) to tetrahydrofolate using NADPH as a coenzyme and is an essential enzyme in all cells. Studies of DHFR provide examples of some of the simplest cases of drug–receptor complexes where the pharmacological action results from an inhibitor molecule binding selectively to this essential enzyme in an invasive cell. We will also discuss several novel applications to other systems taken from the literature in order to illustrate certain aspects of the subject, for example protein–DNA interactions.

In 1981, Jardetzky and Roberts (1) provided a detailed survey of NMR studies of protein–ligand interactions and discussed several systems which had been studied in some depth. These included inhibitor binding to ribonuclease A, staphylococcal nuclease, lysozyme, and dihydrofolate reductase, and several other proteins. The reader is referred to Jardetzky and Roberts

Dihydrofolate **1**

Trimethoprim **3**

Folate **2**

Methotrexate **4**

5 α-amide

6 γ-amide

(1) and other review articles (2–5) for the early literature on these systems. Since that time these proteins have all been studied further, the most notable advances being the sequential assignments of many of their backbone and side chain resonances and the detection of NOEs between the bound ligand and protein resonances. In the past, most of the detailed structural informa- tion about binding sites in protein–ligand complexes has been provided by X-ray crystallographic studies. Earlier NMR studies made their major con- tributions in complementing the X-ray structural data by defining specific interactions and identifying ionization states of interacting groups on the protein and ligand, by monitoring dynamic processes relating to specific interactions within the protein–ligand complexes and, in some cases, by

detecting the presence of multiple conformational states in solution. In each of these latter applications NMR offers considerable advantages over the X-ray method. More recently, NMR methods have been developed for obtaining detailed structural information for small proteins $M_r < 20000$ (see Chapters 4 and 5) and in the best determined NMR structures, resolution comparable with a 2.0 Å resolution X-ray structure has been achieved. While NMR determined structures of more than 100 small proteins have now been reported, relatively few of these have been protein–ligand complexes. However, two examples of protein–DNA complexes have now been well-determined by NMR. These involve the *lac* repressor headpiece (6), and homeodomain proteins (7) binding to the appropriate DNA fragments containing the respective consensus regions for binding. These examples are discussed in Section 7.1. Clearly, these structural methods will be applied more generally to protein–ligand complexes over the next few years.

When a flexible ligand binds to a protein, the process involves a conformational selection of one of the many conformations of the free ligand and a corresponding conformational readjustment of the 'binding site' of the protein to effect the specific binding. The final complex while spending most of its lifetime in a well-defined conformational state will undergo various dynamic processes and some of these are accessible to NMR study. If we are to understand the factors controlling the specificity of the ligand binding it is important to be able to characterize the changes in conformation which occur on binding, to define the specific interactions between the protein and ligand, and to characterize any dynamic processes within the complex.

2. Preparation of the sample and preliminary binding studies

NMR experiments need large quantities of purified protein. The concentrations needed will vary from 10 μM for 1D spectra to up to 5 mM for some 2D experiments. For a protein of M_r 20000 a 0.5 ml sample would require between 1 mg and 50 mg. The ionic strength of the sample will influence the optimum concentration, as high salt reduces the signal-to-noise ratio and higher concentrations will then be beneficial. The highest concentration may not be the optimum one if this causes aggregation and consequent broadening of the signals.

Large quantities of protein are best prepared by cloning the appropriate gene into an over-expressing bacterial cell line. The high concentration of the desired protein in the cell makes the purification procedures easier. Mammalian proteins can also be obtained in this way. Growing the cells on [^{13}C]glucose or [^{15}N]ammonium sulfate produces protein uniformly labelled with ^{13}C or ^{15}N which is very useful for assignment purposes. This requires an organism which grows on minimal media. If one wants to specifically introduce an isotopically labelled amino acid, it is necessary to use an organism which is auxotrophic for that amino acid.

Many NMR experiments require that the sample remains in the spectrometer for several days. It is therefore necessary to establish the best conditions for the stability of the sample, and the range of pH and temperature over which the sample is stable. The ionic strength, the concentration of salt in the buffer, the pH, and the temperature all influence the stability of the sample and need to be considered.

NMR studies of proteins are mostly carried out using aqueous solutions, either H_2O or ^2H_2O. If the protein is amenable to lyophilization, then this is the easiest method of changing from one solvent to the other. However, for some proteins it is necessary to dialyse the sample against the other buffer. Dialysis is also very useful for removing excess free ligand from the sample and for changing one ligand for another when one knows their relative binding constants. Additional details on sample preparation are given in Chapter 2.

In order to design sensible NMR experiments for examining protein–ligand complexes it is useful to know the equilibrium dissociation constants. The binding of a reversible inhibitor, L, to an enzyme, E, results in the formation of a non-covalently bound complex, EL, represented by the equilibrium:

$$E + L \underset{k_{-1}}{\overset{k_{+1}}{\rightleftharpoons}} EL \qquad [1]$$

where the equilibrium dissociation constant K_d is given by:

$$K_d = \frac{[E][L]}{[EL]} = \frac{k_{-1}}{k_{+1}} \qquad [2]$$

3. Analysing the NMR spectra

In order to obtain detailed structural and dynamic information from the NMR spectra of the complexes, the spectral resonances need to be assigned to specific nuclei in the ligand and the protein. Before one can embark on a detailed assignment of the resonances it is first necessary to ascertain whether the bound and free species coexist under conditions of fast or slow exchange on the NMR time scale (see Chapter 6). If a nucleus has chemical shifts of v_B and v_F in the bound and free species respectively, and the lifetime of the complex is long compared with $(v_B - v_F)^{-1}$, then separate signals are seen for the bound and free species: this is designated the *slow exchange* condition. If the lifetime of the complex is short compared with $(v_B - v_F)^{-1}$ then conditions for *fast exchange* prevail and a single observed averaged spectrum is weighted according to the populations and chemical shifts of the two forms (see Section 4.3, Chapter 6).

When the lifetimes are $\sim(\nu_B - \nu_F)^{-1}$ then *intermediate exchange* conditions exist giving rise to broad complex spectra which are difficult to analyse. Clearly it is important to ascertain which exchange regime pertains before further work can be attempted. With typical ^1H chemical shift differences seen between bound and free species in proton spectra (usually less than 1 p.p.m.), an equilibrium dissociation constant of 10^{-3} M or greater will usually result in spectra in fast exchange while dissociation constants of 10^{-6} M or less will almost certainly give rise to spectra in slow exchange. For some ligands where different protons have very different bound shifts, some of the ligand signals can be in fast exchange (small shifts) while others are in slow exchange (large shifts). However, in all cases it is important to determine these exchange regimes experimentally and the methodology for obtaining this information is presented in *Protocol 1*.

Protocol 1. Determination of fast and slow exchange in protein–ligand complexes

1. Determine concentrations of protein and ligand solutions, and if possible the dissociation constant and stoichiometry.

2. Record the spectrum of the protein in the absence of the ligand and also the spectrum of the free ligand.

3. Titrate the enzyme with microlitre volumes of concentrated ligand solution adjusting the sample pH when necessary to keep it constant. Record the ^1H spectra of the protein complexes with 0.2, 0.4, 0.6, 0.8, and 1.0 molar equivalents of ligand.

4. Observe whether the resolved protein signals, (e.g. histidines, high field methyls) are shifting (fast exchange) or changing area (slow exchange).

5. Continue titrating ligand until signals from free ligand can clearly be observed. Look back at the spectra corresponding to lower concentrations to see whether free signals can be observed with hindsight. Determine whether the chemical shift is constant (slow exchange) or changing during the titration (fast exchange).

6. The signals might be broadened due to exchange effects. Record spectra at several temperatures. Sharpening of the ligand signal with increasing temperature indicates intermediate exchange progressing to fast exchange; signal broadening indicates slow exchange. These effects are best seen when the ratio of free to bound ligand in the sample is small (see *Protocols 1* and *2* in Chapter 6).

3.1 Fast exchange

A typical example of a spectrum showing fast exchange behaviour is provided by the ^1H spectrum of the complex of *para*-aminobenzoyl-L-glutamate

(PABG) with dihydrofolate reductase (8). PABG is a fragment of the substrate folate and has a K_d value of 1.2×10^{-3} M. Following *Protocol 1*, increasing amounts of PABG were added to a 1.4 mM solution of dihydrofolate reductase and the ^{1}H spectra recorded after each addition. As the titration proceeds the H2′,6′- and H3′,5′-benzoyl proton signals appear with increasing intensity and with chemical shifts which progressively change towards the chemical shift values for free PABG. The chemical shift data, for the H2′,H6′ protons plotted as a function of the ligand concentration can be analysed with a non-linear regression program to fit eqn 26 in Chapter 6, and thus provides the values of the chemical shifts of the bound ligand and the K_d value.

However, it must be said that NMR is a relatively poor method for determining accurate K_d values and other methods, such as one based on fluorescence spectroscopy, are usually preferred. In order to determine whether or not the fast exchange ligand is binding to a single binding site it is sometimes possible to carry out competition experiments using a more tightly binding ligand which is known to bind to one site. In the example discussed here this could be done by adding an excess of one molar equivalent of a very tightly binding inhibitor (such as methotrexate, $K_d \sim 10^{-10}$ M) which displaces PABG from its binding site and allows one to measure the chemical shifts of the 'free' PABG which result from non-specific binding effects.

If the dissociation rate constant is not sufficiently fast to give true fast exchange conditions but fast enough to give coalesced non-Lorentzian signals for free and bound signals in intermediate exchange, the analysis is more difficult; we have estimated the errors in determined values of K_d and the bound chemical shifts which can result if the assumed fast exchange condition is not fully fulfilled (see Chapter 6). It could be shown that if the observed frequency for the average resonance of the ligand can be measured at very low ligand concentration (for example $[L]_T = 0.1$ mM and $[E]_T = 1$ mM) then a reliable estimate of the bound chemical shift could be obtained (9). However it is usually impossible to see such small ligand signals in ^{1}H spectra in the presence of the complex spectrum from the protein. The photoCIDNP technique can sometimes overcome this problem: for example the H3′, H5′ protons of *p*-aminobenzoyl-L-glutamate, in the presence of DHFR and a laser-excited flavin dye, are strongly polarized thus giving rise to intense signals. The problem can also be addressed by using ^{13}C- or ^{15}N-labelled ligands in combination with appropriate 2D editing experiments to allow only the protons directly bonded to ^{13}C or ^{15}N to be detected (see Section 7.2).

One advantage of working under conditions of fast exchange is that the assignment of bound signals is usually straightforward since the progressively shifting signals monitored during the titration can be related directly to the assigned signals in the free species.

3.2 Slow exchange

Examples of spectra from a complex in slow exchange are shown in *Figure 1*. These are the ^{31}P signals from the 2′-phosphate group of NADP$^+$ in the presence of dihydrofolate reductase as a function of the NADP$^+$ concentration (10). When 1.3 molar equivalents of NADP$^+$ are added to DHFR one observes a signal for bound NADP$^+$ at 3 p.p.m. and a small broad signal at the position of the free NADP$^+$. As increasing amounts of NADP$^+$ are added to the sample, the signal for bound NADP$^+$ remains unchanged while that at the position of free NADP$^+$ increases in intensity and decreases in linewidth. Clearly the exchange between the bound and free NADP$^+$ is sufficiently slow that $\nu_B - \nu_F > (1/\tau_F + 1/\tau_B)$ where τ_F and τ_B are the lifetimes of the free and bound states, and $1/\tau_B = k_{-1}$. The linewidth of the bound ligand signal is independent of the ligand concentration but that of the free ligand signal decreases as the ligand concentration is increased. The exchange rates can be calculated from the observed linewidth contributions (see *Protocol 3* in Chapter 6).

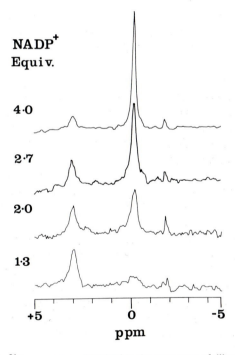

Figure 1. 2′-Phosphate ^{31}P resonance of NADP$^+$ in the presence of dihydrofolate reductase as a function of NADP$^+$ concentration. Microlitre volumes of concentrated NADP$^+$ solution were added to the enzyme sample and the pH readjusted as necessary. The NADP$^+$ concentration is expressed as moles of NADP$^+$ per mole of enzyme. (Reproduced with permission from reference (10).)

4. Assignment of resonances

In order to obtain detailed information about interactions, conformations, ionization states, and dynamic processes within protein–ligand complexes, it is first necessary to assign the protein and ligand resonances. It should be emphasized that the assignments must be carried out on the fully formed complex and preliminary experiments need to be conducted to determine the concentration of free ligand required to saturate the protein.

4.1 Protein resonances

In Chapters 4 and 5 the various methods for assigning protein resonances are outlined in some detail. Using 2D and 3D (and even 4D) NMR methods in combination with ^{2}H-, ^{13}C-, and ^{15}N-labelled proteins it is now possible to obtain almost complete sequential assignments for proteins up to M_r 20000 (11, 12). In our own studies on *L. casei* DHFR, resonance assignments have been obtained for all the 162 amino acids using multi-dimensional NMR techniques on a sample of DHFR uniformly labelled with ^{15}N and ^{13}C. Once the assignments had been made for one complex it proved relatively easy to transfer the assignments to related complexes because of the similarities in the connectivity patterns seen in the 2D spectra of the different complexes. Many of the chemical shifts are not dramatically changed between complexes and it is usually easy to identify those which show substantial shift changes. Assignments can sometimes be transferred between complexes by using 2D exchange spectroscopy to connect resonances in the two complexes: here it is necessary to examine a mixture of two complexes undergoing exchange sufficiently slowly to give separate spectra but sufficiently rapidly to show the exchange cross-peaks (13). In cases of fast exchange between bound and free ligands, assignments of a protein resonance in the bound and free states can be obtained by analysing a binding curve, a plot of the chemical shift of a protein resonance as a function of added ligand concentration. In some cases signal assignments can be transferred to a tightly bound complex (such as DHFR.trimethoprim with signals in slow exchange) from a weakly bound complex (such as DHFR.2,4-diamino pyrimidine with signals in fast exchange). By adding 0.5 equivalents of trimethoprim to a DHFR sample and sufficient free 2,4-diamino pyrimidine to saturate the remaining enzyme, one can obtain a sample containing equal amounts of the two complexes. Using 2D exchange spectroscopy, some resonance assignments can be transferred between the two complexes by observing the exchange cross-peaks. For example, the aromatic protons of Phe49 are very differently shielded in the two complexes and their assignments can be connected by this method.

4.2 Ligand resonances

Assignments of ligand resonances are particularly important because ligand nuclei, of necessity, are well placed to provide direct information about the binding site in the complex. We have seen that for complexes with weakly binding ligands ($K_d > 10^{-3}$ M) fast exchange behaviour is usually observed and the bound chemical shifts can be calculated from the analysis of the binding curves (see Chapter 6). Very tightly binding ligands ($K_d < 10^{-9}$ M) present more of a challenge for the assignment procedures. The classical method is to examine isotopically labelled analogues (^{2}H, ^{3}H, ^{15}N, ^{13}C) in combination with various experimental procedures. For example, deuterated ligands can assist ^{1}H assignments by producing differences between ^{1}H spectra of complexes formed with deuterated and non-deuterated ligands. This is well illustrated by the assignments of the nicotinamide ring protons in complexes of DHFR and NADP$^+$ formed with selectively deuterated NADP$^+$ analogues (see *Figure 2*). A more direct method is to selectively label the ligand with tritium (^{3}H) and then to observe the ^{3}H spectra of the complexes: for example, by examining (^{3}H-7, 3', 5'-) folic acid in its complex with DHFR and NADP$^+$

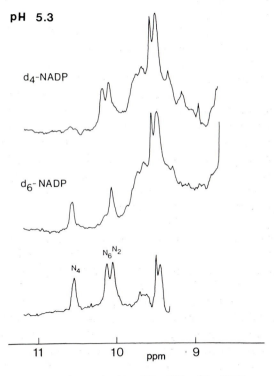

Figure 2. The low field region of the 270 MHz ^{1}H NMR spectrum of the DHFR.folate.NADP$^+$ complex at pH 5.3 (*bottom*). The resonance signal at 10.54 p.p.m. is assigned to the bound nicotinamide ring 4 proton because of the absence of this peak in the spectrum of the corresponding complex with [nicotinamide-4-^{2}H] NADP$^+$ (*top*). The *middle* spectrum from the complex containing [nicotinamide-6-^{2}H] NADP$^+$ assigned the signal at 10.15 p.p.m. to the nicotinamide ring 6 proton in the bound coenzyme. The chemical shifts are referenced to DSS. (Reproduced with permission from reference (58).)

using ^{3}H NMR it was possible to confirm the presence of three different conformational states by monitoring the assigned tritium resonances for the 7,3′,5′-tritium nuclei (the ^{3}H chemical shifts are directly related to the ^{1}H chemical shifts) (14). In the case of ^{15}N- or ^{13}C-labelled ligands their complexes can be examined directly by using ^{15}N or ^{13}C NMR: since only the signals from nuclei at the enriched positions are observed, the assignment problem is trivial. Where there are protons directly attached to ^{15}N or ^{13}C, the opportunity arises for using editing pulse sequences. Heteronuclear multiple quantum coherence (HMQC) experiments allow the attached protons to be detected selectively (they are then characterized by both the ^{1}H and ^{15}N (or ^{13}C) frequencies). These experiments were introduced by Mueller (15) and extended by others (16, 17, 18) to studies of protein–ligand complexes: an example of the use of ^{15}N-edited ligand signals is discussed in Section 7.2. A powerful extension of these studies is to NOESY.HMQC 2D and 3D experiments: these allow selective detection of the NOEs from the ligand protons (on ^{13}C or ^{15}N) to neighbouring protons in the protein (11). This is very important for examining complexes involving large proteins since the normal NOESY spectrum containing all the ligand–protein and protein–protein NOEs is often too complex to analyse.

Complexes formed using less tightly binding ligands ($K_{d} \sim 10^{-6}$ M) have spectra which show slow exchange behaviour, giving separate signals for bound and free species, but have dissociation rate constants sufficiently large to allow transfer of saturation or 2D exchange methods to be used to connect the resonance of the bound and free species. Since the assignments in the free ligand are usually known with certainty, this allows the assignments of the bound species to be made. This method (see Chapter 6) also provides a direct method for measuring the dissociation rate constants. In favourable cases these methods can also be used to assign protein resonances.

Some ligands contain phosphorus and in these cases the resonances from the 100% naturally occurring ^{31}P nucleus can be monitored by NMR. Even in systems where only a few different nuclei are involved, the assignment is not trivial: for example in ^{31}P spectra of NADPH bound to a protein, the two pyrophosphate ^{31}P nuclei can only be assigned unequivocally by resorting to isotopic labelling procedures or to heteronuclear correlation experiments.

When fluorine containing ligands or protein are available, ^{19}F NMR measurements can be used to examine these complexes. The assignment of the ligand fluorine signals is often straightforward since usually only one or two sites are labelled. Fluorine signals from fluorine containing proteins have been assigned by observing the ^{19}F spectra following systematic removal of each fluorine-containing amino acid by site-directed mutagenesis.

5. NMR parameters of the bound species

Once the detailed assignments of the resonances are known for the fully formed complex, the NMR parameters for the various nuclei can be measured and compared for free and bound species in a protein–ligand complex or series of related complexes. Changes in chemical shifts, spin-coupling constants, and relaxation behaviour (including nuclear Overhauser effects) have all been monitored to provide information about such complexes. The chemical shift is the easiest information to acquire but usually the most difficult to interpret. Changes in protein ^{1}H chemical shifts observed on ligand binding can arise either from direct interactions between protein and ligand, or from indirect conformational changes remote from the binding site but induced by the ligand binding. Observations of chemical shift changes on ligand binding indicate those parts of the molecules that are influenced in some way by the binding. If different but related ligands produce similar chemical shift changes for several protons on the protein then one can argue that a particular common fragment of the ligands is binding similarly in the different complexes. We have used such an approach to show that the 2,4-diaminopterin and 2,4-diaminopyrimidine rings of a series of methotrexate and trimethoprim analogues **3–6** are binding similarly to *L. casei* dihydrofolate reductase. In some cases the observed chemical shift changes on ligand binding are noted to be in well-defined parts of the protein and this can help to localize the region of the protein influenced by the interaction. For example, when NADPH binds to the binary complex of DHFR with methotrexate most of the observed chemical shift changes could be explained by movements of the loop comprising residues 13–26 and the α-helix residues 42–49 in the protein structure (13). In some favourable cases, one can attempt a semi-quantitive evaluation of the bound shifts in terms of ring-curent shift models: this is particularly useful when the ring-current shielding of a ligand proton is caused by an aromatic ring in the same ligand and where there is little chance of shielding contributions from other aromatic rings in the protein (19). However, all these various approaches are somewhat qualitative. Chemical shift changes can be best used for determining ionization states or hydrogen bonding states where there is a well documented difference in chemical shift between the two states. Even in such cases, care must be taken to ensure that other shielding effects are not making major contributions. Studies of nuclei other than ^{1}H can be very useful in this regard: for example the ^{15}N chemical shift of the pyrimidine N1 ring nitrogen changes by 80 p.p.m on protonation. Measurements of bound coupling constants offer more promise as a conformational tool since there are well established relationships between three bond vicinal coupling constants and dihedral torsion angles. However, surprisingly few examples of such studies on bound ligands have been reported (20, 21) probably because of the difficulties of obtaining good coupling constant data from the broad lines measured for most protein–ligand complexes. Spin-spin relaxation effects can also lead to the measured values of coupling constants being smaller than the real values which provides a further complication for larger systems ($M_r > 20000$). However, increasing use is being made of spin coupling constant information relating to H^α-H^N and H^α-H^β vicinal coupling constants in proteins: estimates of these are obtained by simulation of the observed cross-peaks in DQF-COSY and HOHAHA-HMQC spectra (22).

Changes in relaxation times at diferent sites in the bound ligand have proved very useful for defining the mobility of different parts of the complex (4, 23, 24). Even from the earliest days, when Jardetzky and Wade Jardetzky (25) showed that different linewidths were observed for different protons in a sulfonamide drug binding to bovine serum albumin, this approach has been valuable.

However, by far the most direct method of obtaining conformational information is by measuring protein–ligand NOEs from 2D NOESY spectra of the complex (see Chapters 4 and 5). Problems of spin diffusion can be overcome by examining the NOESY spectra at a series of mixing times. Using this approach spatial information within the binding site can be obtained and details of the specific interactions deduced.

Other methods of obtaining conformational information rely on introducing paramagnetic probes at defined sites into the complex (transition or lanthanide metal ions or stable free radicals in the form of nitroxide spin labels) and observing their shielding and relaxation effects on neighbouring assigned protons. In favourable cases, distances between the paramagnetic centre and assigned protons have been estimated. While there are many problems associated with such approaches (for example the difficulty of allowing for weak non-specific binding in the case of the paramagnetic ions) useful conformational information can be obtained when the methods are applied with care (26). Particular attention has been focused on studies of enhanced proton relaxation of the water signal from aqueous solutions of metallo enzymes substituted with paramagnetic metal ions. However, attempts to use such measurements to determine hydration numbers are fraught with difficulties (27).

6. Determination of ionization states

Changing the charge state of an ionizable group in a protein or ligand is usually accompanied by large, characteristic changes in shielding of nuclei close to the ionizable group. NMR has thus been used extensively to report on the ionization states of specific groups in proteins and ligands, to measure their pK, values and to detect the changes in pK values which accompany protein–ligand complex formation. X-ray crystallography is unable to provide such direct information about ionization states and clearly this is an area where NMR can make a unique contribution to understanding interactions within the binding site. Reviews of NMR studies of ionizable groups in proteins have been provided by Markley (28), Jardetzky and Roberts (1, 5), and Wüthrich (29).

6.1 Protein ionization states

Ionization effects on histidine residues have been intensely studied even from the earliest days of biological NMR. In ^2H_2O solutions, the ^1H signals from the C2 protons of the imidazole rings of histidines are relatively easy to detect even in the 1D spectra of complex proteins. The pK values of histidines in proteins are typically in the range 4.5 to 8.5 and they can easily be studied by carrying out pH titrations over a pH range accessible for many proteins. Less work has been carried out on other amino acids containing ionizable groups. However, carboxyl group protonation in Asp and Glu residues have been monitored by using ^{13}C NMR of the carboxyl carbon (upfield shifts of 2.5 to 5.0 p.p.m. being caused by protonation (1,2)). Tyrosine phenolic hydroxyl group ionization have also been monitored by measuring ^{13}C chemical shifts. Allerhand and co-workers (30) showed that the Cγ and Cζ ring carbons shift 6 p.p.m. upfield and 10 p.p.m. downfield respectively on deprotonation. Lys and Arg ionizable groups have been less well characterized (1).

Returning to consideration of histidine residues, the signals of the Cε_1 (C2) and Cδ_2 (C4) protons of the imidazole ring are shifted downfield on protonation by 1.0 (7.7 to 8.7). p.p.m. and 0.4 (7.0 and 7.4) p.p.m. respectively.

Because there is usually fast exchange between the protonated and unprotonated species, the ^{1}H signals from the ring protons have averaged chemical shifts weighted according to the populations of the two states. A plot of the chemical shifts against the pH values can be fitted to the Henderson–Hasselbalch equation and thus provide values of the pK and the chemical shifts of the protonated and non-protonated species (see *Figure 3a*).

For some proteins, the histidine titration curves do not conform perfectly to a Henderson–Hasselbalch curve and this indicates the presence of interacting ionizable groups. Such behaviour was seen for two of the histidines in the cAMP receptor protein from *E. coli* and the data could be analysed in terms of two neighbouring interacting His residues with pK values of 6.21 and 7.47 (31) by using equations developed by Schrager and co-workers (32).

L. casei dihydrofolate reductase (M$_r$ 18 400) has seven histidine residues and these have been extensively studied. Five of the histidines are influenced by substrate or inhibitor binding (33). One of the most useful findings from the histidine titration studies is that substrates and inhibitors (for example methotrexate) which contain the L-glutamic acid moiety cause the pK value of His28 to be increased from 6.8 to ~7.8 (see *Figure 3b*). From the crystal structure (see *Figure 4*) it is seen that the γ-carboxylate of the L-glutamic acid moiety of methotrexate is near to His28 and the α-carboxylate is near to the conserved Arg57 residue (34). The pK behaviour of His28 provides a very useful probe for determining whether or not these interactions

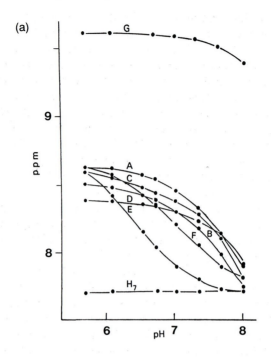

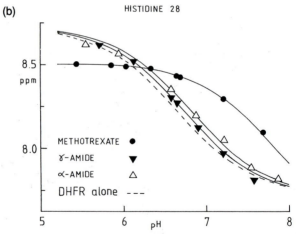

Figure 3. (a) The pH dependence of the chemical shift of histidine imidazole resonances (A–G) for the complex of DHFR with methotrexate α-amide. (b) The pH dependence of the chemical shift of histidine 28 in several dihydrofolate reductase complexes; (●) methotrexate, (▼) methotrexate γ-amide, (△) methotrexate α-amide (---) enzyme alone. Chemical shift scale is in p.p.m. and referenced from DSS.

are present in various complexes. NMR studies of complexes of DHFR with methotrexate **4** and its α-amide **5** and γ-amide **6** analogues have provided interesting results. The binding of the α-amide to DHFR is 100-fold weaker than that of methotrexate whereas the γ-amide binds only nine times more weakly than the parent drug (35). From the histidine titration curves of the complexes it is clear that, in the complex of the γ-amide with DHFR, His28 does not show a large perturbation in pK compared with free DHFR. This result is as expected since the γ-amide is unable to form the ionic interactions with the histidine imidazole group. The complex with the α-amide analogue, however, where the γ-carboxylate is still present, gave an unexpected result; the histidine titration curves for this complex (see *Figure 3a* and *3b*) clearly indicate that once again His28 does not increase its pK. Obviously, the disruption of the interaction between the α-carboxylate group and Arg57 has caused a conformational change which prevents the γ-carboxylate from interacting with His28. More detailed studies of the ¹H chemical shifts in the various complexes using assigned COSY 2D spectra, revealed that the γ-amide complex has a very similar spectrum to the methotrexate complex (except for His28) suggesting that this modification causes only a local perturbation involving the removal of the γ-carboxylate–His28 interaction. The ¹H spectrum of the α-amide complex, however, shows chemical shift changes for the protons of Leu27, Phe49, Leu54, and Leu19 all of which are close to the benzoyl ring of the methotrexate and expected to be influenced by its

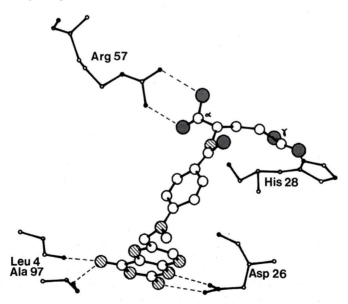

Figure 4. Detail from the X-ray crystal structure of *L. casei* dihydrofolate reductase complex with methotrexate and NADPH indicating the interactions between methotrexate and the protein (from crystal structure data of Matthews and co-workers).

ring-current shielding: this indicates that the disruption of structure caused by introducing the α-amide modification also influences the position of the benzoyl ring in the complex. The NMR findings, indicating that both the α- and γ-carboxylate interactions are affected, thus offer an explanation for why the decrease in binding caused by the α-amide modification is greater than that caused by the γ-amide modification. From the observed changes in binding constant in the different complexes and knowing that both the α- and γ-carboxylate interactions are broken in the α-amide complex, it could be estimated that each interaction is making a similar contribution to the binding energy ($\sim$5.6 kJ mole^{-1}).

6.2 Ligand ionization states

For the case where the ionization state is a protonated species, it is sometimes possible to use NMR to detect directly the proton involved in the protonation. If the protonation is on a nitrogen atom then the use of selective ^{15}N-labelling can provide an unambiguous method of assigning the bonded proton. When inhibitors containing a 2,4-diamino pyrimidine ring (such as the antibacterial drug, trimethoprim) bind to DHFR, NMR evidence clearly shows that the N1 position of the pyrimidine ring is protonated. In a ^{1}H NMR study examining ^{15}N enriched trimethoprim in its complex with DHFR in H_2O solution, a doublet was observed at 14.79 p.p.m. in the spectrum which could be assigned to the N1H proton: the doublet splitting ($\sim$90 Hz) is characteristic of one bond ^{15}N-^{1}H spin coupling and furthermore, the splitting disappears in the ^{1}H spectrum of the DHFR complex formed with non-labelled trimethoprim (36). While this approach is the most direct method of determining the ionization state it is usually easier to monitor chemical shifts of nuclei close to the site of ionization. For example, the ^{15}N chemical shift of the N1 nitrogen has a value characteristic of the protonated species (80 p.p.m. different from the non-protonated species) and can be used to determine the charged state of N1. Similarly one can measure ^{13}C chemical shifts of [^{13}C-2]-trimethoprim to show that the N1 position is protonated in the bound state and to indicate that the pK value for this protonation is displaced by at least two units as a result of complex formation (37). A consideration of the X-ray data strongly suggests that the protonated N1 group is interacting with the carboxylate anion of the conserved Asp26 residue.

Ionization states of phosphate groups can be monitored using ^{31}P NMR and we have used this approach in studies of coenzyme (NADPH or NADP$^+$) binding to DHFR. *Figure 5* shows the plots of ^{31}P chemical shifts of the 2'-phosphate group as a function of pH for complexes of DHFR with oxidized and reduced coenzyme together with the same data for the free coenzymes. While the data for the free coenzymes clearly indicate pK values of 6.1 (NADPH) and 6.4 (NADP$^+$), the ^{31}P chemical shift of the 2'-phosphate group of the coenzymes in the DHFR complexes does not change with pH

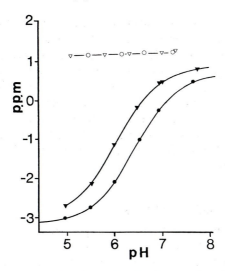

Figure 5. The pH dependence of the chemical shift of the 2'-phosphate ^{31}P resonance of NADP+ (●, ○) and NADPH (▼, ▽). The *open symbols* refer to coenzyme bound to dihydro-folate reductase and the *solid symbols* to free coenzymes. (Reproduced with permission from reference (20).)

over the range 4.5 to 7.5. The bound chemical shifts indicate that the 2'-phosphate is binding in exactly the same manner in complexes with the oxidized and the reduced coenzyme. In each case the phosphate group is binding as the dianionic form with its pK perturbed by at least three units. These studies are relatively straightforward because one is dealing with tightly bound ligands, where the bound and free species exist under conditions of slow exchange. In experiments designed to dissect out the binding contributions of different parts of the coenzyme, fragments of the coenzyme which bind more weakly to DHFR have been studied (38). For some of these, the exchange between bound and free species is rapid. In these cases it is very important to determine fully the pH dependence of the ligand binding to the protein. Such a case is provided by 2'-AMP binding to DHFR where the ^{31}P chemical shift of the 2'-phosphate group varies as a function of the concentration of 2'-AMP in the presence of the enzyme. Binding curves were measured at a series of pH values in the range 4.5 to 7.5 and the curves analysed to give the dissociation constants at the different pH values and the bound shifts. *Figure 6* shows a plot of the pH dependence of the ^{31}P chemical shift of the bound 2'-AMP (curve A) and this indicates that the 2'-AMP is binding in both the monoanionic and the dianionic states. The pK of free 2'-AMP is 6.0 ± 0.05 (see curve B) and there is a shift in pK to 4.8 ± 0.2 on binding to DHFR (curve A). The Dixon plot of the pH dependence of the dissociation constants (curve C) can be satisfactorily explained by this change in pK of the

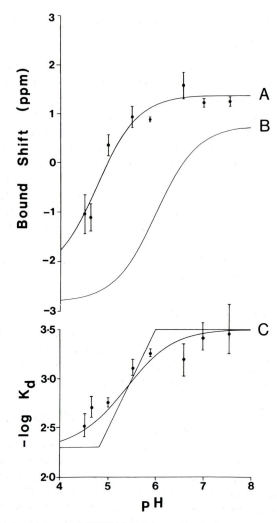

Figure 6. The pH dependence of 2'AMP binding to dihydrofolate reductase. (A) The ^{31}P chemical shift of the 2'-phosphate group of 2'AMP bound to the enzyme as calculated from the binding curves at each pH value. (B) The pH dependence of the ^{31}P chemical shift of free 2'-AMP. (C) The pH dependence of the dissociation constant of the complex. The curves in A and C were calculated using pK values for bound 2' AMP of 4.8 ± 0.2 and for free 2' AMP of 6.0 ± 0.05. (Reproduced with permission from reference (38).)

phosphate group on binding, and therefore there is little or no change in the pK values of groups on the enzyme in the pH range 5 to 7 when 2'-AMP binds to DHFR. The decrease in pK value of 1.2 units indicates that the dianionic form of 2'-AMP binds 16-fold more tightly to the enzyme than does the monoanionic form.

200

7. Determination of conformations of protein– ligand complexes

Ideally, one would like to obtain the conformation of the free species and the conformation of the complex in solution. A good method of achieving this would be to measure intra- and intermolecular NOEs in the complex. However, there are still relatively few examples where this has been achieved using NMR methods, and these are mainly in the area of protein–DNA complexes (6, 7). The general protocols for biomolecular structure determination are discussed later (Chapter 10). First the ^1H assignments are made and then the interproton distance constraints are obtained from NOE measurements and the dihedral angle constraints from vicinal coupling constants and NOEs. Families of structures are then calculated from geometric and covalent structure constraints using distance geometry (DG), distance bounds driven dynamics (DDD), or restrained molecular dynamics (MD) calculations. The final structures are refined using restrained energy minimization (REM) or molecular dynamics calculations (see Chapter 11).

7.1 Protein–DNA complexes

There are two examples of protein–DNA complexes which have been well characterized in solution using NMR methods, namely the *lac* repressor headpiece binding to DNA duplexes (Kaptein and co-workers) (6), and the *Antennapedia* homeodomain protein binding to a DNA 14-mer (Wüthrich and co-workers) (7).

In determining the bound conformations of the components on the complex, intramolecular NOEs and spin coupling constants provide most of the information. In some cases, such as for the complex of the *Antennapedia* homeodomain protein binding to a DNA, the intramolecular NOEs were found to be similar in the free and bound states indicating that the free and bound conformations are very similar. Independent evidence for the similar conformations has been provided by Neri *et al.* (39) who found that the H^{α}-H^N vicinal coupling constants were virtually unchanged for most of the 68 protein residues in the free and bound states. However, amide proton exchange rates for residues 30, 36, and 37 are faster in the complex than in the free protein, which probably indicates some distortion of the protein structure on complex formation. Intermolecular NOEs between nine sets of protons on the DNA and protein were measured for the complex and this was sufficient to obtain a good characterization of the conformation of the complex. This was achieved by docking the protein and DNA in the fixed conformations previously determined by NMR using the measured intermolecular NOEs (translated into distance constraints of 5 Å) in conjunction with an ellipsoid algorithm calculation. The major interactions were found between residues on the elongated second helix of a helix-turn-helix motif in the homeodomain and residues in the major groove of the DNA: additional NOE contacts with the DNA involve a polypeptide loop which precedes the helix-turn-helix motif and also Arg5, the latter interacting with the DNA minor groove. Most of the protons which showed chemical shift changes on binding turn out to correspond to protons located at the protein–DNA interface (7).

7.2 Isotope-editing using labelled ligands

A powerful technique for measuring intra- and intermolecular NOEs in protein–ligand complexes is based on using isotopically labelled (^{13}C or ^{15}N) ligands and examining their complexes using NMR isotope-editing procedures (16, 17). In this way, NOEs involving ligand protons directly attached to ^{13}C or ^{15}N can be selectively detected. *Figure 7* shows a spectrum from an ^{15}N-edited 2D NOE experiment on a pepsin/inhibitor (1:1) complex formed with ^{15}N-labelled inhibitors (17). The spectrum is simple in that it shows only diagonal peaks from the amide protons attached to ^{15}N and NOE cross-peaks between these amide protons. The measured NOEs observed in *Figure 7* could be interpreted by assuming that the inhibitor has an extended backbone conformation when complexed with porcine pepsin (40). Isotope-editing methods have recently been applied (41, 59) to study ^{13}C- and ^{15}N-labelled cyclosporin A bound to cyclophilin: the cyclosporin A protons have been assigned and intramolecular NOE information obtained for free and bound ligand. The bound ligand has all its peptide bonds in the *trans* form, in contrast to free cyclosporin A, which has a *cis* peptide bond between the MeLeu residues at positions 9 and 10.

7.3 Transferred NOEs and bound conformations

The transferred NOE (TNOE) is another useful method for obtaining conformational information about the bound ligand (42–45). In this method, cross-relaxation (NOE) between two protons in the bound ligand is transferred to the free molecule by chemical exchange between bound and free species (see *Figure 8*). The negative NOEs from the bound state can thus be detected by carrying out NOE experiments on the free ligand (in slow exchange), or on the averaged signals for free and bound ligand (in fast exchange). Thus, if one selectively irradiates either the free or the averaged signal of proton S in a 1D experiment then a negative transferred NOE will be observed on the free or averaged signal of proton I and the magnitude of the transferred NOE is proportional to the cross-relaxation rate σ_{IS}^{B} between I and S in the bound state. One should irradiate for sufficiently short times to ensure that the initial rate approximation is fulfilled (see Chapter 4). The method works when

- the chemical exchange rate k is much faster than the relaxation rate of the free proton, ρ_I^F

$$k \geq 10\rho_I^F$$

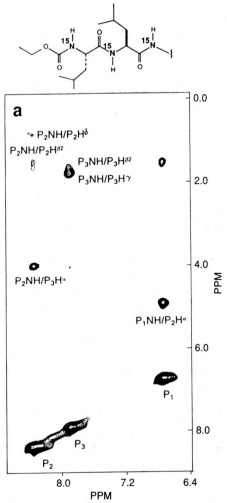

Figure 7. Contour plot of an ^{15}N isotope-edited 2D NOE experiment on a pepsin/inhibitor (1:1) complex. The diagonal peaks corresponding to the three amide protons, P_1–P_3, and various NOE cross-peaks involving these protons are labelled. (Reproduced with permission from reference (40).)

● and $|(1 - p_F)\sigma^B_{IS}| > |p_F\sigma^F_{IS}|$

where p_F is the mole fraction of free ligand and σ^B_{IS} and σ^F_{IS} are the cross-relaxation rates between I and S in the bound and free states respectively. If the bound cross-relaxation rates for two pairs of spins (IS and IT) can be measured, then eqn 3 can be used to estimate the ratio of their internuclear separations. Providing that one of the distances is known, the other can be calculated.

$$\frac{\sigma^B_{IS}}{\sigma^B_{IT}} = \left(\frac{r_{It}}{r_{IS}}\right)^6 \qquad [3]$$

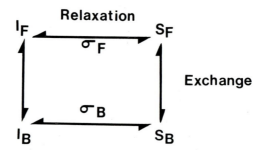

Figure 8. Schematic indication of processes relating to the magnetization of two nuclei I and S in a ligand binding to a protein. (Reproduced with permission from reference (42).)

In practice, the method works very well for large proteins bound to ligands in fast exchange but it is also effective for slowly exchanging ligands if the above conditions can be met. The first observation of such transferred NOE effects for slowly exchanging ligands was in the complex of trimethoprim with a selectively deuterated DHFR sample (19). *Figure 9* shows the simplified aromatic region of its ^{1}H spectrum and indicates the experiments used to detect the transferred NOE effects. Irradiation of the H6 signal from the free ligand is seen to cause a decrease in intensity at the H2′,H6′ signal of the free ligand due to a transferred NOE effect (see *Figure 9a* and *9b*). This result indicates that H6 is close to H2′ (or H6′) in a folded conformation of bound trimethoprim. The use of a partially deuterated protein minimizes the possibility of complications from intramolecular NOE effects from the protein but usually the method is used with unlabelled protein.

The method has also been shown to be useful for determining the glycosidic bond orientations for bound nucleotides. For the complex NADP$^+$.DHFR, a transferred negative NOE was observed between the nicotinamide ribose 1′-proton of free NADP$^+$ and the nicotinamide ring 2 proton of free NADP$^+$ implying an *anti*-conformation for the glycosidic bond in bound NADP$^+$. Gronenborn and Clore (45) have developed the theory for the transferred NOE effect and several workers have exploited the technique, particularly for ligands such as nucleotides binding to proteins under conditions of fast exchange as indicated in *Protocol 2* (for a review see reference 44). Transferred NOE effects can also be detected in 2D NOESY spectra and Clore *et al.* (46) have used this approach to determine the conformation of the tetrapeptide, acetyl-Pro-Ala-Pro-Tyr-NH$_2$, bound to porcine pancreatic elastase. In the 2D NOESY spectra, the only cross-peaks detected were those between the averaged ligand signals (the protein signals being much broader). The cross-peaks had the same sign as the diagonal indicating negative TNOEs from the bound peptide and no spin diffusion effects were observed at a mixing time of 100 msec. Using this method they obtained a set of 23 distance constraints between pairs of protons in the tetrapeptide ligand bound to the protein. They were able to use these constraints, in combination with molecular dynamics

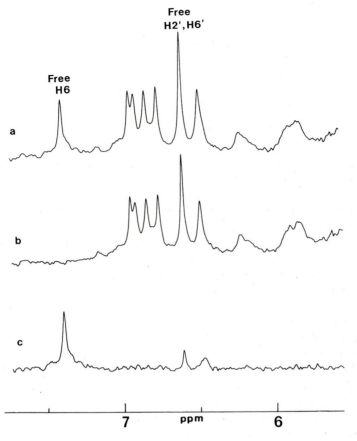

Figure 9. The aromatic region of the 270 MHz ^{1}H NMR spectrum of selectively deuterated dihydrofolate reductase in the presence of 2.7 molar equivalents of trimethoprim; sample temperature 318 K. The resonance at 7.34 p.p.m. arises from the 6 proton, and that at 6.62 p.p.m. from the 2′,6′ protons of free trimethoprim; the other five sharp resonances are those of the tyrosine residues of the protein. (a) Control; (b) irradiation at 7.34 p.p.m.; (c) spectrum (a) minus spectrum (b) (same vertical scale). (Reproduced with permission from reference (19).)

calculations, to define the conformation of the bound ligand. By combining this with some limited crystallographic data they could locate the bound ligand in its protein binding site where it was found to have the reverse orientation to that found in other serine protease–inhibitor complexes.

Protocol 2. Transferred NOE experiment for a protein–ligand complex in fast exchange

1. Assign the ^{1}H signals of the free ligand.

2. In order to check the condition that $(1-p_F)\,\sigma_{IS}^B \geqslant p_F\sigma_{IS}^F$, take a sample of

Protocol 2. *Continued*

free ligand in the absence of protein and selectively irradiate in a systematic manner (every 0.05 p.p.m.) across the spectral region monitoring the intensity of each of the ligand signals: often ligands of M_r 500–1000 show negligible NOE effects particularly if a suitable temperature is chosen.

3. To a 0.1 mM solution of protein add 10 to 30 molar equivalents of ligand.

4. Again irradiate selectively every 0.05 p.p.m. across the spectral region and monitor the intensity for each of the ligand signals.

5. Plot the intensity of each signal as a function of the irradiation frequency.

6. The measured intensity changes are proportional to cross-relaxation rates and can be used with eqn 3 to provide ratios for internuclear distances.

7. In a typical TNOE experiment on the complex of NADP$^+$ bound to yeast glucose-6-phosphate dehydrogenase, the intensity changes observed on the averaged proton signal from the adenine 8 proton were measured when the ribose H1′ and H2′ positions were irradiated (*Figure 2* in reference 44): the results indicated that the glycosidic bond has an *anti*-conformation.

8. Detection of multiple conformations

One of the most useful applications of NMR spectroscopy in the study of protein–ligand complexes is its ability to detect different conformational states in solution. In some cases the different conformations are in slow exchange such that a separate NMR spectrum is observed directly for each conformation (47). In other cases internal inconsistencies between exchange rates measured using different signals provides indirect evidence for multiple conformations (48).

In the ternary complex formed by NADP$^+$ and trimethoprim with DHFR, two conformations of almost equal population have been found to coexist in solution, each form giving a separate NMR spectrum. This was first detected in the histidine C2 proton region of the ^{1}H spectrum where six of the seven His signals appeared as doublets as shown in *Figure 10* (47). Analysis of the changes in lineshape measured as a function of temperature indicated that the conformations are slowly interconverting (6 sec^{-1} at 304 K). By examining ^{13}C spectra of DHFR complexes formed with ^{13}C enriched NADP$^+$ and trimethoprim analogues, and ^{31}P and ^{1}H spectra of non-labelled ligands bound to DHFR it has been possible to detect ligand signals from nuclei in the two different conformations and to use these to estimate the amounts of each form present in the mixture. Complexes formed with analogues of either the coenzyme or trimethoprim with modified structures show changes in the

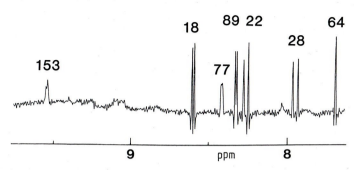

Figure 10. Part of the deconvoluted 500 MHz ^{1}H NMR spectrum of the *L. casei* dihydrofolate reductase.NADP$^+$.trimethoprim complex, showing the C2 proton resonances of the seven histidine residues. (Reproduced with permission from reference (47).)

relative populations of the two forms and the system appears to be a two-state conformational equilibrium which can be 'switched' by the binding of ligands of different structure. We have characterized the structures of the two forms and shown that the major difference between them is in the mode of binding of the nicotinamide ring which in one form (Form I) is within a protein binding pocket and in the other (Form II) extends into solution outside the binding site (49). For Form I the nicotinamide ring protons have very large downfield shifts (0.6 to 1.1 p.p.m.) while those in Form II remain essentially unchanged.

Transferred NOE experiments indicated that the glycosidic bond conformation is different between the two forms, being *anti* in Form I and a *syn/anti* mixture in Form II (49). There are also differences in conformation of the NADP$^+$ pyrophosphate moiety between the two forms which result in characteristically different ^{31}P chemical shifts. In fact, intensity measurements of the ^{31}P NMR spectra provide the most convenient method of estimating the relative populations of the two forms in the different complexes. The conformation of the bound trimethoprim does not appear to change appreciably between the two forms although its environment is modified. Clearly such measurements of conformational mixtures are important when one is considering structure-activity relationships and NMR is the only method which can provide quantitative information about this type of conformational equilibrium in solution.

We have observed multiple conformations in several different complexes of *L. casei* DHFR (involving folate (50) and folinic acid (48)) and others have seen multiple conformations in complexes with *S. faecium* DHFR (51): it seems likely that many other protein–ligand complexes will exist as mixtures of conformations. However such conformations are difficult to detect directly if the conformations are in fast exchange.

9. Dynamic processes in protein–ligand complexes

Studies using NMR relaxation, lineshape analysis, and transfer of magnetization have provided a wide range of dynamic information relating to protein–ligand complexes. The accessible motions range from fast ($> 10^9$ sec^{-1}) small amplitude oscillations of fragments of the complex (available from ^{13}C relaxation measurements) to slow processes (1 to 10^3 sec^{-1}) involved in the rate of dissociation of the complex, rates of breaking and reforming protein–ligand interactions, and rates of flipping of aromatic rings in the bound ligands. A characterization of the various dynamic processes complements the structural information to give a more complete description of the protein–ligand complex. The dynamic processes in dihydrofolate reductase.trimethoprim complexes will be discussed.

9.1 Dissociation rate constants from transfer of saturation studies

In Chapter 6, the use of the transfer of saturation method for measuring dissociation rate constants has been discussed in detail. For example, the method has been used to measure the dissociation rate constant for the trimethoprim.DHFR complex (6 sec^{-1} at 313 K) (19).

9.2 Rapid segmental motions in protein–ligand complexes

Information about rapid molecular motions ($> 10^9$ sec^{-1}) can often be obtained by measuring ^{13}C relaxation times. If the relaxation is completely dipolar involving interactions between the ^{13}C and a directly bonded proton (separated by r_{CH}) and if the motion is isotropic and sufficiently fast to satisfy the extreme narrowing condition then the measured relaxation rate is given by:

$$\frac{1}{T_1^{DD}} = \hbar^2 \gamma_C^2 \gamma_H^2 r_{CH}^{-6} \tau_c \qquad [4]$$

Since all parameters except τ_c, the correlation time, are known, eqn 4 can be used to calculate the correlation time from measured T_1 values under these conditions. The theory has been modified to deal with the non-extreme narrowing condition when $(\omega_H + \omega_C) \approx \tau_c$ and also to allow for multiple correlation times and non-isotropic motions. When a small ligand (typical $\tau_c \approx 10^{-11}$ sec) binds tightly to a macromolecule (typical $\tau_c \approx 10^{-8}$ sec), the bound ligand will show relaxation behaviour corresponding to the correlation time of the macromolecular complex. One can use ^{13}C relaxation time measurements to determine if any parts of the bound ligand are not completely immobilized on binding.

We have measured the ^{13}C relaxation behaviour of the two enriched carbons in [7,4'-OCH$_3$-$^{13}C_2$]-trimethoprim bound to DHFR and found that the calculated correlation time (assuming isotropic motion) is 15.4 ± 1.5 nsec at 295 K which is very close to that expected for the overall tumbling of the protein (52). The relaxation behaviour of the 4'-OCH$_3$ is quite different, having a six-fold slower relaxation rate. Analysis of this data by procedures outlined by Lipari and Szabo (53) provides information about rapid motions about the C7-C1', C4'-O and O-CH$_3$ bonds. The results provide an effective correlation time for the internal motions affecting the 4'-OCH$_3$ (4.3 (± 0.8) psec) and information about an order parameter for the additional motion about the three bonds linking the OCH$_3$ group to C7 (52). The observed order parameter could not be explained simply in terms of one additional motion, corresponding to the methyl group rapidly spinning about its axis, and there must be additional motion on the sub-nanosecond time scale about at least one of the other two bonds: it is impossible to distinguish which bond but if only one bond is involved the amplitude of the rapid motion would be ± 36° while if there is equal motion about both bonds the amplitude would be ± 25°. Other workers have calculated fairly similar estimates for the amplitudes of motions of amino acid side chains from ^{13}C relaxation data.

9.3 Determination of rates of ring flipping

Several workers have pointed out that the aromatic rings of Phe and Tyr residues sometimes take up fixed orientations in a protein such that nuclei at the 2',6' positions (and the 3',5' positions) are non-equivalent due to the asymmetric nature of their environment in the protein. More usually, Phe and Tyr aromatic rings are rapidly flipping, probably by 180° jumps, such that averaged NMR signals are seen for nuclei at the 2',6'- (and 3',5'-) positions. Slow and fast ring flipping has also been observed and characterized in ligands bound to proteins, for example in complexes of methotrexate (54) and trimethoprim analogues (52) with DHFR. Such studies are easier to carry out if isotopically labelled ligands are used. For example, ^{13}C lineshape analysis on the signals from the enriched carbons in [m-methoxy-^{13}C]-brodimoprim bound to DHFR has been used to measure the 'flipping' rates of the benzyl ring (52). (A detailed discussion of lineshape analysis is provided in Chapter 6.) The ^{13}C spectra of this complex recorded at a series of temperatures are shown in *Figure 11*. At 298 K, a single ^{13}C signal is detected for the 3'- and 5'-methoxy carbons. As the temperature is lowered, the signal broadens and below 285 K two separate signals of equal intensity are resolved which become narrower at 278 K. These spectra are characteristic of an exchange

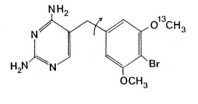

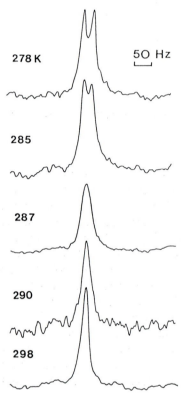

Figure 11. 50.2 MHz ^{13}C NMR spectra of [*m*-methoxy-^{13}C]-brodimoprim with one equivalent of *L. casei* dihydrofolate reductase at a series of temperatures. (Reproduced with permission from reference (52).)

process between two equally populated sites. At low temperatures the benzyl ring is in a fixed orientation on the enzyme with the two *meta*-methoxy groups being in distinct environments in the binding site each having a ^{13}C signal with a different chemical shift. Any exchange process which rapidly interchanges the environments of the 3'- and 5'-methoxy carbons will lead to coalescence of the two signals. The simplest process by which this could be achieved is a 180° flip about the C1'-C4' axis. Similar results were also observed in the complex of the enzyme with trimethoprim. We have analysed the ^{13}C line-

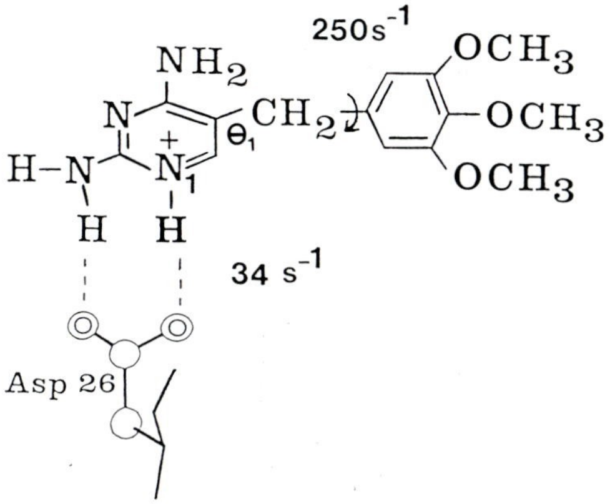

Figure 12. Dynamic processes in complex of trimethoprim with *L. casei* dihydrofolate reductase.

shapes on the basis of McConnell's modification of the Bloch equations and obtained estimates of the rates of 'flipping' and the activation parameters (at 298 K trimethoprim, 250 sec^{-1}; E_a 42 kJ/mol; brodimoprim 140 sec^{-1}, E_a 59 kJ/mol). In both cases these rates are greater than the dissociation rates of the complexes and thus the flipping takes place many times during the lifetime of the intact complex. A consideration of the structure of bound trimethoprim and its analogues (based on X-ray and NMR findings) revealed that in its bound conformation, trimethoprim would find it impossible to undergo a 180° flip of the benzyl ring because of severe steric interactions between the atoms at the pyrimidine C6 and benzyl C2′,6′ positions. In fact, ring 'flipping' can only be achieved by rotating θ_1 by at least 30° (see *Figure 12*) and this in turn requires a substantial conformational change of the protein. Thus the measured rate of flipping is indirectly monitoring transient fluctuations in the conformation of the enzyme structure which are required to allow the flipping to proceed (52).

9.4 Measurements of exchange rates with solvent

Extensive NMR measurements of exchange rates between solvent and labile protons on a protein or ligand have been reported. These are usually based on lineshape analysis or transfer of magnetization methods. The hydrogen exchange can be considered as a two-step process (55) involving first an initial 'opening' of the structure to allow access to the solvent and second the actual chemical exchange process itself. The latter is usually fast in free solution (for an NH proton exchanging with water the pseudo first-order rate constant ~ 10^{-3} sec^{-1} at pH 7) and thus the overall rate constant is often determined by the 'opening' process. Guéron and co-workers (56) have used this method to study such exchange rates for imino protons in many DNA duplexes. Other workers have used this approach to assess changes in the solvent accessibility of NH protons in protein–DNA complexes (7).

It is also possible to measure similar exchange rates in other protein–ligand complexes and we have made such measurements for complexes of trimethoprim analogues with DHFR. Earlier we saw how the N1 proton of bound trimethoprim could be assigned unambiguously using ^{15}N-labelled trimethoprim. On changing the temperature, the lineshape of the N1 proton signal varies due to changes in the exchange rate of this proton with the H_2O solvent. This linewidth data can be analysed to estimate the exchange rate (see Chapter 6). Adding imidazole, an effective catalyst for N-H exchange, to the enzyme–trimethoprim complex, did not perturb the observed exchange process, which indicates that the 'opening' rate, i.e. the rate of structural fluctuation is determining the overall rate of exchange. In this case, the N1H forms a hydrogen bond with the carboxylate group of Asp26 (*Figure 12*) and the measured exchange rate (34 sec^{-1} at 298 K) must involve, at a minimum, the rate of breaking of this hydrogen bonding interaction. Once again a very important interaction in the complex is breaking and reforming at a rate much faster than the dissociation rate. Thus both the pyrimidine ring and the benzyl ring are involved in transient fluctuations in their interactions with the protein during the lifetime of the complex: the rates are summarized on *Figure 12*. When these structural fluctuations take place in close succession they probably form part of a sequence of events leading to dissociation of the complex. It seems likely that the formation and dissociation of a complex involving a flexible ligand takes place via a stepwise 'zipper'-type mechanism involving several such steps (57). In this, an initial binding interaction would take place between the flexible ligand and the protein to form a short-lived complex where rapid conformational readjustments of the ligand and protein would occur to allow further binding interactions to be formed sequentially. Such a mechanism would lead to higher association rate constants than in the case where ligand molecules of the correct conformation bind to the protein in a single-step process. Since dissociation would also proceed by a similar 'zipper' mechanism, the dissociation rate constants would also be increased. These increased rate constants could be of importance in controlling the action of flexible transmitter molecules binding to their receptors.

Acknowledgement

We thank Lindsay Dickinson for processing the manuscript.

References

1. Jardetzky, O. and Roberts, G. C. K. (1981). In *NMR in molecular biology*. Academic Press, London.
2. Feeney, J. (1990). *Biochem. Pharmacol.*, **40**, 141.

3. Handschumacher, R. E. and Armitage, I. M. (ed.) (1990). *NMR methods for elucidating macromolecule–ligand interactions: an approach to drug design.* *Biochemical Pharmacology*, **40**, 1–174.
4. Sheard, B. and Bradbury, E. M. (1970). *Prog. Biophys. Mol. Biol.*, **20**, 187.
5. Roberts, G. C. K. and Jardetzky, O. (1970). *Adv. Protein Chem.*, **24**, 447.
6. Kaptein, R., Boelens, R., and Lamerichs, R. M. J. N. (1989). *Top. Mol. Struct. Structural Biol.*, **10**, 35.
7. Otting, G., Qian, Y. Q., Billeter, M., Müller, M., Affolter, M., Gehring, W. J., and Wüthrich, K. (1990). *EMBO J.*, **9**, 3085.
8. Roberts, G. C. K., Feeney, J., Burgen, A. S. V., Yuferov, V., Dann, J. G., and Bjur, R. (1974). *Biochemistry*, **13**, 5351.
9. Feeney, J., Batchelor, J. G., Albrand, J. P., and Roberts, G. C. K. (1979). *J. Magn. Reson.*, **33**, 519.
10. Hyde, E. I., Birdsall, B., Roberts, G. C. K., Feeney, J., and Burgen, A. S. V. (1980). *Biochemistry*, **19**, 3746.
11. Clore, G. M. and Gronenborn, A. M. (1991). *Prog. NMR Spectrosc.*, **23**, 43.
12. Carr, M. D., Birdsall, B., Frenkiel, T. A., Bauer, C. J., Jiminez-Barbero, J., Polshakov, V. I., McCormick, J. E., Roberts, G. C. K., and Feeney, J. (1991). *Biochemistry*, **30**, 6330.
13. Hammond, S. J., Birdsall, B., Searle, M. S., Roberts, G. C. K., and Feeney, J. (1986). *J. Mol. Biol.*, **188**, 81.
14. Evans, E. A., Warrell, D. C., Elvidge, J. A., and Jones, J. R. (1985). *Handbook of tritium NMR spectroscopy and applications.* J. Wiley and Sons, Chichester.
15. Müeller, L. (1979). *J. Am. Chem. Soc.*, **101**, 4481.
16. Otting, G., Senn, H., Wagner, G., and Wüthrich, K. (1986). *J. Magn. Reson.*, **70**, 500.
17. Fesik, S. W., Zuiderweg, E. R. P., Olejniczak, E. T., and Gampe, R. T. (1990). *Biochem. Pharmacol.*, **40**, 161.
18. Bax, A. and Weiss, M. A. (1987). *J. Magn. Reson.*, **71**, 571.
19. Cayley, P. J., Albrand, J. P., Feeney, J., Roberts, A. S. V., Piper, E. A., and Burgen, A. S. V. (1979). *Biochemistry*, **18**, 3886.
20. Feeney, J., Birdsall, B., Roberts, G. C. K., and Burgen, A. S. V. (1975). *Nature*, **257**, 564.
21. Rodgers, P. and Roberts, G. C. K. (1973). *FEBS Lett.*, **36**, 330.
22. Clore, G. M., Bax, A., and Gronenborn, A. M. (1990). *J. Biomolec. NMR*, **1**, 13.
23. Dwek, R. A. (1973). *Nuclear magnetic resonance in biochemistry: applications to enzyme systems*, Chapter 6. Clarendon Press, Oxford.
24. Casy, A. F. (1971). In *PMR spectroscopy in medicinal and biological chemistry*, pp. 304–27. Academic Press, London and New York.
25. Jardetzky, O. and Wade-Jardetzky, N. G. (1965). *Mol. Pharmacol.*, **1**, 214.
26. Mildvan, A. S. (1989). *FASEB. J.*, **3**, 1705.
27. Burton, D. R., Forsén, S., Karlström, G., and Dwek, R. A. (1979). *Prog. NMR Spectrosc.*, **13**, 1.
28. Markley, J. L. (1975). *Acc. Chem. Res.*, **8**, 70.
29. Wüthrich, K. (1976). *NMR in biological research: peptides and proteins.* North Holland Publishing Co.
30. Wilbur, D. J. and Allerhand, A. (1977). *J. Biol. Chem.*, **252**, 4968.

31. Clore, G. M. and Gronenborn, A. M. (1982). *Biochemistry,* **21,** 4048.
32. Schrager, R. I., Cohen, J. S., Heller, S. R., Sachs, D. H., and Schecter, A. N. (1972). *Biochemistry,* **11,** 541.
33. Birdsall, B., Griffiths, D. V., Roberts, G. C. K., Feeney, J., and Burgen, A. S. V. (1977). *Proc. R. Soc. London, Ser. B,* **196,** 251.
34. Bolin, J. T., Filman, D. J., Matthews, D. A., Hamlin, R. C., and Kraut, J. (1982). *J. Biol. Chem.,* **257,** 13650.
35. Antonjuk, D. J., Birdsall, B., Cheung, H. T. A., Clore, G. M., Feeney, J., Gronenborn, A., Roberts, G. C. K., and Tran, T. Q. (1984). *Br. J. Pharmacol.,* **81,** 309.
36. Bevan, A. W., Roberts, G. C. K., Feeney, J., and Kuyper, L. (1985). *Eur. Biophys. J.,* **11,** 211.
37. Roberts, G. C. K., Feeney, J., Burgen, A. S. V., and Daluge, S. (1981). *FEBS Lett.,* **131,** 85.
38. Birdsall, B., Roberts, G. C. K., Feeney, J., and Burgen, A. S. V. (1977). *FEBS Lett.,* **80,** 313.
39. Neri, D., Otting, G., and Wüthrich, K. (1990). *J. Am. Chem. Soc.,* **112,** 3663.
40. Fesik, S. W., Luly, J. R., Erickson, J. W., and Abad-Zapatero, C. (1988). *Biochemistry,* **27,** 8297.
41. Weber, C., Wider, G., von Freyberg, B., Traber, R., Braun, W., Widmer, H., and Wüthrich, K. (1991). *Biochemistry,* **30,** 6563.
42. Albrand, J. P., Birdsall, B., Feeney, J., Roberts, G. C. K., and Burgen, A. S. V. (1979). *Int. J. Biolog. Macromol.,* **1,** 37.
43. Feeney, J., Birdsall, B., Roberts, G. C. K., and Burgen, A. S. V. (1983). *Biochemistry,* **22,** 628.
44. Gronenborn, A. M. and Clore, G. M. (1990). *Biochem. Pharmacol.,* **40,** 115.
45. Clore, G. M. and Gronenborn, A. M. (1982). *J. Magn. Reson.,* **48,** 402.
46. Clore, G. M., Gronenborn, A. M., Carlsson, G., and Meyer, E. F. (1986). *J. Mol. Biol.,* **190,** 259.
47. Birdsall, B., Bevan, A. W., Pascual, C., Roberts, G. C. K., Feeney, J., Gronenborn, A., and Clore, G. M. (1984). *Biochemistry,* **23,** 4733.
48. Birdsall, B., Hyde, E. I., Burgen, A. S. V., Roberts, G. C. K., and Feeney, J. (1981). *Biochemistry,* **20,** 7186.
49. Gronenborn, A., Birdsall, B., Hyde, E. I., Roberts, G. C. K., Feeney, J., and Burgen, A. S. V. (1981). *Mol. Pharmacol.,* **20,** 145.
50. Birdsall, B., Feeney, J., Tendler, S. J. B., Hammond, S. J., and Roberts, G. C. K. (1989). *Biochemistry,* **28,** 2297.
51. London, R. E., Groff, G. P., and Blakley, R. L. (1979). *Biochem. Biophys. Res. Commun.,* **86,** 779.
52. Searle, M. S., Forster, M. J., Birdsall, B., Roberts, G. C. K., Feeney, J., Cheung, H. T. A., Kompis, I., and Geddes, A. J. (1988). *Proc. Natl Acad. Sci. USA.,* **85,** 3787.
53. Lipari, G. and Szabo, A. (1982). *J. Am. Chem. Soc.,* **104,** 4546.
54. Clore, G. M., Gronenborn, A. M., Birdsall, B., Feeney, J., and Roberts, G. C. K. (1984). *Biochem. J.,* **217,** 659.
55. Englander, W. and Kallenbach, N. (1984). *Q. Rev. Biophys.,* **16,** 521.
56. Leroy, J. L., Kochoyan, M., Huyn-Dinh, T., and Guéron, M. (1988). *J. Mol. Biol.,* **200,** 223.

57. Burgen, A. S. V., Roberts, G. C. K., and Feeney, J. (1975). *Nature,* **253,** 753.
58. Birdsall, B., Gronenborn, A., Hyde, E. I., Clore, G. M., Roberts, G. C. K., Feeney, J., and Burgen, A. S. V. (1982). *Biochemistry,* **21,** 5831.
59. Fesik, S. W., Gampe, R. T. Jr., Holzman, T. F., Egan, D. A., Edalji, R., Luly, J. R., Simmer, R., Helfrich, R., Kishore, V., and Rich, D. H. (1990). *Science,* **250,** 1406.

NMR of nucleic acids; from spectrum to structure

SYBREN S. WIJMENGA, MARGRET M. W. MOOREN,
and CORNELIS W. HILBERS

1. Introduction

Nucleic acids are perhaps the most intensively studied molecules found in nature, which is not too surprising in view of their role as carriers of genetic information. It has been a deep-rooted belief among molecular biologists that structure dictates function and, therefore, these studies have often involved the elucidation of structural aspects of these molecules, be it the linear base sequence or the three-dimensional arrangement of their nucleotide units. In a similar spirit it has been a deep-rooted belief among NMR spectroscopists that 'their' field of spectroscopy should be able to contribute to the elucidation of the structure and dynamics of molecules, in particular, biomolecules, in a way unparalleled by other forms of spectroscopy. Over the last few years this goal has come within reach thanks to the development of multidimensional NMR techniques which provide the necessary spectral resolution for the successful interpretation of the complicated spectra of biomolecules. Triggered off by the development of two-dimensional NMR, an avalanche of NMR studies on nucleic acids has appeared (1). Methods for spectrum interpretation and structure elucidation have been developed, and have been applied in investigations concerning the sequence-dependent structure of double helical fragments, and concerning the influence of mismatches on double helical structure, and in studies on drug–DNA complexes and nucleic acid–protein interactions. This is not meant to say that standard protocols are available which, on the basis of NMR experiments, lead automatically to the detailed description of structures of nucleic acids and their complexes. These topics still form a field of interesting and lively research. However, a number of methods and techniques have become available which can be used to study the structure of nucleic acids in solution. It is the purpose of this chapter to describe in a step-by-step fashion the approaches that can be taken to perform such structural studies. We discuss sample preparation, resonance assignment methods, parameters that determine molecular structure and how to obtain them, and finally, various techniques that may be used to derive molecular structures from the available NMR constraints.

1.1 Specific features of NMR structural information

An important characteristic of the structural information obtained from NMR is that, apart from being indirect, it leads to short-range structural data, i.e. from the J-coupling only torsion angles about single chemical bonds can be determined, and the distances which can be derived from NOE measurements are ≤ 5 Å. In principle it is possible to extend the range of the distance measurements by involving paramagnetic probes, but practical difficulties have precluded their widespread application.

Another typical feature of NMR structural investigations is that the short-range distance information is obtained through the measurement of distances between protons. Thus, apart from torsion angles that are sometimes obtained from J-couplings between protons and heteronuclei (^{13}C, ^{15}N, ^{31}P) and between heteronuclei (^{13}C, ^{31}P), three-dimensional structures derived by means of NMR are essentially 'protonic' pictures. Proton–proton distances can only be used to determine a three-dimensional structure by incorporating knowledge of standard chemical bond lengths and bond angles.

2. Structural parameters

2.1 Nomenclature and definition of structural parameters

2.1.1 Numbering schemes

We shall use the nomenclature agreed upon in the IUPAC/IUB guide-lines (2). The numbering of the atoms in the furanose ring and the five common bases found in DNA and RNA is presented in *Figure 1*. The bases can pair according to different schemes; a number of the most common base pair combinations is presented in *Figure 2*. It is noted that the atoms in the sugar moieties are primed in contrast to the atoms in the bases. The two hydrogens at atom C5′ are labelled with a prime and a double prime. For the interpretation of NMR results it is important to know how these protons are characterized in the molecular frame. The numbering is obtained by looking along the O5′ → C5′ bond in the direction of the mainchain (see *Figure 5, upper right*); going in the counter-clockwise direction the atoms attached to C5′ are: C4′, H5′, and H5″ ((2); this differs from (3)).

2.1.2 Helical parameter and torsion angle description

There are several ways to describe a three-dimensional structure of a molecule. In conformational studies of nucleic acids two methods are in common use, namely, a description in terms of the so-called helix parameters and a description in terms of torsion angles.

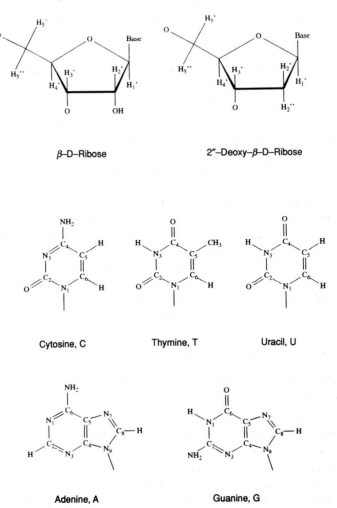

Figure 1. Nomenclature, structures, and standard atom numbering of the β-D-riboses and the five common bases (pyrimidines: C, T, and U, purines: G and A) in DNA and RNA, according to IUPAC/IUB guide-lines (2).

The helix parameters, e.g. helical twist, – rise, are defined in *Figure 3* according to the Cambridge convention (4). It is noted in passing that the six rotational and translational parameters may be different when defined with respect to the local or the global helix axis. A well-known example is the rise in A-type helices: defined with respect to the global helix axis $Dz_g = 2.9$ Å while the distance between stacked base pairs is $Dz_l = 3.4$ Å. (The subscripts g and l indicate the global and local rise respectively.) Several research groups have been engaged in developing programs for the geometrical analysis of

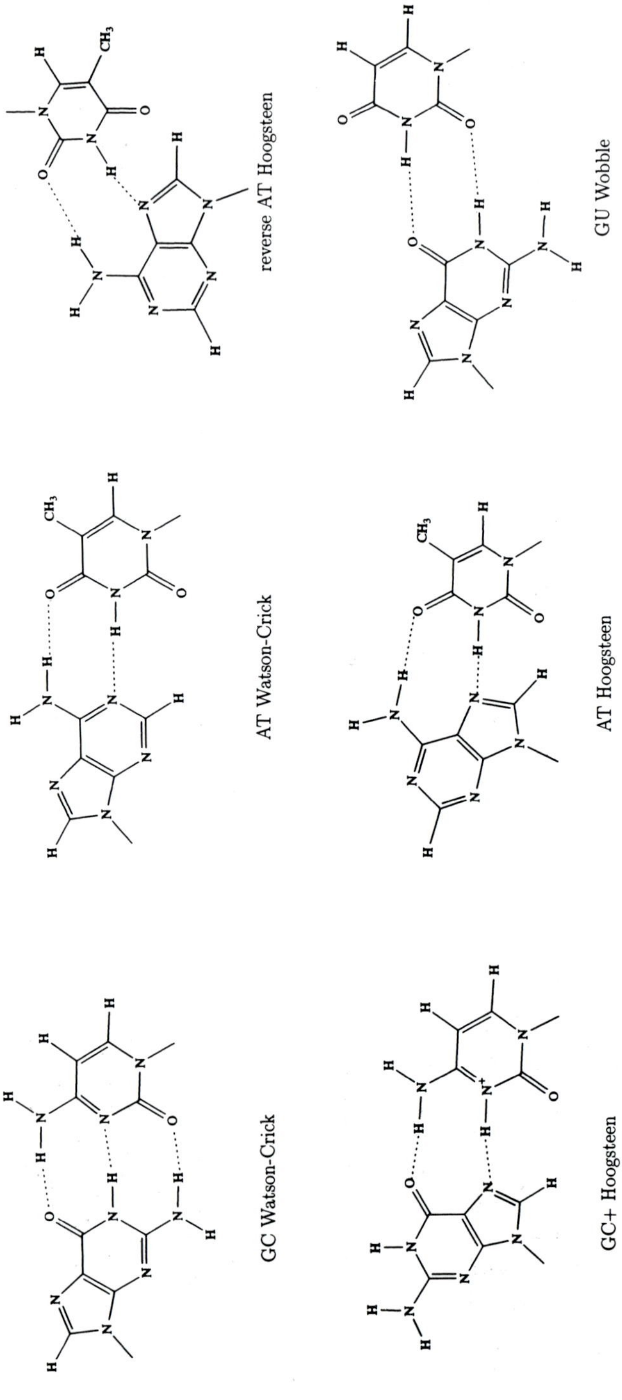

reverse AT Hoogsteen

GU Wobble

AT Watson-Crick

AT Hoogsteen

GC Watson-Crick

GC+ Hoogsteen

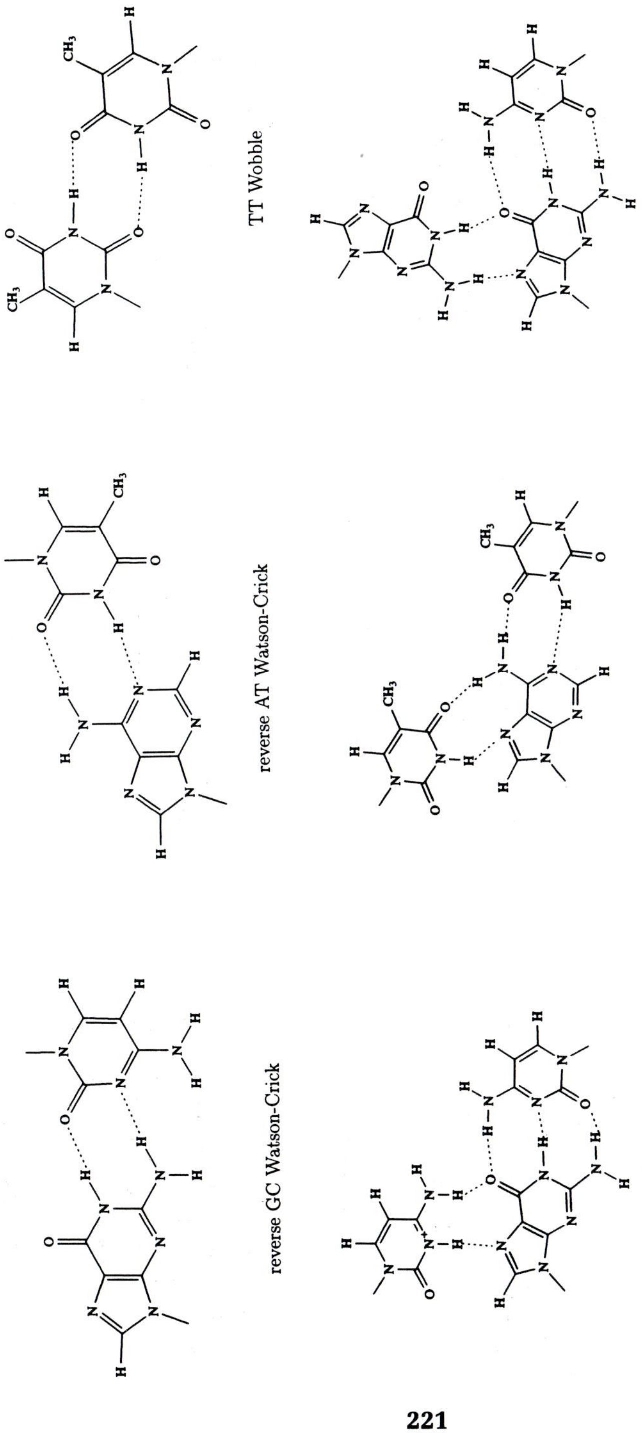

Figure 2. Possible base pairs and base triplets occurring in DNA and RNA. In RNA the uracil base (U) is found instead of thymine (T).

221

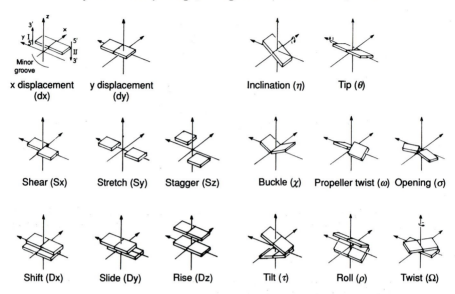

Figure 3. Definitions of nucleic acid structure parameters, agreed upon at the EMBO workshop on DNA Curvature and Bending (4). The first three columns present the definitions of the various translations involving two bases of a pair (*upper two rows*) or two successive base pairs (*bottom row*). The last three columns present the definitions of the various rotations of two bases of a pair (*upper two rows*) or two successive base pairs (*bottom row*). In the *top row* the motions of the two bases are coordinated, while in the *middle row* their motions are opposed. The standard coordinate frame is defined in the *upper left corner*. (Diekman *et al.* (1989), *EMBO J.*, **8**, 1.)

nucleic acid chains. These programs are available from Dr R. Lavery (4), Dr D. H. Soumpasis (4), Dr E. von Kitzing and Dr S. Diekman, Dr Manju Bansal (4), and Dr R. Dickerson (4).

The torsion angles that define the structure of a nucleotide unit are the backbone torsion angles α, β, γ, δ, ε, and ζ, the endocyclic torsion angles determining the furanose ring conformation, ν_0 to ν_4, and finally, the glycosidic torsion angle χ, between the base and the sugar ring. Alternatively, the furanose conformation is described in terms of the phase angle of pseudorotation, P, and the puckering amplitude, Φ_m, (see next section). The nomenclature and definitions of the torsion angles and atoms of the nucleotide unit are presented in *Figure 4* and *Table 1*. The value of the torsion angles is defined via the so-called Newman projections; examples of such projections, with the corresponding sign convention, for the staggered conformations *gauche*-minus (g^-), *gauche*-plus (g^+), and *trans* (*t*) are given in *Figure 5*. As an example, the average values of the torsion angles found in A-RNA, A-, B-, and Z-DNA (3) are given in *Table 2*.

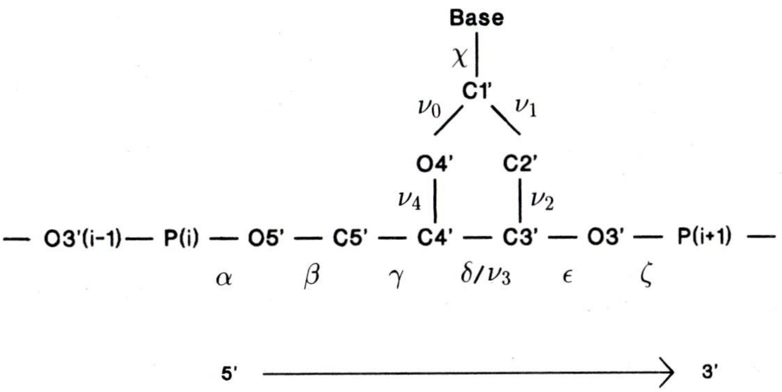

Figure 4. Definition of the torsion angles in the sugar phosphate backbone: α, β, γ, δ, ε, and ζ, the glycosidic torsion angle χ, and the endocyclic torsion angles ν_0–ν_4 in the sugar ring.

Table 1. Definition of torsion angles

Sugar phosphate backbone torsion angle	Atoms defining torsion angle
α	O3'—P —O5'—C5'
β	P —O5'—C5'—C4'
γ	O5'—C5'—C4'—C3'
δ	C5'—C4'—C3'—O3'
ε	C4'—C3'—O3'—P
ζ	C3'—O3'—P —O5'
χ (Py)	O4'—C1'—N1—C2
χ (Pu)	O4'—C1'—N9—C4
Endocyclic sugar torsion angle	
ν_0	C4'—O4'—C1'—C2'
ν_1	O4'—C1'—C2'—C3'
ν_2	C1'—C2'—C3'—C4'
ν_3	C2'—C3'—C4'—O4'
ν_4	C3'—C4'—O4'—C1'

2.2 Description of the furanose ring

The description of the structure of the sugar ring deserves separate attention. The five endocyclic torsion angles defining the sugar conformation are not all independent from one another. The conformation of non-planar five-membered cyclic molecules can be described by the concept of pseudorotation (5). In this approach the endocyclic torsion angles defined in *Figure 4* and

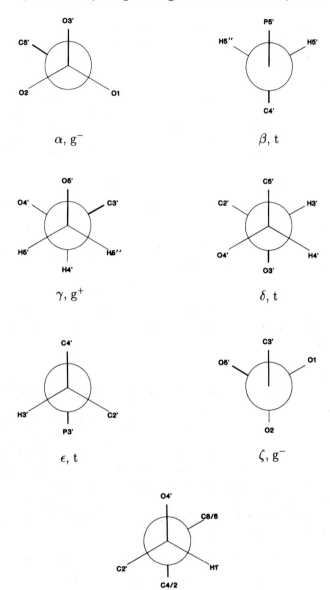

χ, anti

Figure 5. Newman projections of the torsion angles α, β, γ, δ, ε, and ζ, shown in the staggered conformations, *gauche* minus (g⁻), *gauche* plus (g⁺), and *trans* (t), which are typically found in B-DNA. The sign convention of the torsion angles is as follows (2): a positive torsion angle is obtained after a clockwise turn of the bond containing the front atom (*thick line*) about the central bond for it to eclipse the bond to the back (*thick line*) regardless of the end from which the system is viewed. The glycosidic torsion angle χ is called *anti* when χ = 180 ± 90° and *syn* when χ = 0 ± 90° (2).

224

Table 2. Average values of backbone torsion angles and sugar conformations in polynucleotides

Polynucleotide	α	β	γ	δ	ϵ	ζ	χ	Sugar
A-RNA	−68	178	54	82	−153	−71	−158	N-type
A-DNA	−50	172	41	79	−146	−78	−154	N-type
B-DNA	−46	−147	36	157	155	−96	−98	S-type
Z-DNA G [a]	52	−153	178	76	−72	102	89	N-type
Z-DNA C [b]	−110	−168	54	147	−103	−91	−159	S-type

[a] Guanine residue in Z-DNA.
[b] Cytosine residue in Z-DNA.

Table 1) are interrelated via two independent parameters, P and Φ_m, which characterize the conformation of the five-membered ring (5,6):

$$\nu_j = \Phi_m \cos(P + 144(j - 2)) \qquad [1]$$

In this equation $j = 0$ to 4, P is the phase angle of pseudorotation, and Φ_m is the puckering amplitude. The backbone torsion angle δ is related to ν_3 according to (1):

$$\delta = \nu_3 + 125° \qquad [2]$$

(The value to be added to ν_3 depends somewhat on the value of δ; for a more accurate definition *Figure 4.6* in (3) can be consulted.)

Eqn 1 is, however, only applicable to equilateral five-membered rings such as cyclopentane. In heterocyclic rings with varying bond distances and angles, adjustments have to be introduced to account for these 'non-idealities'.

For practical reasons the equations, in which the necessary corrections for the non-ideality of the furanose ring have been incorporated, are given for the torsion angles, ϕ_{ij}, between the protons in the sugar ring (7, 8). As will be discussed in Section 5.2 these torsion angles can be determined from the J-couplings between these protons.

$$\phi_{1'2'} = 121.4 + 1.03\Phi_m\cos (P - 144) \qquad [3a]$$

$$\phi_{1'2''} = 0.9 + 1.02\Phi_m\cos (P - 144) \qquad [3b]$$

$$\phi_{2'3'} = 2.4 + 1.06\Phi_m\cos (P) \qquad [3c]$$

$$\phi_{2''3'} = 122.9 + 1.06\Phi_m\cos (P) \qquad [3d]$$

$$\phi_{3'4'} = -124.0 + 1.09\Phi_m\cos (P + 144) \qquad [3e]$$

In practice the value of the pucker amplitude may vary between 34° and 42°. A good impression of how the sugar conformation varies is obtained when the phase angle of pseudorotation is allowed to traverse a complete cycle of 360°.

This is demonstrated in the pseudorotation wheel in *Figure 6*. In most practical situations the sugars are found to adopt P-values between 0° and 36°, and between 144° and 180°. The first category of conformations are called N-type, the second category S-type conformations (they occur in the Northern and Southern hemisphere of the pseudorotation cycle respectively). As a result of changes in the phase angle of pseudorotation the distance between the protons in the sugar moiety may change. This is indicated in *Figure 7*, which may be consulted in the analysis of the sugar conformation discussed in Section 5.3.

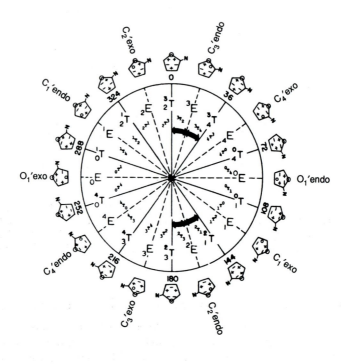

Figure 6. Pseudorotation cycle of the furanose ring, showing the relation between the phase angle of pseudorotation (P) and the envelope (E) and twist (T) notations and the endo and exo notations. Preferred P-values are concentrated in two parts of the wheel: P = 0°–36° (C3'endo; N-type) and P = 144°–180° (C2'endo; S-type). (Altona and Sundaralingam. (1972). *J. Am. Chem. Soc.*, **94**, 8205; Copyright by the American Chemical Society.) The puckering of the most common conformations, C2'endo and the C3'endo conformation, is shown *below*.

226

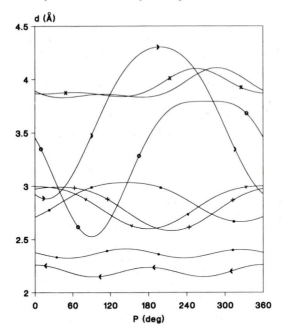

Figure 7. Intranucleotide distances between for hydrogen atoms in the deoxyribose ring versus the pseudorotation phase angle P. The amplitude of pseudorotation ϕ_m was kept at 35°. The following distances are plotted: $d_i(1';2'')$ (−<−), $d_i(2';3')$ (−·−), $d_i(1';4')$ (−○−). $d_i(2';3')$ (−v−), d_i (2″;4′) (−>−), $d_i(2';4')$ (−−−), d_i (1′;3′) (−x−), $d_i(1';2')$ (−*−), and $d_i(3';4')$ (−+−).

2.3 Description of the mononucleotide conformation

Once the phase angle of pseudorotation and the pucker amplitude of the furanose ring are known, the conformation of a mononucleotide can be derived by determining the value of the glycosidic torsion angle, χ, as will be discussed in Section 5.4. This can be achieved by measuring the distances between the sugar protons and the base ring protons of the mononucleotide. The distance between the ring H8 (in purines) or the ring H6 (in pyrimidines) and the sugar H1′ protons does not depend on the sugar conformation, but only on the glycosidic torsion angle. The latter dependency is presented in *Figure 8*. The distances between these ring protons and the other sugar protons depend on both the puckering of the sugar and the value of the glycosidic torsion angle. This is outlined in *Figure A1* in the Appendix.

As has been pointed out, the success of NMR structural studies of nucleic acids relies heavily on the possibility of interpreting their 2D-NOESY spectra. Only short-range distances (in practice ⩽ 5 Å) between protons in proton pairs will give rise to observable NOE cross-peaks in these spectra. The plots in *Figure A1* provide for a quick overview of those conformations of a

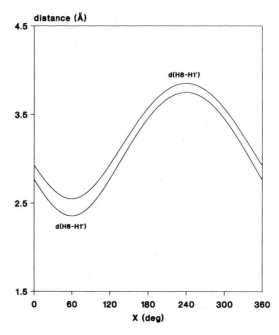

Figure 8. Intranucleotide distances between aromatic protons H6 or H8 and the sugar proton H1′ versus the torsion angle χ.

mononucleotide which will lead to important NOE connectivities in the NMR spectra and can thus be analysed by this approach.

2.4 Distances in secondary elements of nucleic acid structures

NOE effects also play an important role in the sequential analysis of the NMR spectra (see Section 4.2.2). Thus, the availability of short internucleotide proton–proton distances (≤ 5 Å) is required to be able to successfully perform sequential assignments of the proton resonances. In this analysis the standard nucleic acid secondary structural elements play a crucial role. We have therefore summarized the relevant internucleotide distances in A-, B-, and Z-type helices (see *Figures A2*). To facilitate further discussions we introduce a shorthand notation for the distances between protons *l* and *r*, analogous to that which has proven to be very useful in proteins (Chapters 4, 10, 11)(9):

$d_i(l;r)$ for intranucleotide distances, e.g. $d_i(8;2')$
$d_s(l;r)$ for internucleotide distances, e.g. $d_s(1';6)$

In the latter case we follow the convention that *l* corresponds with the proton in the 5'-nucleotide and r with the proton in the 3'-nucleotide. For the methyl protons of thymidines l or r is indicated by the letter M. Sometimes we use two symbols for l or r, for instance in $d_s(3';5/M)$. This signifies the sequential distance between H3' in the 5'-nucleotide and H5 or the methyl protons in the 3'-cytosine or thymine respectively.

Cross-strand distances are defined as follows:

$d_{ci}(l;r)$ for distances within base paired nucleotides, e.g. $d_{ci}(T\text{-}NH3;A\text{-}NH_26)$

$d_{cs}(l;r)_p$ for distances between adjacent base paired nucleotides, e.g.
$\quad\quad d_{cs}(2,1')_{3'}$

The symbols NH and NH_2 represent imino- and amino-protons, respectively. To avoid ambiguities in the characterization of the sequential cross-strand distances (d_{cs}) the directionality of the connectivity has to be indicated. Consider two adjacent base pairs, and define 5'- and 3'-nucleotides. It is easy to see that cross-strand sequential distances are only found between the two 3'-nucleotides or between the two 5'-nucleotides. This is indicated by the last subscript p, which is either 3' or 5', and which suffices to characterize this distance. Thus, in the example above the cross-strand distance between H2 and H1' of the 3'-nucleotides of adjacent base pairs in A-DNA is indicated (*Figure A2a*).

A few comments on some of the differences between A- and B-type helices are in order. Examination of *Figure A2a* and *A2b* shows that for the sequential analysis the distances $d_s(2';6/8)$ and $d_s(2'';6/8)$ are important. They are quite different, however, in the A- and B-type helices. While in B-DNA $d_s(2'';6/8)$ is relatively short (2.2 Å) and $d_s(2';6/8)$ is relatively long (4.0 Å), the opposite is true in A-DNA. This is a direct consequence of the fact that in idealized A-type helices the sugars adopt N-type conformations while in B-type helices S-puckered sugars are found. It is noted in passing that the distance $d_s(1';6/8)$, which is also important in the sequential analysis of the NOESY spectra, is moderately different in A-type (4.6 Å) and B-type (3.6 Å) helices, but other d_s values do differ between the types of helix.

Examination of *Figure A2* also shows that the cross-strand distances show interesting differences for the different types of helix. These have not been employed systematically in the analysis of nucleic acid spectra, but certainly are interesting in providing a further definition of nucleic acid conformation.

3. Recording of spectra

Spectrum interpretation and structure determination require the recording of spectra of H_2O as well as of ^2H_2O solutions. The procedure for obtaining suitable NMR samples is outlined in *Protocol 1*. The concentration necessary for two-dimensional (or higher-order) NMR experiments is 1 to 3 mM in nucleic acid. This leads, with the presently available technology, to measuring times of 8 to 24 hours, depending on the type of experiment (homo or hetero 2D experiments) for 2D experiments, and 100 to 120 hours for 3D NMR experiments. Salt concentrations (NaCl or KCl) may vary from a few millimolar to 0.2 M concentrations in standard experiments, depending on the particular requirements; on the average 0.1 M concentrations are used.

Protocol 1. Preparation of H_2O or ^2H_2O samples

1. Obtain oligonucleotide in suitable form, for the preparation of H_2O or ^2H_2O solutions, by three or four cycles of freeze-drying it from H_2O or 99.8% ^2H_2O to remove volatile components possibly present in the preparations. The safest way of storing oligonucleotide samples over a longer period of time is to desalt and freeze-dry the samples, and subsequently deep-freeze them.

2. Obtain buffer solution by weighing the required amounts of buffer salts into H_2O or ^2H_2O; for the preparation of ^2H_2O buffers it is necessary to first freeze-dry the buffer salts three times from ^2H_2O or use deuterated salts. Most often a phosphate buffer is used.

3. Dissolve oligonucleotide in 0.5 ml of the required buffer, (e.g. 10 mM phosphate, 0.1 M NaCl, 0.05 mM EDTA, pH 7.0).

4. Adjust pH to the desired value.

Recommendations

(a) Prepare ^2H_2O samples under a nitrogen or argon atmosphere to avoid the uptake of oxygen and water from the air.

(b) Heat the sample up to about 80°C and quickly cool it in ice to get rid of concatamers and other aggregates which may have possibly formed during freeze-drying.

(c) To prevent line-broadening by paramagnetic ions some EDTA may be added to the buffer (0.05 mM is generally sufficient).

(d) To limit hydrogen exchange of imino protons a high buffer concentration should be avoided as it may catalyse the exchange. Phosphate buffer is recommended because of its favourable exchange characteristics (10). The actual determination of the exchange of imino protons is complicated by the fact that a three-site exchange problem has to be handled. A method has been worked out by Guéron and collaborators. The imino proton exchange and by inference the base pair lifetimes are determined from the imino proton linewidths or their T_2 relaxation times as a function of the buffer concentration. We refer to the original literature for a more detailed description (10, 11).

The most important type of experiments are summarized in *Table 3* where they are ordered according to the type of correlations which become manifest in the spectra:

(a) Pulse sequences that correlate proton signals via J-couplings, i.e. COSY and TOCSY-type experiments.

(b) Pulse sequences that correlate signals of protons that are in close proximity, i.e. NOESY and ROESY-type experiments.

(c) Pulse sequences that correlate the signal of a heteronucleus to that of its J-coupled proton(s), HETCOR-type experiments.

Descriptions of the most important of those experiments, and discussions of practical aspects of their use, are given in Chapter 3; here we shall concentrate on comments specific to work with nucleic acids.

3.1 Water suppression

In NMR studies of nucleic acids, be it 1D or 2D spectroscopy, the investigation of exchangeable protons (e.g. imino protons), and hence the use of H_2O solutions, plays an important role. The techniques to overcome the dynamic range problem which such solutions present are discussed in Chapter 3. The simplest of these techniques is presaturation of the water signal. However, for most studies of imino and amino protons in nucleic acids this approach can not be used. The relatively fast exchange of these protons with water leads to (partial) saturation of their resonances by transfer of saturation (see Chapter 6) from the water resonance. In order to observe these protons, selective excitation or selective observe pulses have to be applied. The most straightforward approach is the application of the so-called $1\bar{1}$ (or jump-return) or $1\bar{3}3\bar{1}$ observe pulses as used in NOE experiments (12). The DANTE pulse sequence can also be used as an observe pulse (13). In combination with these semi-selective pulses digital shift accumulation, DSA, can be used to further suppress the water signal (14). In our experience, the $1\bar{1}$ pulse is quite sufficient for most applications performed on modern NMR spectrometers. Finally, we mention the technique involving the use of pulsed gradient fields which has emerged recently and promises to be quite valuable for studying H_2O solutions (15).

3.2 NOESY and ROESY spectra

As mentioned in Chapters 3 and 6, the cross-peaks observed in NOESY spectra can be generated through two separate mechanisms: dipolar relaxation and chemical exchange (16, 17). The magnetization transfer takes place during the mixing period, τ_m, of the pulse sequence and can be described by a set of coupled differential equations which are more fully discussed in Chapter 11. Here we consider the combined effect of the relaxation and exchange processes from the practical point of view of obtaining the best spectra. For

Table 3. Summary of the 2D NMR experiments that are currently important for the analysis of nucleic acids

Name	Application	Comment	Pulse sequence
Proton–proton J-correlation techniques			
COSY normal with delay	Homonuclear shift Correlation of spin systems Detection of small J-couplings	Fast routine technique No pure phase	
DQF-COSY	Identification of spin systems, determination of J-couplings	Standard technique	
β-COSY	Relative sign of J	Pure phase low intensity diagonal alternatives	
E.COSY	Accurate measurement of J	High resolution, good suppression of diagonal signals	
P.E.COSY	Accurate measurement of J	Simpler method	(57)
ω_1-COSY	Determination of J	Enhanced resolution of J-couplings in ω_1 direction	(21)
TOCSY (HOHAHA)	In-phase multiplets	Transfer across total J-coupled spin network possible, cross-peaks appear also for large molecules.	(22) (58)
Proton–proton correlation through NOE and/or exchange			
NOESY	Distance information or exchange rates	Besides COSY most important	
MINSY	NOESY with suppression of spin diffusion pathways		(18) (59)
ROESY	Separation of chemical and NOE information in medium sized molecules	Suppression of J peaks necessary	(60)
Heteronuclear 2D NMR techniques			
H,X-COSY	Heteronuclear shift correlation	Phase-sensitive	(61)
X,H-COSY	Heteronuclear shift correlation via detection of proton	Phase-sensitive Very sensitive technique	(23)
Solvent suppression techniques			
Presaturation		Most commonly used	
11	Read pulse	Simplest form	(16)
1331	Read pulse	Slightly better suppression	(62)

Table 3. *Continued*

Name	Application	Comment	Pulse sequence
DANTE	Read or excitation pulse		(14)
DSA	Digital shift accumulation, solvent suppression, used in addition to other techniques	Leads to rather strong baseline artefacts	

T1 and T2 measurements

Inversion recovery	T1 measurement		
CPMG	T2 measurement		

3D NMR experiments

Name	Application	Comment	Pulse sequence
TOCSY-NOESY	Additional separation of proton resonances with respect to 2D	Provides both J-coupling and NOE information in one experiment	(30)
NOESY-NOESY	Additional separation of resonances	Provides additional information on spin diffusion	(31)

An excellent overview of modern NMR experiments is presented in (19), which can be consulted for references to original papers for most of the experiments listed here. The references listed in the last column are to applications in nucleic acid studies (see also text).

practical purposes it is convenient to take the apparent longitudinal relaxation times, T_1, as a yard stick. For short mixing times ($\ll T_1$) the cross-peak intensities increase linearly with the mixing time, for longer mixing times ($> T_1$) magnetization is spread out over the whole system of interconnected spins and the cross-peak intensities start to decrease again due to relaxation leaks (spin diffusion). In recording NOESY spectra it is thus important to choose the correct mixing time in order to obtain sufficiently intense cross-peaks; similar considerations apply to 1D NOE difference spectra. Since the basic geometry of the DNA duplex is known (in contrast to the situation in proteins), it is possible to calculate the expected NOEs and to use these calculations as a guide to the choice of the optimum mixing time. In *Figure 9a–d*, a number of NOE curves are shown for exchangeable protons for different values of the exchange rate k and different τ_c values. It is also very simple and helpful to obtain, in advance, a rough estimate of the T_1 values. For a mixing time of the length of the average T_1 values the diagonal peaks will be reduced to approximately one third of their value at $\tau_m = 0$ and the NOE cross-peaks will be close to their maximum intensities. As can be seen in *Figure 9a–d* a high exchange rate markedly reduces the NOE cross-peak intensity (note the different vertical scales). Recording spectra at low

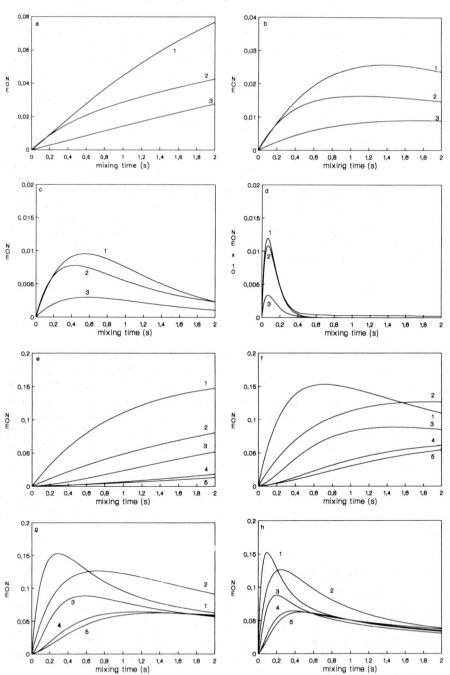

Figure 9. NOE build-up curves for exchangeable and non-exchangeable protons. NOE curves were calculated from the relaxation matrix (see Chapter 11). Relaxation matrix elements σ_{ij} were calculated from the distance matrix elements r_{ij} according to $\sigma_{ij} = 5.69 \times 10^4$ (sec^{-2} nm^6) $\times \tau_c$ (sec)/r_{ij}^{-6} (nm^{-6}), which is correct for $\omega\tau_c \gg 1$. (a–d) Exchangeable protons. The distance matrix was set-up for the duplex d5'(T$_1$C$_2$T$_3$)3'.3' (A$_3$G$_2$A$_1$)5' and contains all distances between protons within a 5 Å radius of G$_2$-NH1, assuming a B-DNA conformation. To account for exchange to H$_2$O, an identical exchange rate constant was added to each diagonal element, except for the non-exchangeable protons in the matrix. The rotational correlation time, τ_c, is 2 nsec. In each panel build-up curves are shown for NOEs from G$_2$-NH1 to A$_1$-NH6 (1), T$_1$-NH3 (2) and A$_3$-NH6 (3), for exchange rates of 1.78×10^{-6} sec^{-1} (a), 0.6 sec^{-1} (b), 1.78 sec^{-1} (c), and 15.1 sec^{-1} (d), respectively. Notice the very low NOE intensities in *Figure 9b, c,* and *d,* corresponding with imino proton lifetimes found in practice (10 msec to 1 sec). (e–h) Non-exchangeable protons. NOE cross-peak intensities were calculated here for a system of 24 inter-connected protons, involving all relevant (< 5 Å) intra- and interresidue distances around C$_2$-H6 in the duplex mentioned above. The curves were calculated for rotational correlation times of 0.2 nsec (e), 0.8 nsec (f), 2 nsec (g), and 6 nsec (h) and represent the NOE intensities of C$_2$-H6 to H2' (1), H5 (2), H2" (3), H1' (4) and H3' (5) of residue 2. For comparison it is noted that a τ_c value of approximately 2 nsec is expected for a hairpin of 15 nucleotides.

---

temperature may help to suppress exchange between imino and water protons and lead to better NOESY spectra.

In *Figure 9e–h* similar curves are shown for cross-peaks between non-exchangeable protons. In a similar fashion they may provide a rough estimate of the mixing time to be used and also an estimate of the amount of spin diffusion, when magnetization is transferred from one proton to another not directly but via an intermediate. For example, transfer of magnetization from the base proton H6 to H3' of the sugar often occurs via the H2'. In B-DNA $d_i(6;3') > 4$ Å while $d_i(6;2')$ and $d_i(2';3')$ are both ≈ 2.2 Å, providing an efficient spin diffusion pathway. In order to block a spin diffusion pathway one can saturate the intermediate protons, as is done in a MINSY experiment (18); in this example one could saturate the H2' and H2" protons, suppressing the spin diffusion pathway, via H2', for transfer of magnetization from H6 to H3'.

3.3 J-correlated spectra

In J-correlation spectroscopy two types of experiments can be distinguished: experiments which provide information about directly coupled spins and experiments which connect spins which are part of a J-coupled network in which individual spins are not necessarily coupled to all other spins in this network. The COSY experiment belongs to the first, the TOCSY experiment to the second category.

As discussed in Chapter 3, the DQF-COSY is now the standard COSY experiment for correlating J-coupled spins and as such invaluable for distinguishing networks of coupled spins for assignment purposes. The fine-structure of the cross-peaks in phase-sensitive COSY spectra is characterized by a multiplet pattern, which for the weak coupling limit is simple to interpret, and from which J-couplings constants can be obtained (9, 19).

In principle, J-couplings can be extracted directly from these multiplets, but in practice one is confronted with two problems which may hamper the interpretation, the finite linewidth and/or the limited digital resolution. The latter problem occurs mainly for the F1-dimension and may be ameliorated by using a variant of the COSY experiment, the ω_1-scaled DQF-COSY experiment (20), in which the chemical shift evolution is scaled down by a chosen factor whereas the evolution due to the J-couplings is unaffected. In this way sometimes the fine-structure of the H1'-H2' and H1-H2'' cross-peaks can be made visible (21), allowing for a stereospecific assignment of these protons, particularly in S-type sugars. The first problem, i.e. finite linewidth, has a profound effect on the peak separation in a multiplet. In order to obtain reliable coupling constants from the multiplet, the linewidths should be a factor of two smaller than the coupling constants involved (9). For N-type sugars *all* antiphase peaks in a H1'-H2' multiplet tend to cancel by virtue of the small $J_{1'2'}$ (2 Hz), while for the H1'-H2'' multiplet and for the multiplets of S-type sugars only partial cancellation may occur.

Simplification of multiplet structure can be achieved by recording E-COSY spectra (19) (or the improved version P.E.COSY spectra (19)) or β-COSY spectra (small flip-angle COSY spectra (19)). In the simplified multiplets there is less cancellation and coupling constants are more readily measured (see also Chapter 10). For an N-type sugar the antiphase peaks at the edge of the H1'-H2'' multiplet lie relatively free so that both $J_{1'2'}$ and $J_{1'2''}$ can be determined more reliably, while for S-type sugars the same applies for the H1'-H2'' multiplet; in fact, $J_{1'2'}$ can now be determined even when the linewidth is of the same order as $J_{1'2'}$.

The additional cross-peaks in the TOCSY or HOHAHA (19, 56) spectra provide a means of recognizing spins belonging to the same spin system and in this way appreciably aid in the spectrum interpretation. The coherence transfer functions for the deoxyribose system are shown in *Figure 10* (22). These plots can be used to determine a suitable mixing time for a particular experiment; in many instances a mixing time of 100 milliseconds will provide excellent transfer.

For the determination of nucleic acid structures it is essential that information is available on the conformation of the sugar phosphate backbone. In this respect it is important to be able to study ^{31}P and ^{13}C nuclei in addition to protons. Proton observe heteronuclear correlation experiments, such as HMQC, are commonly used for this purpose (see Chapter 3). A particular member of this class of experiments has been proposed by Sklenar *et al.* (23) for the study of ^{31}P-^1H correlation, which allows for the determination of the J_{1H31P} couplings from the antiphase multiplet patterns of the ensuing cross-peaks.

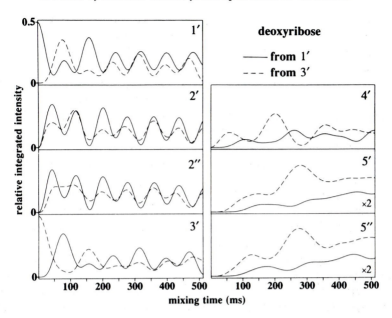

Figure 10. Coherence transfer during TOCSY mixing times for the deoxyribose system, starting from the H1′ proton (*solid lines*) and from the H3′ proton (*dashed lines*). The vertical scale of the autocorrelation data for the H1′ and H3′ protons has been compressed by a factor of two, while the vertical scale of the data for the H5′ and H5″ protons has been multiplied by a factor of two. The J-coupling constants employed in the calculations were $J_{1'2'} = 7.7$ Hz, $J_{1'2''} = 6.4$ Hz, $J_{2'2''} = -14.1$ Hz, $J_{2'3'} = 6.9$ Hz, $J_{2''3'} = 2.7$ Hz, $J_{3'4'} = 3.3$ Hz, $J_{4'5'} = 2.1$ Hz, $J_{4'5''} = 4.4$ Hz, and $J_{5'5''} = -11.8$ Hz. (Cavanagh *et al.* (1990). *J. Magn. Res.*, **87**, 110. Copyright Academic Press.)

4. Spectrum interpretation

4.1 Distribution of resonances

The extent to which a structure determination can be carried out depends on the possibility of assigning particular resonances or cross-peaks to certain nuclei or pairs of nuclei in the molecule. This is aided by our knowledge of the distribution of the resonances in the spectrum, e.g., the imino proton resonances are always found around 14 p.p.m., while the methyl proton resonances are found around 1 p.p.m. downfield from the reference signal DSS (2,2-dimethyl-2-silapentane-5-sulfonate). A compilation of the distribution of the ^{1}H resonances and ^{13}C resonances of nucleic acid molecules is presented in *Figure 11a* and *b*.

4.1.1 Exchangeable protons

As can be seen in *Figure 11a* the imino proton resonances scatter over an appreciable spectral region. An obvious reason for this behaviour is the large number of different combinations of hydrogen bonded base pairs that can be

237

a.

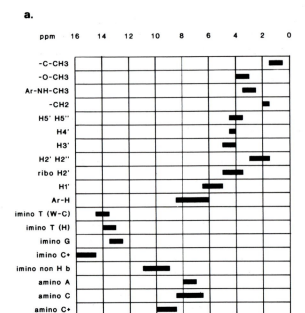

b.

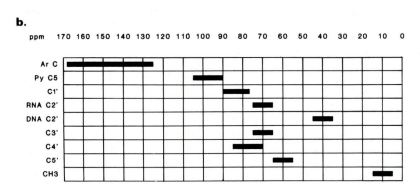

Figure 11. (a) Approximate ^{1}H chemical shift ranges in DNA and RNA fragments, relative to DSS. −C−C−CH3, −O−CH3, and Ar−NH−CH3: methyl protons attached to various groups; H2′ H2″, H3′, H4′, and H5′ H5″ refer to sugar protons in DNA; ribo H2′: β-D-ribose H2′ proton. (The H3′, H4′, H5′, and H5″ protons of RNA have the same chemical shift range as ribo H2′); H1′: H1′ proton in RNA and DNA; Ar-H: aromatic proton; imino T (W−C) and imino T (H): thymine imino proton involved in Watson−Crick and in Hoogsteen base pairing; imino G: guanine imino proton; imino C+: protonated cytosine imino; imino non H b: imino proton of either G, T, or U not involved in hydrogen bonding; amino A, amino C, and amino C+: amino proton of an adenine, cytosine, and protonated cytosine residue, respectively; the hydrogen bonded amino proton generally resonates downfield of the non-hydrogen bonded amino proton. (b) Approximate ^{13}C chemical shift ranges in DNA and RNA fragments, relative to TMS. Ar C: aromatic carbon; Py C5: pyrimidine C5 carbon; C1′, C3′, C4′, C5′, and CH3: the C1′, C3′, C4′, C5′, and methyl carbon in DNA and RNA, respectively; RNA C2′: C2′ carbon in RNA; DNA C2′: C2′ carbon in DNA.

formed (see *Figure 2*). If not hydrogen bonded, the imino protons of G and T (or U) resonate around 10 p.p.m. Upon hydrogen bonding, the resonances shift downfield, the largest downfield shifts being found for positively charged bases. A case in point is the position of the imino proton resonance of the protonated cytosine occurring in the Hoogsteen base pair of a CCG base triple (see *Figure 2*), which is found around 15 p.p.m. The imino proton resonances of Watson–Crick A-T (or U) pairs are found between 14.5 and 13.0 p.p.m., while those of G-C pairs fall between 13.5 and 11.5 p.p.m. Ring-current shifts are probably the main cause for the spread observed for the resonance positions of particular types of imino protons (24), although other phenomena such as chemical exchange may also contribute. In all of the examples mentioned, imino protons are hydrogen bonded to an imino nitro-gen. In other type of base pairs, e.g. the wobble G-T (or U) or T-T base pairs the imino proton is hydrogen bonded to a carbonyl oxygen. In this situation imino proton resonances are observed between 12.0 and 10.0 p.p.m.

The behaviour of amino proton resonances is similar to that of the imino protons. The position of amino proton resonances involved in hydrogen bonding in Watson–Crick base pairs is between 6.5 and 9.0 p.p.m., while the amino proton resonances of protonated cytosines in GC Hoogsteen base pairs are shifted downfield to 9.5 p.p.m. Hydrogen bonding of amino protons to phosphate groups has also been observed, e.g. in duplexes with protonated A-A pairs, with the amino proton resonance at ≈ 8 p.p.m. (25).

4.1.2 Non-exchangeable protons

The non-exchangeable protons resonate in narrower spectral regions (see *Figure 11a*). For DNA the resonances of the different type of protons are reasonably well separated on the chemical shift scale except for the H4′ and H5′/5″ resonances. For RNA the situation is somewhat less favourable; the introduction of the hydroxyl group on the C2′ sugar atom moves the H2′ resonances downfield such that the H2′, H3′, H4′, and H5′/H5″ all resonate in one rather narrow spectral region which causes appreciable overlap.

For deoxyribonucleotides the H2′, H2″, and the H5′, H5″ resonances fall in the same spectral region (see *Figure 11a*) and it is important to be able to distinguish between the two resonances of each of these methylene groups. In the practice of spectral assignment two rules of thumb are frequently used:

- H2′ resonates upfield from H2″.

- H5′ resonates downfield from H5″ (Shugar and Remin's rule; (26)).

Although application of these rules often leads to correct answers they do not have general validity. For instance, the relative positions of the H2′ and H2″ resonances are often reversed for 3′-terminal residues in double helical frag-ments. Such inversions may also be observed in non-helical regions, e.g. in the hairpin formed by d(ATCCTA-TTTA-TAGGAT) the relative positions of the H5′ and H5″ resonances of the third thymidine in the loop region are inverted (27). Methods for stereospecific assignment of these resonances are however available (see Section 4.2.3) so that we do not have to rely entirely on these empirical rules.

The resonance position of a particular proton in a molecule is in essence determined by its chemical environment. For the analysis of protein spectra this notion is of little use, since *a priori* not much can be said about the local protein structure. However, in the DNA (or RNA) double helix the direct environment of any particular proton is determined by the nucleotide sequence which is also a determinant of the local conformation. Therefore, to explain the distribution of resonance positions of particular protons in double helices, attempts have been made to relate their chemical shifts within the framework of the base sequence in which they are incorporated. The large data set available for the resonance positions of the non-exchangeable protons in B-type double helices formed by deoxyribo-oligonucleotides lends itself to such an approach.

In an analysis of over 600 nucleotides, the resonance positions of the non-exchangeable protons (except for H5′, H5″, and H2) of the central nucleotide of a trinucleotide in a B-type helix, were examined (28). Most of the resonance positions turn out to be influenced mainly by one base. For the H5 of cytosine and the CH_3 of thymine it is the neighbouring base on the 5′-side and for H1′ the one on the 3′-side. The positions of the remaining proton resonances, i.e. H2′, H2″, H3′, H4′, and H6/H8, are predominantly determined by the base of the central nucleotide itself. The chemical shift distribution profiles observed for the non-exchangeable protons are shown in *Figure 12*. For H2′, H2″, H3′, H4′, and H6/H8 the distribution of the resonances is indicated for the residue to which they belong, for the H5 and CH_3 resonances this is shown as a function of the residue at the 5′-side and for the H1′ resonance as a function of the residue at the 3′-side. It can be seen from *Figure 12* that these chemical shift distribution profiles are relatively broad, their widths varying between 0.3 and 0.6 p.p.m. Thus, there must be other factors involved than just the influence of the bases discussed above. Further analysis showed that the distribution of the chemical shifts of the protons of the central residue in the data set for an individual trinucleotide, e.g. ATA, TAT, CGC, or GCG, is generally not narrower than that of the total data set. The spread in these resonance positions can in part be attributed to the different experimental conditions at which the measurements were performed and in part to long-range effects involving more remote sites in the DNA duplex.

Figure 12. Chemical shift distribution profiles for non-exchangeable protons of nucleotides in regular double stranded B-type DNA oligomers. The chemical shifts are of protons belonging to the central residue of a trinucleotide. The distributions reflect the influence of one of the residues, i.e. the central (a–e), the 3′ flanking (f), and the 5′ flanking (g, h) residue on the proton of interest, which is indicated in the *upper right corner* of the plots. The notation 3′ or 5′ in f–h indicates the absence of a 3′ or 5′ flanking nucleotide.

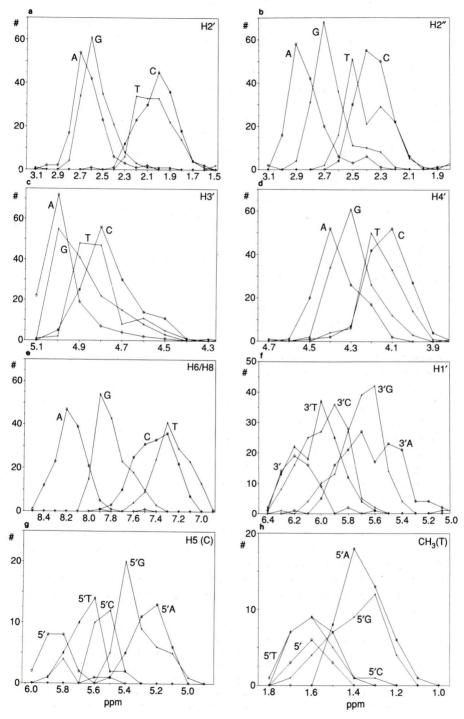

4.1.3 ^{13}C resonances

Figure 11b provides an overview of the ^{13}C chemical shifts found for nucleic acids. A large number of factors may influence the resonance positions of the carbon atoms. For a detailed discussion of the relation between the chemical shifts, base pairing, and stacking in nucleic acids we refer to (24). Here we just want to point out the interesting fact that ribose and deoxyribose C5′ resonances are well resolved from those of the other sugar carbons. This is of value for the assignment of both ^{1}H and ^{13}C resonances through HETCOR experiments which make use of selective excitation.

4.1.4 ^{31}P resonances

The chemical shift distribution profiles of the ^{31}P resonances of the phosphates in nucleic acids show that two classes of signals can be distinguished. Resonances from the sugar phosphate backbone of more or less regular A- and B-type helices are found in a region of 0 to 1.2 p.p.m. (with respect to external 15% aqueous H_3PO_4). Outside this region, i.e. 4 p.p.m. up- or downfield, one may find additional resonances. These signals arise from monoester phosphate groups at the ends of the nucleic acid chains or from diester phosphate groups which have conformations deviating significantly from those in regular A- or B-type helices or which are involved in specific interactions. The monoester phosphate signals are shifted downfield from 0 p.p.m. depending on their state of protonation. For its non-protonated state the ^{31}P resonance is at 3.5 p.p.m., protonation shifts the signal upfield.

NMR studies of model systems and theoretical calculations (24, 29) have shown that the ^{31}P chemical shifts of phosphate groups are sensitive to the conformation of the sugar phosphate backbone, the ionization state of the phosphate group, to complex formation with metal ions, and to the character and structure of chemical substituents bound to the phosphate oxygens. Experimentally it is rather difficult to discriminate between different effects, e.g. the ^{31}P chemical shift is quite sensitive to changes in the O—P—O bond angle, but also alters as a function of the values of the torsion angles ζ and α. Yet some general remarks with respect to the observed shifts can be made with some practical consequences for structure determination. The first concerns diester phosphate resonances which are shifted about 1 to 3 p.p.m. downfield. These resonances have been observed in tRNAs, in complexes of drugs with double stranded DNA, and in Z-DNA (29). These resonances are assigned to phosphates in which ζ and α occur in the combination *g* and *t* or *vice versa* in the combination *t* and *g* respectively. In A-type or B-type helices, whose ^{31}P resonances are found between 0 and 1.2 p.p.m., the torsion angles ζ and α are thought to be both in the *gauche* domain, be it g^+ or g^-. This spread in positions has been interpreted as a demonstration of the sequence or positional dependence of the conformation of the local phosphate backbone leading to subtle changes in ζ and α. Reliable rules for the interpretation of the shifts are, however, not available. The origin of the resonances upfield from 1.2 p.p.m. is less clear. In our laboratory, we have found that hydrogen bonding of amino groups to diester phosphates leads to an upfield shift of 1.8 p.p.m. of the ^{31}P resonances; further experiments are necessary to unravel the mechanisms underlying the observed shifts.

4.2 Assignment strategies

The procedure for the interpretation of nucleic acid spectra, like that of protein spectra (Chapter 4), can be divided into two steps: first the assignment of the signals to particular residues (mononucleotides) and/or base pairs, secondly the sequential assignment, in which signals are attributed to specific residues in the nucleic acid sequence.

4.2.1 Residue-specific assignments

In the first step the connectivity patterns of individual networks of J-coupled spins are traced by means of COSY and TOCSY experiments. Obviously, the majority of these spin systems are formed by the sugar protons. Given sufficient resolution, the connectivity patterns arising from each of the different sugars can be identified in the spectrum (*Figure 13*). In the same manner the

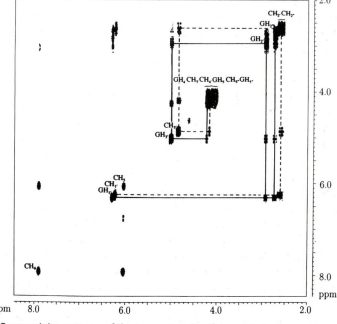

Figure 13. Connectivity patterns of the two networks of J-coupled sugar protons of the G and C nucleotides in a COSY spectrum of the circular dinucleotide d < CpGp >. Starting from the H1'-H2' cross-peak and ending at the cross-peaks between H4'-H5'/H5" it is possible to identify all sugar ring protons.

cytosine H5 and H6 resonances and their corresponding connectivities can be found. Moreover, it is our experience that the cross-peaks between the H6 and methyl resonances of thymines can be detected in COSY (or TOCSY) spectra. Given the absence of (sufficiently large) J-couplings between the sugar protons and the ring protons of the base i.e. between the sugar H1′ and base H6 or H8, the signals of a particular sugar and those of its corresponding base need to be connected via NOEs. Apart from the above-mentioned J-coupled spins the remaining protons of the base (or base pair) also have to be connected by NOEs. These connections are thus made, in effect, as part of the sequential assignment procedure described in the following section.

4.2.2 Sequential assignment

Once the connectivity patterns of individual mononucleotides have been identified in the spectrum the next step is to connect these patterns so as to obtain sequential assignments. Ideally, this would be performed by connecting the spin systems of the mononucleotides to each other by 'through-bond' connectivities, i.e. by the J-coupling of spins present in neighbouring residues. This is indeed possible, by means of hetero correlation experiments, through the coupling of the H5′ and/or H5″ to the neighbouring phosphorous atom, and in turn via the coupling of this phosphorous atom to the H3′ of the adjacent sugar. The method has so far, however, only been successful for small molecules (9), because the ^{1}H $-^{31}$P couplings are usually small and the ^{31}P signals are often badly resolved. Therefore, in the majority of studies, sequential assignments have been obtained through NOE experiments. As for proteins, the distances of the most abundant secondary structure elements have served to set-up the assignment procedures; for nucleic acids these are the A- and B-type double helices. It is convenient to distinguish between two categories of protons:

- the imino and amino protons in the inner part of the double helix
- the base ring protons H6 and H8, and the sugar protons H1′, H2′, and H2″

Sometimes, the non-exchangeable protons on adjacent bases or base pairs are used in the sequential assignment procedure, but depending on the base(pair) sequence the distances between these protons may vary to such an extent that oftentimes the chain of connectivities will be disrupted. In this respect these protons are therefore of limited value.

i. Exchangeable protons

The distances between the imino protons of adjacent base pairs in A- and B-type helices are (except for an U-A and a T-A stack in A-type helices) less than or equal to 5 Å (see *Figure A2*), so that assignment of these protons can be achieved via a chain of connectivities provided a suitable starting point or a unique sequence is available. This approach has been successfully applied to molecules as large as tRNAs (1). In *Figure 14* a NOESY spectrum is pre-

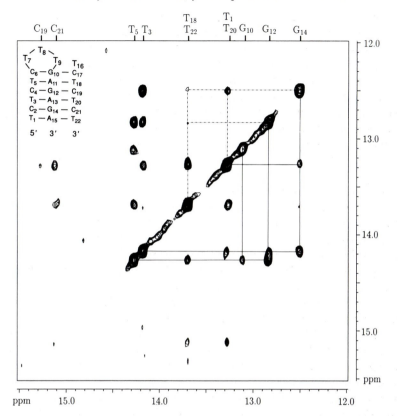

Figure 14. Demonstration of the assignment procedure in a NOESY spectrum of a DNA triplex formed by the hairpin d5'($T_1C_2T_3C_4T_5C_6$-$T_7T_8T_9$-$G_{10}A_{11}G_{12}A_{13}G_{14}A_{15}$)3' and the oligonucleotide d5'($T_{16}C_{17}T_{18}C_{19}T_{20}C_{21}T_{22}$)3'. A sequential walk can be made between the imino protons involved in Watson–Crick hydrogen bonding in the hairpin as is shown below the diagonal, starting at the imino proton resonance position of residue T_1, via the imino proton of G_{14}, T_3, G_{12}, and T_5, ending at the imino proton resonance position of residue G_{10}. Note that the incompletely resolved T_3-G_{12} and G_{12}-T_5 cross-peaks (due limited digital resolution in F1) are resolved above the diagonal. Above the diagonal the sequential walk is shown between the imino protons of the Hoogsteen base paired residues. The latter involve the purine strands of the hairpin stem and the pyrimidine oligonucleotide. Starting at the resonance position of the imino proton of residue G_{12} the NOE walk proceeds to the resonance position of the imino proton of residue G_{14}.

sented in which the assignment procedure is demonstrated. The cytosine amino protons can be incorporated in this scheme, but the G-NH$_2$ protons are less suitable, because it turns out that even in GC base pairs the G-NH$_2$ group may 'flip' around the C-N axis, smearing out the amino proton resonances.

ii. Non-exchangeable protons
This second category of protons has proven to be most useful for sequential assignments. Two connectivity pathways (indicated by the distances in *Figure*

A2) are commonly used for this purpose, namely those formed by series of cross-peaks between (H6 or H8) and H1′ resonances and pathways formed by cross-peaks between (H6 or H8) and (H2′ or H2″) resonances. The method can best be illustrated by an example, given in *Figure 15*, which shows the NOE connectivities between the aromatic ring proton resonance and the H1′ resonances of the sequence 5′GTTCCA-AAC-TGGAAC3′, part of the replication origin of the genome of the M13 phage. This partially complementary sequence forms a hairpin under the conditions of the NMR experiment. The sequential assignment procedure is outlined in *Protocol 2*.

Protocol 2. Sequential assignment procedure

1. Search for a unique cross-peak at which to start the analysis. The H6/H8 resonance of the 5′-terminal residue is connected only to the H1′ of its own sugar and therefore exhibits only one cross-peak, in contrast to other residues in a regular A-type or B-type helix. In this particular example it is the cross-peak indicated by an *arrow* at 7.89 p.p.m. along the F2-direction (*Figure 15A*).

2. Search for a cross-peak which connects the H1′ resonance of the terminal residue to the H6/H8 resonance of the following residue in the sequence, in our example T2, along the horizontal direction (F2).

3. From the established position of the H6/H8 ring proton of the second residue search for the position of its H1′ resonance along the vertical direction.

   In the example of *Figure 15* this seems difficult because there are three candidates for these cross-peaks. The two peaks upfield of 6.0 p.p.m. (F1-direction) can however be disregarded because they correspond to cytosine H5 not to H1′ resonances. This leaves the broad cross-peak between 6.1 and 6.2 p.p.m. (F1-direction).

4. Continue with the same procedure by searching for the cross-peak between the H1′ resonance of the second residue and the H6/H8 position of the third residue (in our case T3).

   In our example this poses a problem because the H6 resonances of T2 and T3 are overlapping. Sometimes this problem can be alleviated by repeating the measurement at a different temperature or at different salt concentrations. Here it was possible to proceed with the analysis because T3-H1′ should be connected to C4-H6; the cross-peak at F2 = 7.52 p.p.m. is the only candidate because the isolated resonance at this position exhibits a cross-peak to an H5 resonance which is confirmed in the DQF-COSY spectrum (not shown).

   From here on the analysis can be continued to the H1′ resonance of the second residue in the loop. At some positions the cross-peaks may be rather weak, but combining this H6/H8 → H1′ analysis with the sequential analysis of the H6/H8 to H2'/H2″ resonances, which proceeds in an exactly analogous way, an unambiguous result is obtained.

5. When a discontinuity in the sequential analysis occurs, try to find a new starting point.

    In the present example this was possible by starting at the 3' end (see 3' *arrow* in *Figure 17A*), and continuing to C9-H6. From this analysis it is concluded that the backbone conformation between residue A8 and C9 deviates from the regular B-type helix such that the sequential analysis can not 'bridge' this step.

As an aid in this assignment process, COSY and/or TOCSY spectra can be used to identify H6–H5 cross-peaks from cytidines and H6-methyl connectivities from thymidines to distinguish between the H6 positions of the two residue-types. In addition, the chemical shift profiles shown in *Figure 12* provide a guide-line for the approximate resonance positions of the non-exchangeable protons.

The routes illustrated in *Figure 15A* and *B* provide the assignments of the base protons, as well as those of the H1', H2', and H2″ protons of the furanose ring. As indicated above COSY spectra can be used to assign the other furanose ring protons.

iii. Connectivities between exchangeable and non-exchangeable protons

As we have seen in the above example the exchangeable imino and amino protons represent one and the non-exchangeable H6/H8 and H1', H2'/H2″ protons represent a second route for sequential assignment. They can be linked together and this may sometimes aid in the analysis of more complicated nucleic acid molecules. For example, the G-NH1 resonance can be connected to the C-H5 resonance of the cytidine which forms a base pair with the guanine, via the amino protons of the cytidine base. Subsequently, through the connectivity of the C-H5 with the C-H6 resonance of the cytidine, the G-NH1 resonance is linked to the backbone proton signals.

The A-H2 resonances can very often be identified through their NOE connectivities with the resonance from the T(U)NH3 proton in the same base pair. The distance between these two protons is 2.8 Å. Assignment of the A-H2 resonances is important in situations where interstrand cross-peaks are seen between these A-H2 resonances and H1' resonances, since the latter help to establish the details of the conformation of the double helix. Situations for which such connectivities are expected can be found in *Figure A2*.

4.2.3 Stereospecific assignment of H2' and H2″ as well as H5' and H5″ resonances

As mentioned in Section 4.1.2 the assignment of these resonances is very often based on their relative position in the spectrum. For irregular

A.

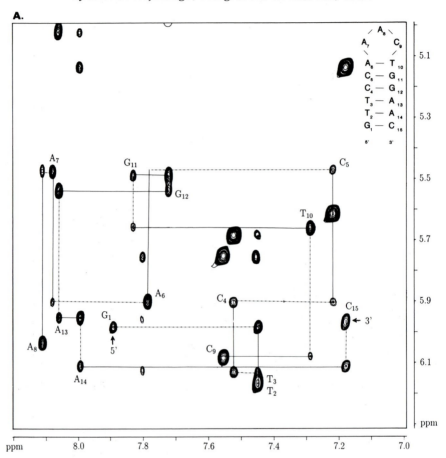

Figure 15. (A) Part of a 600 MHz phase-sensitive NOESY spectrum of the hairpin d(GTTCCA-AAC-TGGAAC), dissolved in ^2H_2O, exhibiting cross-peaks between aromatic H6/H8 ring protons and sugar H1' protons. The mixing time was 0.3 sec. The sequential analysis of the cross-peaks, as outlined in the text, is indicated. The *solid lines* represent the intraresidue connectivities, the *dashed lines* the sequential connectivities. (B) Cross-peaks between aromatic H6/H8 protons and sugar H2'/H2'' protons. *Solid lines* represent intraresidue connectivities, the *dashed lines* the sequential connectivities. The interpretation of the spectrum is left as an exercise for the reader.

structures, however, the availability of a more reliable procedure remains necessary.

i. H2' and H2'' resonances

The H2' and H2'' signals can be separately identified on the basis of the intensity of their NOEs to H1', provided spin diffusion effects are negligible or can be accounted for. Under these conditions the H1'-H2'' cross-peaks will

B.

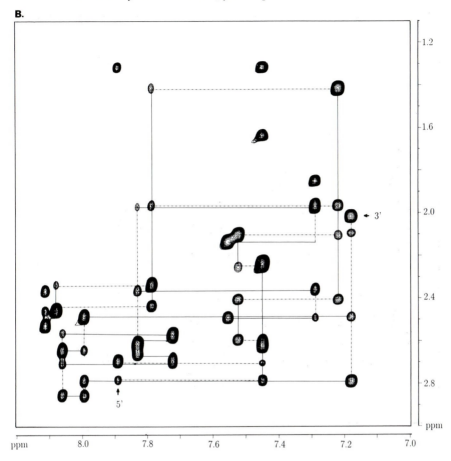

be more intense than the H1'-H2' cross-peaks. This is basically true through-out the pseudorotation cycle of the sugar moiety. Analysis of the J-couplings $J_{1'2'}$ and $J_{1'2''}$ may aid in the stereospecific assignment even when the sugar is involved in an N $\longleftrightarrow$ S equilibrium. When the time a sugar spends in the S-conformation is smaller than 50%, $J_{1'2'}$ will always be smaller than $J_{1'2''}$. In the pure N-conformer $J_{1'2'}$ will be almost negligible. On the other hand, if a sugar spends more than 70% of the time in the S-state $J_{1'2'} > J_{1'2''}$ (see Section 5.2.1).

ii. H5' and H5" resonances

The stereospecific assignment of H5' and H5" resonances is intimately con-nected to the determination of the torsion angle γ. It can be achieved by combining measured $J_{4'5'}$ and $J_{4'5''}$ couplings with the intensities of NOE cross-peaks between the resonances of the proton pairs H3'-H5', H3'-H5", H4'-

H5′, and H4′-H5″. This is closely analogous to the procedures used for stereospecific assignment of $C_\beta H_2$ resonances in the amino acid side chains of proteins (see Chapter 10). The procedure can best be illustrated by making use of *Figure 16*. If both coupling constants $J_{4'5'}$ and $J_{4'5''}$ are small, (i.e. ≈ 2–3 Hz) and we assume that only the normal staggered conformations can be populated, then the torsion angle γ corresponds to a g^+ conformation. In this situation, since $d_i(3';5'') \approx 2.4$ Å and $d_i(3';5') \approx 3.8$ Å, the NOE between H3′ and H5″ will be much more intense than that between H3′ and H5′, and this establishes the identification of the H5″ resonance. The γ^t and γ^- state can be distinguished from the γ^+ state because then either or both $J_{4'5'}$ or $J_{4'5''}$ become large (see *Figure 16*). The intensity of the NOE cross-peaks can help us distinguish between these two states. For the γ^t conformer $d_i(3';5') \approx d_i(3';5'') \approx 2.6$ Å which gives rise to reasonably intense NOE cross-peaks. Then the proton pair with the large J-coupling will be H4′-H5″. In the γ^- conformation the intensity of the H3′-H5″ and H3′-H5″ NOE cross-peaks is

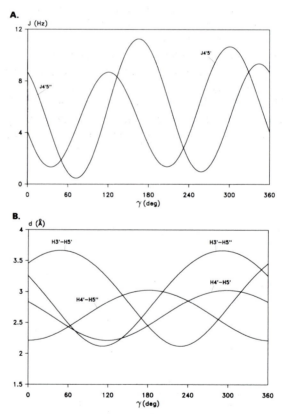

Figure 16. (A) $J_{4'5'}$ and $J_{4'5''}$ coupling constants calculated as a function of the torsion angle γ on the basis of the Karplus relations (*Table 4*). (B) ^1H–^1H distances $d_i(3';5')$, $d_i(3';5'')$, $d_i(4';5')$, and $d_i(4';5'')$ as a function of torsion angle γ.

again quite different, the most intense cross-peak belonging to the H3'-H5' pair.

If there is a rapid equilibrium (on the NMR time scale) between the three staggered conformations the situation obviously becomes more complicated. The data available so far indicate that in the majority of cases the γ^- state is hardly populated. Then the average value of $J_{4'5''}$ is always larger than that of $J_{4'5'}$ except when the γ^+ population is larger than 85% (see *Figure 17*). If there are no indications to the contrary, e.g. no unusual chemical shifts, one often falls back on Remin and Shugar's rule for the stereospecific assignment of H5' and H5'' signals (Section 4.1.2) (26).

4.3 Assignment via 3D NMR

The assignment procedure described in the preceding sections can in most cases be extended to involve the H4' resonances. Only occasionally, (e.g. in 'non-regular' loop regions; 27) can the H5' and H5'' signals be assigned, since the region containing the H4', H5', and H5'' resonances is so crowded. The resulting overlap is worsened by the often small chemical shift differences between the H4', H5', and H5'' resonances, which places the cross-peaks close to the diagonal. These problems may, to a large extent, be alleviated by homonuclear three-dimensional NMR experiments. So far, 3D NMR applications to nucleic acids have been very limited (30, 31), and the isotopic labelling procedures available for proteins (see Chapter 5), which have facilitated heteronuclear 3D NMR studies of these molecules, are just becoming available for nucleic acids, but the preparation of these isotope-enriched molecules is much more involved than that of proteins.

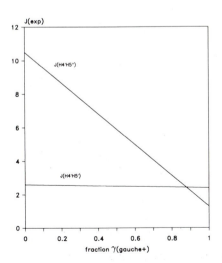

Figure 17. Plot of $J_{4'5'}$ and $J_{4'5''}$ as a function of the fraction γ^+, assuming that the γ^- state is not populated. $J_{4'5''}$ is larger than $J_{4'5'}$ if the fraction γ^+ is less than 85%.

Homonuclear three-dimensional experiments can be considered as a combination of two two-dimensional NMR experiments. Here we will examine the TOCSY-NOESY (30) and the NOESY-NOESY experiments (31); the former is a combination of a two-dimensional TOCSY and a two-dimensional NOESY experiment, the latter a combination of two two-dimensional NOESY experiments. It has not been generally recognized that in these spectra repeating connectivity patterns occur which, when used properly, may facilitate their interpretation. We will illustrate this for the TOCSY-NOESY experiment. The transfer of coherence and magnetization during the TOCSY-NOESY pulse sequence is schematized in *Figure 18*. After labelling of the resonance during the t_1-period, transfer of coherence of protons J-coupled to a particular proton (in our example the H2″) takes place during the spin-lock period. Then, after labelling in the t_2-period magnetization is transferred during the NOESY mixing period to other protons in the molecule (in our particular example from H2″ to other protons).

We first examine the F_1F_3-plane at the position of the H2″ resonance along the F_2-axis. The connectivities observed there arise solely from coherence transfer during the spin-lock period (there is of course no magnetization transfer for the same proton H2″ → H2″ during the NOESY mixing time). This set of connectivities therefore falls on the so-called TOCSY line. When NOE transfer from the H2″ to a different proton does take place during the NOESY mixing period then the same set of connectivities gets labelled with the resonance frequency of this third proton. Hence at the position of this third proton, along the F_3-axis, we expect to observe a repeat of the connectivities present at the TOCSY line. This is illustrated in *Figure 18* and an experimental example (the spectrum of the hairpin formed by the sequence GTTCCA-AAC-TGGAAC, part of the replication origin of the M13 phage) is presented in *Figure 19*. Similarly, it is possible to observe NOE connectivities between H3′ on one hand and H4′, H5′, and/or H5″ protons on the other hand, which are unresolved in 2D spectra, because in the 3D spectrum they can be labelled with H1′ resonance frequencies and as a result fall in a relatively empty region of the spectrum. This is demonstrated in *Figure 20* where the F_1F_3-plane is shown which crosses the F_2-axis at the H3′ resonance position of residue A14. During the TOCSY period the H3′ is, among other protons, connected to H1′. The resulting coherence is translated into magnetization, labelled with both the H3′ and H1′ resonance frequencies, that is transferred to, for instance, the H5′ position in the NOESY mixing period. Hence, a cross-peak is observed representing the H3′-H5′ NOE connectivity at the F_1-position of the H1′ proton. A similar analysis can be developed for the 3D NOESY-NOESY spectrum. Thus, the 3D approach provides additional means for interpreting the spectrum and, hence, additional constraints for structure determination.

a.

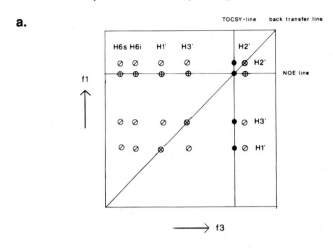

b.

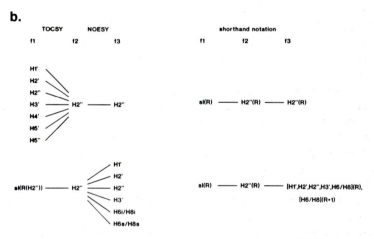

Figure 18. Coherence-magnetization transfer scheme of a TOCSY-NOESY experiment for an oligonucleotide. (a) F1F3 plane which is assumed to cross the F2 axis at the position of a H2″ resonance of sugar residue R in an oligonucleotide. (b) Shorthand notation, summarizing the transfer routes, leading to the cross-peaks in the F1F3 plane shown in (a).

5. Local conformational analysis

In our view it is important to have a good understanding to what extent torsion angle and distance information are able to define the structure of the building blocks of nucleic acid molecules, i.e. the conformation of the furanose ring and the mononucleotide. This facilitates the assessment of the importance of particular distance or torsion angle constraints in the structure determination procedure, and also the assessment of the quality of the overall structures determined along the lines described in Section 6. Thus, in this section we discuss which structural constraints determine the pseudorotation

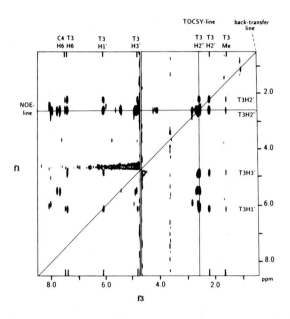

Figure 19. F1F3 plane from a TOCSY-NOESY spectrum of the hairpin 5′d(G$_1$T$_2$T$_3$C$_4$C$_5$A$_6$-A$_7$A$_8$C$_9$-T$_{10}$G$_{11}$G$_{12}$A$_{13}$C$_{15}$)3′, through the H2″ resonance position of residue T$_3$. Note that the spin-lock pattern of residue T$_3$ is repeated at different F3 positions (see text). (Mooren *et al.* (1991). *J. Magn. Res.*, **94**, 101; Copyright Academic Press.)

parameters of the sugar moiety and the value of the torsion angle χ, which together define the mononucleotide structure, and then consider the determination of the backbone torsion angles which are important in constraining the conformational space available to a dinucleotide. Throughout, attention will be paid to the influence of internal mobility on the determination of the values of these parameters.

5.1 NOEs

NOE experiments provide information on interproton distance constraints (1, 9, 16, 17). It has been shown that in structural studies of proteins the quantity of these NOE constraints is more important than their quality, in the sense that it suffices to interpret strong, medium, and weak NOEs to correspond with distance constraints 2.0 to 2.5, 2.5 to 3.5, and 3.5 to 5.0 Å respectively. For nucleic acids, however, it is necessary to derive more accurate interproton distances from the volumes of the NOE cross-peaks. Approaches which take account of spin diffusion effects, such as the full relaxation matrix treatment discussed in Chapters 10 and 11, have been developed and a range of different programs (IRMA (32), MARDIGRAS (33), NO2DI(34)), which incorporate these have been developed to deduce

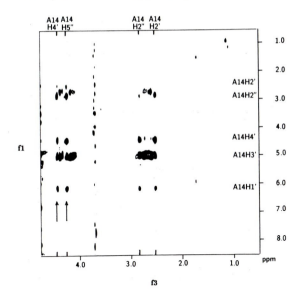

Figure 20. F1F3 plane from a TOCSY-NOESY spectrum of 5'd(GTTCCA-AAC-TGGAAC)3' (numbering is the same as in *Figure 19*) through the H3' resonance position of residue A_{14} which shows that it is possible to observe NOE connectivities between H3' on the one hand and H4', H5' and/or H5" protons on the other hand. This is possible because in the 3D spectrum the signals of the latter can be labelled with H1' resonance frequencies, as indicated with *two arrows*.

reliable distance information from NOE spectra. An outline of this approach will be presented here; further details are given in Chapters 10 and 11.

The time-dependence of the magnetization of the individual spins in the NOESY can be described by a set of differential equations which, in matrix form, are denoted the relaxation matrix. The off-diagonal elements of this matrix, known as the cross-relaxation rate constants, σ_{ij}, determine the cross-peak intensity. For short mixing times:

$$I_{ij} = \sigma_{ij} \tau_m + \sum_{i=1}^{n} \sigma_{ik}\sigma_{kj}\tau_m^2 + \ldots \qquad [4]$$

(where a power series expansion of the exponential dependence is used). If the series can be truncated after the first term, the NOE intensity increases linearly with τ_m and is proportional to $1/r^6_{ij}$. Such calculations of distances are said to make use of the 'isolated spin pair approximation' (ISPA) (16,17). In practice, it is often very hard if not impossible to decide whether the limit of short mixing (which rules out contributions from spin diffusion) is applicable in a particular experimental situation. Thus, to be able to extract reliable distances from the NOE experiment, the values of the elements of the relaxation matrix must be determined. These can in principle be derived from

255

the matrix defining the NOE intensities. In practice, mainly as a result of peak overlap and incomplete assignment of the spectrum, the NOE matrix will be incomplete and the relaxation matrix can not be derived in a straightforward manner. This problem can be approached as follows. A model for the molecule under study is used to simulate a 'theoretical' NOESY spectrum. This trial model can be derived from distances deduced from the available NOESY data using the isolated spin pair approximation or it can be purely hypothetical, e.g. a pure B-type helix. The NOEs in the 'theoretical' matrix are scaled with respect to the available experimental values and subsequently in this matrix those elements are replaced for which experimental values are available. In this way, a hybrid NOE matrix is constructed which is complete and from which the elements of the relaxation matrix can be calculated. This, and hence the interproton distances can be iteratively refined by several different approaches including IRMA or related methods (32), and MARDIGRAS (33).

A problem which may occur in the complete relaxation matrix treatment is the generation of negative eigenvalues from the hybrid NOE matrix, which hampers the calculation of the relaxation rate constants. To avoid this, methods have been proposed which only consider relatively small cross-relaxation networks involving a subset of the total set of protons of the nucleic acid molecule (34, 35).

Experience gained so far with these methods for distance determination is that the outcome is not very sensitive to the initial distances introduced into the calculations. The accuracy of the derived distances, in so far as corrections for spin diffusion are being made, will depend, however, on the experimental data used in the calculations. It is important to use data obtained at mixing times, where the equalizing effect of spin diffusion is not yet too pronounced. If the NOE peak intensities are tending to a common value insufficient information is retained to determine distances reliably.

In the situation that anisotropic tumbling occurs, we do not expect that for nucleic acid molecules which can be subjected to NMR studies the NOE enhancements are significantly affected compared to the situation in which isotropic tumbling predominates. This can be gleaned from *Figure 2.7* in (16), where it is shown for an axially symmetric molecule how the NOE effect depends on the ratio of the rotational correlation time perpendicular and parallel to the symmetry axis of the molecule. When this figure is consulted it should be kept in mind that for short oligonucleotides, up to about ten base pairs (small molecules consisting of a few nucleotides are excluded from consideration), the hydrodynamic radii parallel and perpendicular to the symmetry axis of the molecule are of the same magnitude (≈ 30 Å). Larger (longer) duplexes do have an axial ratio above unity but this does not affect the NOE enhancements because then $\omega\tau_c \gg 1$, certainly at 500 or 600 MHz.

So far, it has been assumed that the considered molecules are rigid. There are strong indications, however, that nucleic acid molecules in solution may be quite flexible and that intramolecular conformational interconversions may occur. For example, the furanose ring may interconvert between N- and S-type conformations and a methyl group may rapidly rotate around its axis. We have to consider, therefore, how this affects the determination of the interproton distances from the observed NOE intensities. The following division can be made according to the time scales involved in the interconversion processes (16, 17; see also Chapter 6). Descending the list corresponds to an increasing rate of intramolecular motion.

(a) Interconversion is slow on the chemical shift time scale; the NMR spectrum is a superposition of the NMR spectra of the different conformers and the NOEs pertain to these different conformers. Two cases can be distinguished:

  i. interconversion is slow compared to NOE buildup ($\approx T_1$), averaging occurs over NOE enhancements

  ii. interconversion is faster than T_1; averaging occurs over relaxation rates.

(b) Interconversion is fast compared to the chemical shift differences; one averaged spectrum will be observed for the different conformers present in solution. Again two situations can be distinguished.

  i. Interconversion between different conformers is slow compared to the overall rotational correlation time of the molecule. The distances in the elements of the relaxation matrix, obtained from the observed NOE effects, are now $<1/r^6>$ in which $< >$ denotes the averaging over the allowed conformations. Thus, for a two-site exchange problem:

$$I_{ij} = f_a I_{ij}^a + f_b I_{ij}^b \qquad [5]$$

  where f_a and f_b denote the fractions and I_{ij}^a and I_{ij}^b the NOE intensities of conformers a and b. Furanose interconversions, i.e. N- to S-type transfers, probably fall into this exchange category, and this approach has been used below to predict the effect of the N–S interconversion on the NOE effect observed for, e.g. the H3′-H6 proton pair.

  ii. Interconversion is faster than the overall tumbling time of the molecule. Rotation of methyl groups around their symmetry axis is considered to fall into this category. The distance constraints in the elements of the relaxation matrix, obtained from the observed NOE effects, are now $<1/r^3>^2$ where $< >$ again denotes the average over the allowed conformations. Moreover, when the internal rotation is much faster than the overall rotation a correction factor is required for the elements in the relaxation matrix (16, 36). Corrections accounting for these effects on methyl protons have been incorporated in most of the above-mentioned programs as a model independent correction factor (37).

5.2 J-couplings

For nucleic acids, J-coupling information is especially important since it is difficult to characterize the structure of certain elements of these molecules by NOEs alone. In the translation of vicinal J-couplings to torsion angles use is made of the celebrated Karplus equation and its descendants. *Table 4* summarizes the Karplus equations (38) which are of particular value in nucleic acid research. These are empirical relationships but they have been tested extensively and have proven to be reliable in structure determination, although this is not to say that they could not be improved. Examination of *Table 5* shows that the backbone angles β, γ, and ε as well as the furanose ring are amenable to J-coupling analysis. The torsion angles α and ζ and the glycosidic torsion angle χ can not be characterized in this way because the necessary, coupling spins are missing. Because of internal molecular mobility individual torsion angles may, during the course of time, adopt different staggered conformations. If the motion is fast with respect to the values of the J-coupling constants, the weighted average of the J-couplings of the different conformers will be observed in the NMR spectrum. This effect will be discussed below in relation to the determination of the individual torsion angles. We mention that Altona's review (39) is still an important source of information on the subject matter discussed in this section.

5.3 Conformation of the furanose ring

5.3.1 The furanose conformation as reflected through J-couplings

The phase angle of pseudorotation, P, and the pucker amplitude, Φ_m, which characterize the conformation of the furanose ring (Section 2.2) are related to the endocyclic torsion angles (see eqn 1) which in turn can be correlated with the vicinal proton–proton coupling constants in the sugar ring via the EOS-Karplus equation in *Table 4*. Knowledge of two independent J-couplings permits the determination of the conformation of a rigid furanose ring. It can be seen from *Figure 21* that the coupling constants $J_{1'2'}$, $J_{2''3'}$, and $J_{3'4'}$ are good indicators of the sugar conformation. For the most important values of P, i.e. for P = 0°–36° and P = 150°–190° (N- and S-conformations respectively), the variation of the J-couplings is no more than 1 or 2 Hz when Φ_m is varied within the practically allowed boundaries (35°–42°). The changes in the coupling constants, in particular the afore-mentioned $J_{1'2'}$, $J_{2''3'}$, and $J_{3'4'}$, are much larger for variations in P, e.g. for N-type sugars $J_{1'2'}$ is small, ≈ 1 Hz, while for S-type sugars $J_{1'2'}$ is large, ≈ 10 Hz. The way in which the J-couplings $J_{1'2'}$ and $J_{3'4'}$ change in tandem upon traversing the full pseudorotation cycle is shown in *Figure 22*. The lines represent conformations around the conformation wheel, while the points indicate measured coupling constants for different oligonucleotides. It is clear that no points are found on the lines,

Table 4. Relations between torsion angles and J-couplings

Karplus equations

$$^3J_{HH} = P1 \cos^2 \phi + P2 \cos \phi + P3 + \sum_{i=1}^{4} \Delta\chi_i (P4 + P5 \cos^2 (\xi_i\phi + P6 \mid \Delta\chi_i \mid)) + \Delta J$$

$$\Delta\chi_i = \Delta\chi_{i,\alpha} - P7 \sum_{j=1}^{3} \Delta\chi_{ij,\beta}$$

$$^3J(HCOP) = 15.3 \cos^2 \phi - 6.2 \cos \phi + 1.5$$

$$^3J(CCOP) = 8.0 \cos^2 \phi - 3.4 \cos \phi + 0.5$$

$\Delta\chi_{i,\alpha}, \Delta\chi_{ij,\beta}$: Huggins electronegativity difference between α or β substituent and H; with $\Delta\chi_{i,\alpha}, \Delta\chi_{ij,\beta} = 1.3$ (O), 0.4 (C), 0.85 (N), -0.05 (P).

ξ_i: relative orientation factor of substituents. In Newman projection of $H_\alpha S_{1\alpha}S_{3\alpha}$-$C_1$-$C_2$-$S_{2\alpha}S_{4\alpha}H_b$: $S_{1\alpha}$ has positive position, $\xi_i = +1$, if H_α-C_1-$S_{i\alpha} = 120°$, negative position, $\xi_i = -1$, if angle $= -120°$.

ΔJ: $-2.0\cos(P-234)$ for $144° < P < 324°$ for J'2''
$-0.5\cos(P-288)$ for $198° < P < 360°$ and $0° < P < 18°$ for J2'3''
in all other cases $\Delta J = 0.0$

ϕ: torsion angle (°)

Numerical values of P in Karplus equations depending on number of substituents

Number of substituents	P1	P2	P3	P4	P5	P6	P7
2 α substituents	13.89	−0.96	0	1.02	−3.40	14.9	0.24
3 α substituents	13.22	−0.99	0	0.87	−2.46	19.9	0.00
4 α substituents	13.24	−0.91	0	0.53	−2.41	15.5	0.19
Overall	13.70	−0.73	0	0.56	−2.47	16.9	0.14

i.e. for sugar conformations on the pseudorotation cycle between the N- and the S-conformers (see *Figure 6*). This a reflection of the fact that the furanose rings occur only in either N- or S-conformations, and that due to rapid inter-conversion between the two forms average J-couplings are being observed (1). Thus,

$$J_{obs} = f_N J_N + f_s J_s \qquad [6]$$

with J_{obs} being the observed J-coupling, f_N and f_s the fraction of N- and S-conformers respectively, and J_N and J_S the J-coupling in the N- and S-conformer respectively.

Table 5. J-couplings between nuclei, monitoring backbone torsion angles

Torsion angle	J-coupled nuclei			
α	—	—	—	—
β	C4'P5	H5'P5	H5"P5	—
γ	H4'H5'	H4'H5"	C3'H5'[a]	C3'H5"[a]
δ[b]	H3'H4'	C5'H3'[a]	C5'C2'[a]	—
ε	H3'P3	C4'P3	C2'P3	—
ζ	—	—	—	—
χ[c]	—	—	—	—

[a] These torsion angles have as yet not been used.
[b] The torsion angle δ determines the sugar pucker which can also be determined via J-couplings between other sugar protons (*vida supra*).
[c] The J-coupling between H1' and C2/C4 or between H1' and C6/C8 may be a future candidate to determine the torsion angle χ.

Under these conditions, if it is possible to measure all five averaged coupling constants of the sugar ring the conformational parameters characterizing the two states can be deduced, since in this situation five unknowns have to be determined (P_N, Φ_{mN}, P_S, Φ_{mS}, and K, i.e. the pseudorotation parameters of the N- and the S-forms, and the equilibrium constant describing the N–S equilibrium). It should be noted that in this situation where one adjusts five conformational parameters to five experimental parameters, excellent fits may be obtained with physically unrealistic conformational parameters. One way around this problem is to constrain the pseudorotation parameters of the less abundant conformer. In practice, it is often not feasible to measure all five of the individual coupling constants. However, it may still be possible to make an estimate of the percentage of N- and S-conformers and of their conformational parameters. In particular, sums of coupling constants may be very useful because it is often easier to deduce these, rather than individual coupling constants, from the spectra or from the cross-peaks. The sum of the coupling constants of a certain proton is just the distance between the outer peaks of its multiplet signal or cross-peak. This idea has been exploited by Rinkel and Altona (40) who have shown that the three sums of coupling constants:

$$\Sigma 2' = J_{1'2'} + J_{2'3'} + J_{2'2''} \qquad [7a]$$

$$\Sigma 2'' = J_{1'2''} + J_{2''3'} + J_{2'2''} \qquad [7b]$$

$$\Sigma 3' = J_{2'3'} + J_{2''3'} + J_{3'4'} \qquad [7c]$$

in combination with the individual coupling constants $J_{1'2'}$ and $J_{1'2''}$ have the same information content as the five individual constants $J_{1'2'}$, $J_{1'2''}$, $J_{2'3'}$, $J_{2''3'}$,

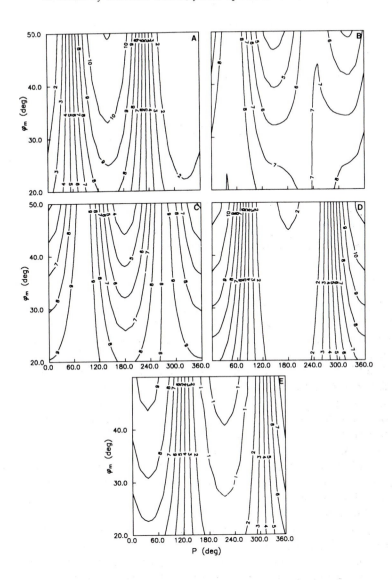

Figure 21. Contour lines of ^1H-^1H coupling constants (Hz) in the sugar ring as a function of P and Φ_m. The coupling constants were calculated with the aid of the EOS-Karplus equation and corrected for the Barfield transmission effect (*Table 4*). (A) $J_{1'2'}$, (B) $J_{1'2''}$, (C) $J_{2'3'}$, (D) $J_{2''3'}$, and (E) $J_{3'4'}$.

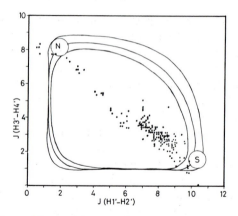

Figure 22. Combinations of $J_{1'2'}$ and $J_{3'4'}$ coupling constants reported for a wide variety of nucleic acids at various temperatures. The combinations of $J_{1'2'}$ and $J_{3'4'}$ expected for pure N-type and S-type conformers are indicated as *encircled regions*. The *solid lines* represent the combinations of the two coupling constants as the full pseudorotation circle would be traversed. The lines were calculated for pucker amplitudes $\Phi_m = 30°$, $35°$, and $40°$ going from the inner to the outer curve. (v.d. Ven and Hilbers. (1988). *Eur. J. Biochem.*, **178**, 1; Copyright Federation of European Biochemical Societies.)

and $J_{3'4'}$. This information can be used in a graphical determination of the pseudorotation parameters (see following Section 5.3.2). If the system is not conformationally pure then an estimate is first made of the fraction of S-conformers present for a particular sugar, using one of several empirical relationships. An excellent determination of the fraction of S-type conformers can be performed with the expression:

$$f_s = (31.5 - \Sigma 2'')/10.9 \qquad [8]$$

In case $\Sigma 2''$ is not available, e.g. when the H2' and H2'' have (nearly) the same chemical shift, then estimates can be obtained from the equations:

$$f_s = (16.9 - J_{2'3'} - J_{2''3'})/8.9 \qquad [9]$$

and/or

$$f_s = (\Sigma 1' - 9.8)/5.9 \qquad [10]$$

Estimates of the fraction of S-conformers are required to be able to calculate the conformational parameters of the S- and the N-conformers (see Section 5.3.2).

5.3.2 Determination of the furanose conformation through J-couplings; practical aspects

For nucleic acid fragments consisting of about 20 nucleotides (i.e. $M_r \approx 6000$) the H1' resonances are usually reasonably well resolved and hence from this spectrum the $J_{1'2'}$ and $J_{1'2''}$ coupling constants can be obtained. This is most easily done by simulating the spectrum; spectrum simulation programs are readily available, e.g. the Bruker program PANIC, which allows the simulation of spectra of spin systems of up to nine spins. As input these programs need the chemical shifts of the different H1' spins, the J-coupling constants $J_{1'2'}$, $J_{1'2''}$, $J_{2'2''}$, and the linewidth. At this stage of the analysis an estimate of the chemical shift values will normally be available from the sequential analysis of the spectrum; a first impression of the values of the coupling constants can be obtained from a first-order J-coupling analysis of the H1' multiplets as described in textbooks on NMR. In an iterative procedure the program adjusts the input parameters to new values so as to obtain an optimal fit between the experimental and simulated spectrum.

In the following step one tries to get information about the other J-couplings as well. For smaller molecules this again can be attempted by analysing one-dimensional spectra, but for the molecules considered here this is usually not feasible and one has to take recourse to the analysis of cross-peaks in 2D spectra. This can be performed by a computer simulation of the fine-structure of the cross-peaks, of for instance COSY-type spectra, as a function of the values of the coupling constants involved (SPHINX, 41). An example of this approach is shown in *Figure 23* for cross-peaks between H1' and H2'/2'', and between H2'/2'' and H3'. By comparing the fine-structure of the experimental with the simulated patterns individual J-couplings can be estimated. The fine-structure of the cross-peaks also lends itself to the determination of sums of coupling constants. Sums of coupling constants can sometimes be determined directly from the distance between the outer peaks of resonance multiplets in a 1D or from the fine-structure of the cross-peaks in a 2D spectrum.

If the J-coupling analysis has been completed (as far as possible) the conformational analysis of the sugar ring can be started. We first consider the graphical method introduced by Rinkel and Altona (40) and then we will briefly describe a computer based approach.

The graphical method, which so far is only available for deoxyribose sugars, makes use of *Figures 24* and *25*. In *Figure 24* the predicted values of the coupling constants and sums of coupling constants are plotted as a function of the phase angle, P, while in *Figure 25* predicted values of coupling constants and sums of coupling constants are plotted as a function of the percentage of S-type conformers. The analysis is started by fitting the available J-coupling information to the curves in *Figure 24*. Thus, for a single P-value the experimental data should fall on or close to the curves. An exact fit might not be obtained, because the curves have been drawn for one pucker amplitude, $\Phi_m = 35°$ and the actual value may deviate somewhat. If no acceptable fit between the experimental and simulated coupling constants can be obtained, the analysis is continued on the basis that the sugars exist as a mixture of rapidly interconverting N- and S-conformers. This mixture is analysed using the curves in *Figure 25*, which were calculated on the basis of

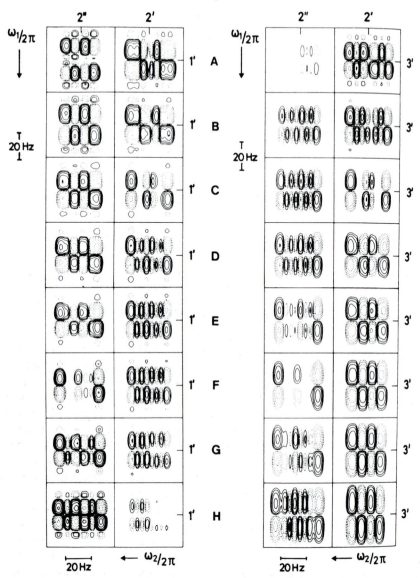

Figure 23. Simulated DQF-COSY cross-peaks of a deoxyribose proton spin system. The H1'-H2', H1'-H2'', H1'-H3', and H2'-H3' cross-peaks are shown. (A) Coupling constants correspond to the pure C2'endo conformation (S-type); (B)–(G) rapid interconversion between S- and N-type sugars with relative populations of (C2'endo)/(C3'endo) of 8:2, 6:4, 5:5, 4:6, 3:7, and 2:8, respectively; (H): pure C3'endo conformation (N-type). In the simulation strong coupling is taken into account only between H2' and H2'' protons. A chemical shift difference of 100 Hz was used. The linewidth was set at 4 Hz. $J_{2'2''} = -14.1$ Hz, $J_{H3'-^{31}P} = 5.8$ Hz, and the values for the other coupling constants (in Hz) are: $J_{1'2'} = 9.5$, $J_{1'2''} = 5.8$, $J_{2'3'} = 5.5$, $J_{2''3'} = 1.3$, $J_{3'4'} = 1.0$ for C2'endo (S-type), and $J_{1'2'} = 1.5$, $J_{1'2''} = 7.7$, $J_{2'3'} = 7.2$, $J_{2''3'} = 9.7$, $J_{3'4'} = 7.8$ for C3'endo (N-type). (Widmer and Wüthrich. (1987). *J. Magn. Res.*, **74**, 316; Copyright Academic Press.)

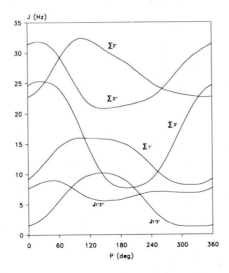

Figure 24. Calculated coupling constants and sums of coupling constants in deoxyribo-furanose rings plotted versus the phase angle of pseudorotation, P. The pucker amplitude, Φ_m, was kept at 35°. $\Sigma 1' = J_{1'2'} + J_{1'2''}$, $\Sigma 2' = J_{1'2'} + J_{2'3'} + J_{2'2''}$, $\Sigma 2'' = J_{1'2''} + J_{2'3''} + J_{2'2''}$, and $\Sigma 3' = J_{2'3'} + J_{2''3'} J_{3'4'}$. For $J_{2'2''}$ a value of -14 Hz was used.

the assumption that the N-conformer ($\Phi_m = 35°/37°$ and $P_N = 9°$) is the minor component, as is the case in the majority of B-DNA fragments. The procedure is now continued as follows. First, an initial estimate of the percentage of S-conformers is made, using eqn 8 or 10; this is refined by fitting the experimental coupling data to simulated curves (*Figure 25*) to determine the best %S-value. The value of P_s is then deduced from $\Sigma 2'$ and/or $\Sigma 3'$, which are the most sensitive measures of P_s. Usually, the experimental values fall within the range $117° < P < 189°$ and the actual value of P_s can be obtained by interpolation. If the H2' and H2" resonances have (nearly) the same chemical shift, the analysis can be started with the aid of eqn 9 and 10. More details on the application of the method can be found in (40).

An alternative, computer based approach, in which no assumption about the conformation of the minor component is required, has been used as well (MARC; 27). In this method a set of torsion angles is generated on the basis of eqn 3 by systematically varying P and Φ_m in steps of, e.g. 5°. These torsion angles are then used to derive the corresponding values of the J-coupling constants between the sugar protons through the application of the EOS-Karplus equation (see *Table 4*). At this point a set of five coupling constants is obtained for each generated P, Φ_m pair, i.e. for each corresponding sugar conformation. A set of J-couplings is then generated for different ratios of N- and S-conformers. (These J-couplings are the weighted average of the coupling constants in each conformation; see eqn 6.) Five coupling constants obtained for each combination of the five parameters P_N, Φ_{mN}, P_S, Φ_{mS}, and f_S. The program then tests each line for the occurrence of J-couplings which

265

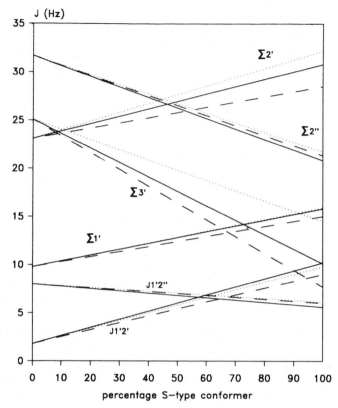

Figure 25. Predicted coupling constants and sums of coupling constants plotted versus the percentage S-type conformer for deoxyribofuranose ring in case a fast equilibrium exists between N- and S-type conformers. The pseudorotation parameters for the N-conformer are, $P_N = 9°$, and $\Phi_{m,N} = 35°$. The pseudorotation parameters for the S-conformer are, $P_s = 189°$ (*dashed line*), $P_s = 144°$ (*solid line*), $P_s = 117°$ (*dotted line*), and $\Phi_{m,s} = 35°$. Use has been made of eqn 6.

match the experimental J-coupling constraints. For instance, when $J1'2' = 10 \pm 0.5$ Hz, and $\Sigma 2' = 24 \pm 1$ Hz are the experimental constraints used as input, all values P_N, Φ_{mN}, P_S, Φ_{mS}, and f_S satisfying these constraints are generated.

5.3.3 The furanose conformation as reflected through NOE effects

Attempts have been made to determine the sugar conformation on the basis of NOE distance constraints between the sugar protons (1). In our experience and that of others analysis of the J-couplings is more suited for the description of the furanose ring. This follows from an examination of *Figure 7*, where the distances between the protons in the sugar ring have been plotted as a function of the phase angle of pseudorotation. It is clear that the variations of the distances between the protons as a function of P is, except for $d_i(1';4')$ and $d_i(2'';4')$, less than 0.3 Å. This is less than the accuracy of a distance determination expected from NOE measurements ($\pm$ 0.2 Å or worse). The distances $d_i(1';4')$ and $d_i(2'';4')$ do vary sufficiently to convey information concerning the sugar conformation. The distance $d_i(1';4')$ is of little use, however, for distinguishing between N and S, since it is about equal for the N- and S-conformers. Only in the situation that the sugar would adopt a conformation intermediate between an N- and S-type pucker, i.e. P $\approx$ 90°, it could be monitored by H1'–H4' connectivities in the NOE spectrum. This leaves $d_i(2'';4')$ as a measure of P (at a given Φ_m), but the determination of this distance is particularly sensitive to spin diffusion effects. If these can be accounted for, H2''–H4' connectivities may provide useful information, whether the sugar adopts a rigid N- or S-conformation, but obviously only for deoxyribose sugars.

Another distance that has been reported to be sensitive to the sugar conformation is $d_i(3';6)$ and $d_i(3';8)$, when the base is in an *anti*-orientation (see *Figure A1a* and *b*). Indeed, with the sugar in the N-conformation the distance between H3' and the ring H6 or H8 becomes short, i.e. between 2 and 3 Å, in contrast to the situation where the sugar has adopted an S-conformation and the distance between H3' and the ring H6 or H8 is between 4.5 and 5 Å. Since these effects are intimately connected with the determination of χ a more detailed discussion is given in Section 5.4.

These conclusions are only valid for rigid sugar conformations. When there is a rapid interconversion between the N- and the S-puckers the situation changes markedly, as we have seen in connection with the J-coupling analysis of the sugar conformation. This is demonstrated in *Figure 26*, where average $d_i(2'';4')$ distances are presented as derived from NOEs as a function of the N/S ratio and χ for a rapid equilibrium between the sugar conformations. As can be seen, to determine with any accuracy the value of N/S requires an extremely accurate NOE measurement, except in case the sugar pucker is purely of the S-type. Hence, it is not possible to determine the N/S ratio from sugar proton–proton distances alone. A similar conclusion holds for the H6-H3' and H8-H3' pairs (*Figure 27*). As little as $\approx$ 10% of N-conformers leads to an intense NOE for the H3'-H6 pair, rendering such an observation rather useless as a tool for the determination of either the N/S equilibrium or the sugar conformation (see also Section 5.4).

5.4 Rotation about the χ angle

The distance $d_i(6/8;1')$ is basically independent of the sugar pucker and H6/H8-H1' connectivities therefore are obvious candidates for estimating the torsion angle χ (*Figure 8*). In practice, however, it is not possible, given the experimental accuracy of the distance determination ($\pm$ 0.2 Å), to determine

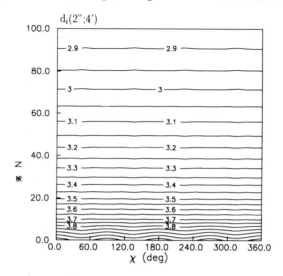

Figure 26. Contours of $d_i(2''; 4')$ plotted as a function of N/S and χ. The average distance $d_i(2''; 4')$ is shown for an equilibrium between N- and S-type sugar conformations, with $P_N = 10°$, $P_s = 165°$, and $\Phi_m = 35°$. Interconversion between N- and S-type sugars is assumed to be slower than the molecular tumbling time τ_c, in which case NOEs are averaged according to eqn 5, Section 5.1. The average distance $d_i(2''; 4')$ is, therefore, calculated according to $d_i(2''; 4') = \{(\%N/(d_i(2''; 4')_N)^6) + ((1 - \%N)/(d_i(2''; 4')_s)^6)\}^{-1/6}$. The N- as well as the S-type sugar are assumed to have the same χ values.

χ more precisely than $\pm 50°$. Rather, $d_i(6/8;1')$ may be used to establish its global range, i.e. whether χ is in the *syn-* or *anti-*domain.

The combination of the distances $d_i(6/8;2')$ and $d_i(6/8;2'')$ do, however, provide a good measure for the value of the glycosidic torsion angle. These distances depend strongly on χ, in particular in the range $\chi = 180°$ to $300°$ and $\chi = 30°$ to $90°$, domains which comprise the χ-values most abundant in nucleic acids. Moreover, these distances hardly depend on P for either N-type or S-type sugars (*Figure A1a* and b).

As mentioned in Section 5.3.3, it is possible to constrain the phase angle of pseudorotation, P, when $d_i(3';6/8)$ can be estimated. In fact the combination of $d_i(6/8;2'/2'')$ and $d_i(3';6/8)$ serves to constrain both the value of P and that of χ. An example is presented in *Figure 28* in which, for a practical situation, (P,χ) combinations are sketched allowing for an uncertainty in the distances $d_i(6;2'/2'')$ and $d_i(6;3')$ of ± 0.4 Å. It is seen that χ is better constrained than P. Note that in this calculation a value $\Phi_m = 35°$ was assumed.

When an N/S equilibrium exists, an analogous figure can be drawn, with N/S replacing P. *Figure 29* thus illustrates that it is possible to determine both the N/S ratio and χ from these distance constraints, but with limited accuracy in so far as the value of N/S is concerned. Taking a reasonable value for the

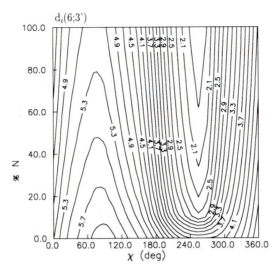

Figure 27. Contours of $d_i(6; 3')$ plotted as a function of N/S and χ. The averaged distance $d_i(6; 3')$ was calculated along the lines indicated in the legend to *Figure 26* for an equilibrium between N- and S-type sugar conformations. The same value for torsion angle χ is assumed for the N- and S-type sugar. In practice χ-*anti* is about 206° in N-type sugars and about 260° in S-type sugars, and when this difference is taken into account the contour plot does not change significantly.

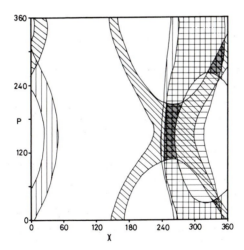

Figure 28. Combination of contours for $d_i(6; 3')$ and $d_i(6; 2'/2'')$ as a function of P and χ. Taken together these distances confine the conformation space of nucleosides. Shown is the situation as it would occur for the canonical B-conformation using uncertainties of ± 0.2 Å in the distances, $d_i(6; 2') = 2$ Å (*horizontal shading*), $d_i(6; 2'') = 3.5$ Å (*vertical shading*), and $d_i(6; 3') = 4$ Å (*diagonal shading*). The *dark region* represents the allowed values for P and χ. (Copyright, Federation of European Biochemical Societies.)

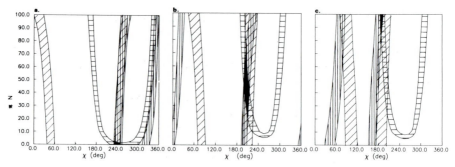

Figure 29. Combination of contours of $d_i(6; 2'/2'')$ and $d_i(6; 3')$ as a function of N/S and χ. Average distances were calculated as indicated in the legends to *Figures 26* and *27*. (a) Contours typical for B-type DNA distances. The upper and lower bounds for the distances were set at $+ 0.2$ Å and $- 0.2$ Å from the median distances, $d_i(2'; 6) = 2.2$ Å (*vertical shading*), $d_i(2''; 6) = 3.6$ Å (*diagonal shading*), and $d_i(3'; 6) = 3.9$ Å (*horizontal shading*). The *dark region* represents the area to which N/S and χ are constrained (this applies also to (b) and (e)). As can be seen the allowed values for P and χ both lie in a rather narrow range. (b) Contours for average distances typical for a 1:1 ratio of A- and B-type DNA. As in (a) the distance ranges are ± 0.2 Å. The average distances used are $d_i(2'; 6) = 3.1$ Å (*vertical shading*), $d_i(2''; 6) = 4.2$ Å (*diagonal shading*), and $d_i(3'; 6) = 3.0$ Å (*horizontal shading*). As can be seen these distances do still determine N/S ratio albeit with limited accuracy (20 to 80%), but the torsion angle χ remains well determined (200°–220°). (c) Contours typical for A-DNA distances; distances used are, $d_i(2'; 6) = 4.0$ Å (*vertical shading*), $d_i(2''; 6) = 4.5$ Å (*diagonal shading*), and $d_i(3'; 6) = 3.0$ Å (*horizontal shading*), with distance ranges of ± 0.2 Å. As can be seen the allowed values of both torsion angle χ and N/S ratio lie in a rather narrow region.

error in the distance determination, ± 0.2 Å, the angle χ can be determined from these distances with an uncertainty of $\pm 10°$. The uncertainty in the N/S ratio depends on the value of the N/S ratio; the boundaries of the nearly pure N- and S-conformations can be defined quite well, but for average distances corresponding to a 1:1 ratio, the uncertainty in the estimated value of N/S may increase to $\pm 30\%$. We note in passing that knowledge of the other intranucleotide distances does not help to further limit the range of values for χ and N/S. An uncertainty in the distance larger than ± 0.2 Å degrades the possibility to establish the value of N/S; an uncertainty of ± 0.5 Å makes it virtually impossible to determine N/S, except in the case of an almost pure S-type sugar.

If, however, the N/S ratio is available from J-coupling analysis the torsion angle χ can be determined with reasonable accuracy from $d_i(6/8;3')$ and/or $d_i(6/8;2'/2'')$.

5.5 The backbone angles

5.5.1 Conformation around O5'-C5'

The value of the torsion angle $\beta(P\text{-}O5'\text{-}C5'\text{-}C4')$ can be determined by making use of the coupling constants $J_{H5'-31P}$, $J_{H5''-31P}$ and $J_{13C4'-31P}$. The Karplus equations valid for these coupling constants are summarized in *Table 4* (42–44). The lowest energy conformation is the β^t form which is observed in regular double helices of RNA and DNA. For the β^t form a relatively large value, ≈ 10.9 Hz, for $J_{13C4'-31P}$ will be observed. Note that in the right-handed helix both β and ε are in the *trans*-domain; this leads to comparable splittings of the C4′ signal by both the $J_{13C4'-31P5'}$ and $J_{13C4'-31P3'}$ coupling constants, and it may be difficult to decide which is which.

Deviations from the ideal $\beta^t = 180°$ are indicated by the inequivalence of the values of $J_{H5'-31P}$ and $J_{H5''-31P}$. For the β^t form the values of the of $J_{H5'-31P}$ and $J_{H5''-31P}$ coupling constants are rather small, i.e. ≈ 3 Hz (see *Figure 30*). When β adopts a g^+ or a g^- conformation these coupling constants may increase to about 23 Hz. This will be manifest in 1D ^1H spectra and in the cross-peaks in, for instance, double quantum filtered COSY spectra if the H5′ and H5″ signals are well enough resolved. To be able to discriminate between the presence of β-rotamers with g^+ or g^- conformations, it is necessary that stereospecific assignments of the H5′ and H5″ resonances are available; this can be achieved following the procedure outlined in Section 4.2.3. If all three staggered conformers contribute to the observed J-couplings the percentage of the β^t conformer can be estimated with the aid of eqn 11 (42):

$$f_t = (25.5 - J_{H5'-31P} - J_{H5''-31P})/20.5 \qquad [11]$$

Information about β can also be obtained from the four bond coupling constant $J_{H4'-31P}$. If the atoms P-O5′-C5′-C4′-H4′ lie in the same plane and form a W-shaped conformation, $J_{H4'-31P}$ couplings of about 3 Hz will be

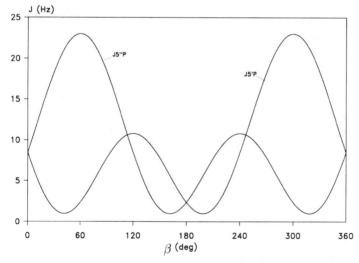

Figure 30. J-coupling constants calculated for $J_{H5'-31P}$ and $J_{H5''-31P}$ as a function of the torsion angle β. The coupling constants were calculated by means of the Karplus equations given for $^1H-^{31}P$ couplings (*Table 4*).

271

observed. This is the case if β is in the *t*- and γ is in the g^+-domain as is observed in regular B-type helices. In practice this information will only be available when the H4′ resonances are well enough resolved from neighbouring resonances.

5.5.2 Conformation around C5′-C4′

The torsion angle γ(O5′-C5′-C4′-C3′) may occur in the three staggered conformations g^+, g^-, and *t*, although the g^- conformation is much the least abundant. As discussed in Section 4.2.3, the stereospecific assignment of the H5′ and H5″ resonances is intimately interwoven with the determination of γ and, provided that backbone conformation is rigid, the determination of one leads to the other. That is to say it follows which staggered conformation about C5′-C4′ is present. To obtain precise values for this torsion angle the coupling constants $J_{4'5'}$ and $J_{4'5''}$ must be accurately known.

If for some reasons H5′ and H5″ can not be stereospecifically assigned, or if a rapid equilibrium exists between the three staggered conformations, then the sum of the coupling constants $\Sigma 4' = J_{4'5'} + J_{4'5''}$ can be used to estimate the percentage of the population present in the g^+ form with the help of the empirical relationship (39):

$$f_{g+} = (13.3 - \Sigma 4')/9.7 \qquad [12]$$

It follows from this equation that if $\Sigma 4'$ exceeds 12.5 Hz the dominant rotamer is either γ^- or γ^t, and if $\Sigma 4'$ is smaller than 4 Hz it is γ^+. Precise determination of γ from $\Sigma 4'$ in the γ^+ domain is not feasible, however, because the variation of $\Sigma 4'$ as a function of γ is too slight in the region $40° < \gamma < 60°$ (39).

If one can use Remin and Shugar's rule (compare with Section 4.1.2), the population of the different γ rotamers can be estimated from the measured J-couplings using the equations (44):

$$J_{4'5'} = f_{g+} J'_{g+} + f_t J'_t + f_{g-} J'_{g-} \qquad [13a]$$

$$J_{4'5''} = f_{g+} J''_{g+} + f_t J''_t + f_{g-} J''_{g-} \qquad [13b]$$

$$f_{g+} + f_t + f_{g-} = 1 \qquad [13c]$$

The f_i indicate the fractions of the different staggered rotamers; the primed J'_i represent the $J_{4'5'}$ in the different staggered conformers, the double primed J''_i the $J_{4'5''}$ in these conformers.

As followed from the discussion on the stereospecific assignment of the H5′ and H5″ resonances, Overhauser effects may also be used in the determination of γ. The $d_i(4';5'/5'')$ and $d_i(3';5'/5'')$ distances depend on this torsion angle as indicated in *Figure 16* and can thus be used to establish γ. In addition, $d_i(6/8;5')$ and $d_i(6/8;5'')$ should be considered. These distances do not strongly depend on the values of P and χ, provided that the latter parameters adopt their standard values, but are essentially determined by γ. For instance, as can be gleaned from *Figure 31A* and *B*, when γ falls in the g^+ domain $d_i(5'';6)$ = 4.3 Å both for P = 10° and χ = 180°, and for P = 165° and χ = 240°. When γ falls in the *trans*-domain $d_i(5'';6)$ = 2.5 Å for the same values of P and χ. Thus $d_i(5';6/8)$ and $d_i(5'';6/8)$ can be used to investigate γ even if P and χ are not exactly known.

If an equilibrium exists between the staggered conformations of γ^+ and γ^t (from here on we assume that the γ^- conformation can be disregarded) average distances can be calculated for the afore-mentioned proton pairs. In *Figure 31*, the average distances of $d_i(5'/5'';6)$ and $d_i(3';5'/5'')$ are shown as a function of the fraction γ^+ and P or χ. For standard values of P and χ, if (averaged) NOEs indicate that either $d_i(5';6)$ or $d_i(5'';6)$ (averaged values) is shorter than 3.0 Å, this can only be the distance between H5'' and H6. Thus, a stereospecific assignment can then be made for the H5' and H5'' resonances.

Is it possible to derive the fraction of the γ^+ conformation from observed NOE effects, given that these lead to ± 0.2 Å uncertainties in the distance determination? We first notice that this is not possible from averaged $d_i(3';5'5'')$, since these distances are not sufficiently sensitive to changes in the γ^+ fraction (*Figure 31A*). A similar conclusion applies when only one of the distances $d_i(5';6)$ *or* $d_i(5'';6)$ can be deduced. Only when $d_i(5'';6)$ is large (4.1 ± 0.2 Å) is the γ^+ fraction confined to the 95–100% region. In all other situations the fraction of γ^+ can not be established.

On the other hand, when both averaged distances $d_i(5';6)$ and $d_i(5'';6)$ can be determined the situation is more favourable. This is best illustrated with an example. If we assume that $d_i(5';6)$ = 4.2 ± 0.2 Å and $d_i(5'';6)$ = 2.5 ± 0.2 Å and superimpose the plots of the left- and right-hand side of *Figure 31B*, we find that γ^+ is present between 0 and 15% of the time for both P = 10° and 165°, and for χ values between 170° and 260°. A similar situation is found for the other extreme, i.e. distances can be found that confine the γ^+ fraction close to 100%. Thus, it can be established from NOEs whether γ is either in the g^+ or the *trans*-domain (provided P and χ fall in their standard domains). When the distances $d_i(5';6)$ and $d_i(5'';6)$ do not match these conditions it is no longer possible to measure the fraction of γ^+ with any precision. For instance, if it is found that the average distances $d_i(5';6)$ = 3.9 ± 0.2 Å and $d_i(5'';6)$ = 2.9 ± 0.2 Å (assuming P = 10° or 165° and χ is covering the region 170° –260°) a fraction of γ^+ anywhere between 20–80% will be found to match these distances. Confining χ to the region 170°–200° does not significantly improve matters.

We finally note that if the fraction γ^+ is available from J-coupling data (*Figure 17*; eqn 13) and the value of χ is known, the distances $d_i(5';6)$ and/or $d_i(5'';6)$ constrain the value of P considerably. For example, for a fraction γ^+ = 50%, χ = 240° and $d_i(5';6)$ = 4.0 ± 0.2 Å we find P = 163 ± 12° or P = 11 ± 8°.

A.

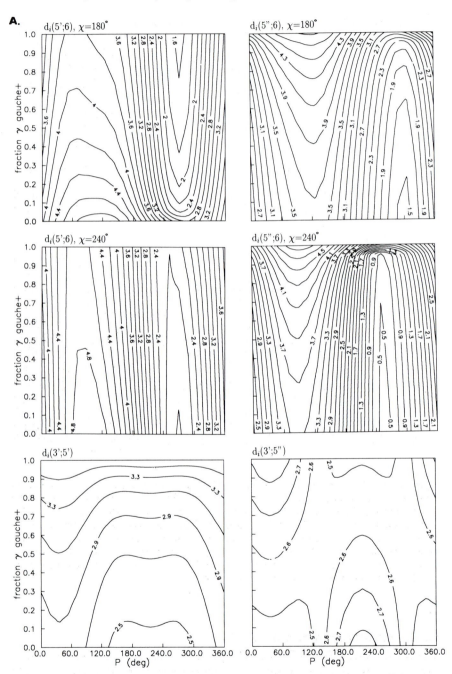

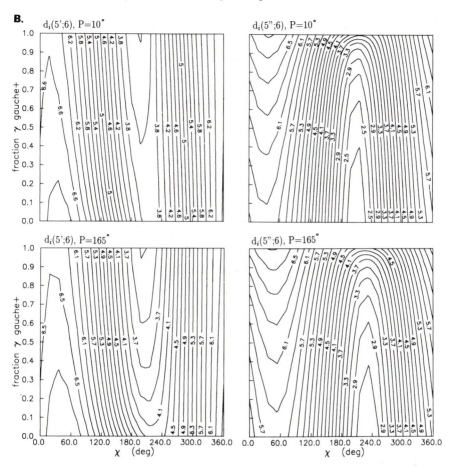

Figure 31. (A) Contours of $d_i(5'/5''; 6)$ and $d_i(3'; 5'/5'')$ plotted as a function of the fraction γ *gauche+* and the sugar pucker P (and in case of $d_i(5'/5''; 6)$ for different values of the torsion angle χ). The average distances $d_i(5'/5''; 6)$ and $d_i (3'; 5'/5'')$ were calculated as indicated in the legend to *Figure 26* for a rapid interconversion between the rotamers γ *gauche+* and γ *trans*. The γ *gauche−* population is assumed zero as it is found in practice to be hardly populated. (B) Contours of $d_i(5'/5''; 6)$ plotted as a function of the fraction γ *gauche+* and the torsion angle χ, for P = 10° and P = 165°, assuming a rapid interconversion between the rotamers γ *gauche+* and γ *trans*.

5.5.3 Conformation around C4′-C3′

The torsion angle δ(C5′-C4′-C3′-O3′) is determined by the pseudorotation parameters of the furanose ring. Once these have been established δ can be calculated with the aid of eqn 1 and 2; δ varies from 82° to 89° for N-type sugars and from 137° to 157° for S-type sugars.

5.5.4 Conformation around C3′-O3′

The torsion angle ε (C4′-C3′-O3′-P) can be monitored by three J-couplings, JH3′-^{31}P, J^{13}C2′-^{31}P and J^{13}C4′^{31}P. The Karplus equations applicable to these heteronuclear couplings are presented in *Table 4*; the most recent parametrization is given (43). As for β and γ one might consider the three staggered conformations g^+, g^-, and t in the conformational analysis of ε. It is usually assumed, however, that the g^+ conformation can be disregarded. This conformation has not been found in X-ray diffraction studies of nucleic acids, and energy calculations of oligonucleotides also support this conclusion (39, 45). There are experimental as well as theoretical indications that the value of ε in a nucleotide is correlated with the conformation of the sugar moiety. For N-type sugar conformations ε tends to adopt a *trans* conformation in DNA as well as in RNA nucleotides; the $ε^-$, N combination seems to be forbidden. For S-type sugars both $ε^-$ and $ε^t$ conformers are found. Furthermore, NMR experiment have indicated that the values for $ε^t$,S and $ε^t$,N may differ from one another ($ε^t$,S ≈ 192° and $ε^t$,N ≈ 212° (46)).

6. Determination of the overall structure of nucleic acids

An understanding of the three-dimensional folding in nucleic acids, of their sequence-dependent structure, and of the mechanism of interaction of double and single stranded nucleic acids with proteins and drug molecules requires an insight in the finer details of nucleic acid structure. Therefore attempts have been made to proceed a step further and use the information discussed in Section 5, extended with so-called holonomic constraints such as bond lengths and bond angles, to derive the total structure of nucleic acid molecules in detail.

For the computation of the structure of biomolecules from NMR constraints different methods have been developed, as discussed in Chapter 11:

- restrained molecular dynamics methods (rMD) (47–51) for which different programs are available, (e.g. QUANTA, XPLOR, BIOGRAPH, DISCOVER, GROMOS, AMBER)
- distance space methods (DG) which use the embed algorithm (52)
- programs which use this algorithm are DISGEO and D-SPACE (52)
- variable target algorithms which work in torsion angle space; examples of programs which use this algorithm are DISMAN and DIANA (9, 53)

It is also possible in special circumstances to apply grid search procedures to derive structures from NMR constraints. The advantage of this approach is that the sampling of the conformational space allowed by the NMR constraints can be well defined, that it is exhaustive and that it is transparent, in the sense that it is clear to what extent the NMR constraints determine the structure. As an example we mention the derivation of the loop structure of the hairpin formed by the partly self-complementary sequence d(ATCCTA-TTTA-TAGGAT). In the grid search procedure (program HELIX; 27) conformers of the loop region, first for dinucleotides, then for trinucleotides, and finally for the tetranucleotide, are generated by systematically varying all torsion angles in all possible combinations. Only those conformers which fulfil the NMR constraints are retained. An example of the family of trinucleotide structures that is obtained in this manner is presented in *Figure 32*. A disadvantage of the method is that it is very demanding in terms of computer time, in particular when the step size in the sampling procedure is small. In practice, this approach will therefore be restricted to small oligonucleotides, e.g. four nucleotides in the mentioned example.

In this respect the afore-mentioned algorithms are much more powerful, but their sampling properties are less transparent, and in case of the rMD applications the force field could play an influential role in the final structure determination. For proteins, these methods have been extensively tested and are believed to yield reliable results. For nucleic acids the situation is somewhat less advanced, particularly because most of the nucleic acid molecules studied by NMR do not have the globular structures found for proteins. Thus, in double helices no long-range distance constraints are available from NMR, making the definition of the overall structure difficult. Moreover, since the overall double helical conformation is often known, the questions asked

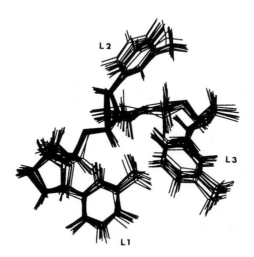

Figure 32. Superposition of twelve representative structures of the L1-L2-L3 segment of the hairpin d5'($A_1T_2C_3C_4T_5A_6$-$T_{L1}T_{L2}T_{L3}A_{L4}$-$T_6A_5G_4G_3A_2T_1$)3' obtained with the aid of a conformational analysis algorithm which makes use of a grid search procedure. The root mean square deviation between the structures was 0.5 Å. (Blommers *et al.* (1991). *Eur. J. Biochem.*, **201**, 33; Copyright Federation of European Biochemical Societies.)

about nucleic acids are concerned with the finer details of the structure putting higher demands on the performance of these algorithms.

6.1 Restrained molecular dynamics

The approach used by different investigators (32, 33, 47–50) to determine nucleic acid structures from NMR constraints by means of rMD are quite similar. To the force field used in the rMD program the NMR constraints, from NOES and J-couplings, are added as pseudo-potentials, commonly square-well potentials (see Chapter 11).

The protocol used for the structure determination of nucleic acids by means of rMD differs somewhat among different investigators, but common features can be recognized (see also Chapter 11).

(a) At the start of the calculations the NMR constraints are kept at low values and during the first 10 psec of the rMD run they are gradually increased to their final values. The values used for the final constraint force constants varies from 25 to 200 kcal/mol.Å^2. The temperature at which the influence of the constraint force constants is introduced is an additional variable. Some groups increase the values of the NMR constraints when the 'sample' has been brought to higher temperatures (e.g. 400 K), while others raise the temperature (e.g. to 600 K) after the final values of the NMR constraints have been introduced.

(b) After equilibration at the high temperature the 'sample' is cooled down to 300 K, often slowly (4 psec), to avoid trapping in local minima, and restrained MD is continued, during 10 to 20 psec, with the maximum constraint force constants. The coordinates of the structures generated during the last 5 or 6 psec are averaged to yield the final rMD structures, which are sometimes subjected to a further round of energy minimization.

An example of a result obtained with the IRMA procedure is shown in *Figure 33*. Two superimposed structures are presented for the double helix formed by the self-complementary sequence d(GCGATCGC). For one structure, B-DNA was used as a starting structure in the iterative relaxation approach, for the other structure A-DNA was used as a starting structure. The effect of fast local motion (internal mobility) was included in the calculations via the introduction of generalized correction factors (see Section 5.1; reference 37).

6.2 The embed algorithm

The embed algorithm, which was originally designated distance geometry algorithm (DG)(52), starts from a distance bounds matrix containing upper and lower bounds of the holonomic distance constraints together with the NMR distance constraints, smoothed by the application of triangle inequalities. The algorithm generates different distance matrices by choosing random

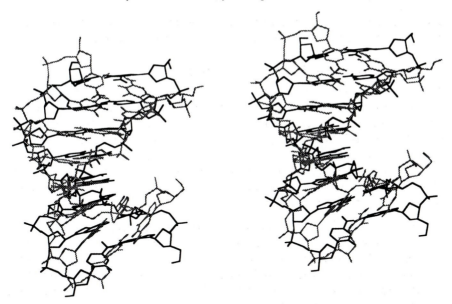

Figure 33. Stereopair of two superimposed structures of the octamer d(GCGATCGC). (GCGATCGC), obtained by application of the IRMA procedure. The structure drawn in *solid lines* represents the final structure derived from a B-DNA starting structure, the structure drawn in *dotted lines* represents the one derived from A-DNA. (T. M. G. Koning. (1991). Thesis, University of Utrecht, with permission.)

distances between the lower and upper bounds for each matrix element. Each of these matrices are then embedded in three-dimensional space to yield starting structures which will generally no longer satisfy the distance constraints. These structures are therefore refined (by conjugate gradient methods) to match these constraints. Experience gained so far shows that a family of structures is obtained with fairly small root mean square deviations.

In the earlier studies performed in this way the input parameters were in essence derived by the isolated spin pair approximation. Recent investigations have shown, however, that it is necessary to refine the DG structures by back-calculations of the NOESY spectra and/or simulated annealing (54). Sometimes the last step in the iterative back-calculations is followed by restrained molecular dynamics calculations, which are performed to search for additional energy minima (54).

Comparison of the different procedures used by different investigators shows that there is considerable similarity in their approaches. Some authors make use of NMR distance constraints (determined for example by a method such as MARDIGRAS or NO2DI (see Section 5.1)) and subject these constraints afterwards to rMD to obtain a nucleic acid structure (33), while others combine the derivation of the distance constraints with rMD (32) or DG calculations (54) in an iterative manner. This approach yields at the same time distance constraints and structures satisfying these distance constraints; examples are the IRMA procedure (see Section 5.1; 32) and the DG calculations described above (54). We mention that, to our knowledge, the variable target algorithm has so far not been used in nucleic acid structure determination.

6.3 Conclusions

At this point we return to the question posed in the beginning of this section, namely, to what extent have these methods been successful in providing detailed structural information on nucleic acids. In trying to answer this question a few aspects of the applied procedures should be considered. It has been mentioned already, that the sampling properties of the distance geometry algorithms are not particularly transparent. This is exemplified by the recent demonstration that due to its sampling properties the standard embed algorithm leads to extended structures (55). These results may well explain the occurrence of the ladder-like DNA structures found after the embedding step in the DG calculations. One has tried to remedy this problem by borrowing techniques (simulated annealing) from MD. It has also been shown that it may be worthwhile to conduct quenched dynamics calculations after the iterative DG calculations, involving the NOE back-calculations, have been finished. This suggests that the MD techniques may have more favourable sampling properties.

The use of rMD in the structure derivation of nucleic acids has raised the question whether and to what extent the resulting structures are determined by the NMR constraints or by the MD force field. The issue is of particular importance since the aim of various NMR studies is to unravel the relation between nucleotide sequence and local helical structure in DNA. The matter is complicated by the fact that the NMR data themselves are sequence-dependent, e.g. depending on sequences cytosine H5 and adenine H2 connectivities contribute or do not contribute to the definition of the helix structure.

In the present discussion it has been tacitly assumed that the nucleic acid molecule in solution can be characterized by 'a' structure. This only valid if the molecule samples a small conformational space. Conformational mobility of nucleic acids, involving large displacements of the atoms from their equilibrium position, has been documented, however, in the literature. In the foregoing sections the influence of internal mobility on the measured values of some of the distance parameters and torsion angle constraints has been discussed. The question how the relaxation matrix approach or related methods, discussed in the above, perform in the presence of internal motion has only been touched upon in the literature. These topics are still a matter of debate as well as of ongoing research.

References

1. van de Ven, F. J. M. and Hilbers, C. W. (1988). *Eur. J. Biochem.*, **178**, 1.
2. IUPAC/IUB JCBN (1983). *Eur. J. Biochem.*, **131**, 9.
3. Saenger, W. (1984). *Principles of nucleic acid structure.* Springer Verlag, Heidelberg.
4. Diekman, S. (1989). *EMBO J.*, **8**, 1.
5. Altona, C. and Sundaralingam, M. (1972). *J. Am. Chem. Soc.*, **94**, 8205.
6. Altona, C., Geise, H. J., and Romers, C. (1968). *Tetrahedron*, **24**, 13.
7. de Leeuw, H. P. M., Haasnoot, C. A. G., and Altona, C. (1980). *Isr. J. Chem.*, **20**, 108.
8. Haasnoot, C. A. G., de Leeuw, F. A. A. M., de Leeuw, H. P. M., and Altona, C. (1981). *Org. Magn. Reson.*, **15**, 43.
9. Wüthrich, K. (1986). *NMR of proteins and nucleic acids.* Wiley, New York.
10. Leroy, J. L., Broseta, D., and Gueron, M. (1985). *J. Mol. Biol.*, **184**, 165.
11. Leroy, J. L., Bolo, N., Figueroa, N., Plateau, P., and Gueron, M. (1985). *J. Biomol. Struct. Dyn.*, **2**, 915.
12. Hore, P. J. (1985). *J. Magn. Reson.*, **64**, 38.
13. Morris, G. A. and Freeman, R. (1978). *J. Magn. Reson.*, **29**, 433.
14. Haasnoot, C. A. G. and Hilbers, C. W. (1983). *Biopolymers*, **22**, 1259.
15. Moonen, C. T. W. and van Zijl, P. C. M. (1990). *J. Magn. Reson.*, **88**, 28.
16. Neuhaus, D. and Williamson, M. (1989). *The nuclear Overhauser effect in structural and conformational analysis.* VCH Publishers Inc., New York.
17. Ernst, R. R., Bodenhausen, G., and Wokaun, A. (1987). *Principles of nuclear magnetic resonance in one and two dimensions.* Clarendon Press, Oxford.
18. Massefski, W. Jr. and Redfield, A. G. (1988). *J. Magn. Reson.*, **78**, 150.
19. Kessler, H., Gehrke, M., and Griesinger, C. (1988). *Angew. Chem. Int. Ed. Engl.*, **27**, 490, and references therein.
20. Hosur, R. V. (1990). *Prog. NMR Spectrosc.*, **22**, 1. Rance, M., Wagner, G., Sørensen, O. W., Wüthrich, K., and Ernst, R. R. (1984). *J. Magn. Reson.*, **59**, 250.
21. Chazin, W. J., Wüthrich, K., Hyberts, S., Rance, M., Denny, W. A., and Leupin, W. (1986). *J. Mol. Biol.*, **190**, 439.
22. Cavanagh, J., Chazin, W. J., and Rance, M. (1990). *J. Magn. Reson.*, **87**, 110.
23. Sklenar, V., Miyashiro, H., Zon, G., Miles, H. T., and Bax, A. (1986). *FEBS Lett.*, **208**, 94.
24. Giessner-Prettre, C. and Pullman, B. (1987). *Q. Rev. Biophys.*, **20**, 113, and references therein.
25. Hilbers, C. W. (1979). In *Biological applications of magnetic resonance* (ed. R. G. Shulman). Academic Press, New York.
26. Remin, M. and Shugar, D. (1972). *Biochem. Biophys. Res. Commun.*, **48**, 636.
27. Blommers, M. J. J., van de Ven, F. J. M., van der Marel, G. A., van Boom, J. H., and Hilbers, C. W. (1991). *Eur. J. Biochem.*, **201**, 33.
28. van de Ven, F. J. M. and Hilbers, C. W. (1988). *Nucleic Acids Res.*, **16**, 5713.
29. Gorenstein, D. G. (1981). *Annu. Rev. Biophys. Bioeng.*, **10**, 355.
30. Mooren, M. M. W., Hilbers, C. W., van der Marel, G. A., van Boom, J. H., and Wijmenga, S. S. (1991). *J. Magn. Reson.*, **94**, 101.

31. Boelens, R., Vuister, G. W., Koning, T. M. G., and Kaptein, R. (1989). *J. Am. Chem. Soc.*, **111**, 8525.
32. Boelens, R., Koning, T. M. G., and Kaptein, R. (1988). *J. Mol. Struct.*, **173**, 299. Boelens, R., Koning, T. M. G., van der Marel, G. A., van Boom, J. H., and Kaptein, R. (1989). *J. Magn. Reson.*, **82**, 290. Post, C. B., Meadows, R. P., and Gorenstein, D. G. (1990). *J. Am. Chem. Soc.*, **112**, 6796.
33. Gochin, M. and James, T. L. (1990). *Biochemistry*, **29**, 11172. Gochin, M. and James, T. L. (1990). *Biochemistry*, **29**, 11161. Borgias, B. A. and James, T. L. (1990). *J. Magn. Reson.*, **87**, 475.
34. van de Ven, F. J. M., Schouten, R., Blommers, M. J. J., and Hilbers, C. W. (1991). *J. Magn. Reson.*, **94**, 140.
35. Koehl, P. and Lefevre, J.-F. (1990). *J. Magn. Reson.*, **86**, 565.
36. Tropp, J. (1980). *J. Chem. Phys.*, **72**, 6035.
37. Lipari, G. and Szabo, A. (1982). *J. Am. Chem. Soc.*, **104**, 4546.
38. Haasnoot, C. A. G., de Leeuw, F. A. A. M., and Altona, C. (1980). *Tetrahedron*, **36**, 2783.
39. Altona, C. (1982). *Recl. Trav. Chim. Pays Bas*, **101**, 413.
40. Rinkel, L. J. and Altona, C. (1987). *J. Biomolec. Struct. Dyn.*, **4**, 621.
41. Widmer, H. and Wüthrich, K. (1987). *J. Magn. Reson.*, **74**, 316.
42. Lankhorst, P. P., Haasnoot, C. A. G., Erkelens, C., and Altona, C. (1984). *J. Biomolec. Struct. Dyn.*, **1**, 1387.
43. Mooren, M. M. W., van der Marel, G. A., van Boom, J. H., and Hilbers, C. W. *Biochemistry*, submitted.
44. Haasnoot, C. A. G., de Leeuw, F. A. A. M., de Leeuw, H. P. M., and Altona, C. (1979). *Recl. Trav. Chim. Pays Bas*, **98**, 576.
45. Pearlman, D. A. and Kim, S.-H. (1986). *J. Biomolec. Struct. Dyn.*, **4**, 49.
46. Lankhorst, P. P. (1984). Thesis, State University Leiden, Leiden, The Netherlands.
47. Scheek, R. M., van Gunsteren, W. F., and Kaptein, R. (1989). In *Methods in Enzymology* (ed. N. J. Oppenheimer and T. L. James), Vol. 177 B, pp. 204–18. Academic Press, San Diego.
48. Clore, G. M., Oschkinat, H., McLaughlin, L. W., Benseler, F., Scalfi Happ, C., Happ, E., and Gronenborn, A. M. (1988). *Biochemistry*, **27**, 4185.
49. Gronenborn, A. M. and Clore, G. M. (1989). *Biochemistry*, **28**, 5978.
50. Nillson, L., Clore, G. M., Gronenborn, A. M., Brunger, A. T., and Karplus, M. (1986). *J. Mol. Biol.*, **188**, 455.
51. Van Gunsteren, W. F. and Berendsen, H. J. (1990). *Angew. Chem.*, **102**, 1020.
52. Kuntz, I. D., Thomason, J. F., and Oshiro, C. M. (1989). In *Methods in Enzymology* (ed. N. J. Oppenheimer and T. L. James), Vol. 177 B, pp. 159–204. Academic Press, San Diego.
53. Wüthrich, K. (1989). In *Methods in Enzymology* (ed. N. J. Oppenheimer and T. L. James), Vol. 177 B, pp. 125–31. Academic Press, San Diego.
54. Pardi, A., Hare, D., and Wang, C. (1988). *Proc. Natl Acad. Sci. U.S.A.*, **85**, 8785. Banks, K. M., Hare, D. R., and Reid, B. R. (1989). *Biochemistry*, **28**, 6996. Metzler, W. J., Wang, C., Kitchen, D. B., Levy, R. M., and Pardi, A. (1990). *J. Mol. Biol.*, **214**, 711.
55. Havel, T. (1990). *Biopolymers*, **29**, 1565.
56. Griesinger, C., Otting, G., Wüthrich, K., and Ernst, R. R. (1988). *J. Am. Chem. Soc.*, **110**, 7870.

57. Macaya, R. F., and Feigon, J. (1992). *J. Am. Chem. Soc.*, **114**, 781.
58. Nikonowicz, E. P. and Gorenstein, D. G. (1990). *Biochemistry*, **29**, 8845.
59. Mooren, M. W., Pulleyblank, D. E., Wijmenga, S. S., Blommers, M. J. J., and Hilbers, C. W. (1990). *Nucleic. Acids Res.*, **18** (22), 6523.
60. Bauer, C. J., Frenkiel, T. A., and Lane, A. (1990). *J. Magn. Reson.*, **87**, 144.
61. Roongta, V. A., Jones, C. R., and Gorenstein, D. G. (1990). *Biochemistry*, **29**, 5245.
62. Rajagopal, P. and Feigon, J. (1989). *Biochemistry*, **28**, 7859.

Appendix

In this appendix we summarize, in the form of a set of contour plots, *Figure A1*, distances which are important in defining the structure of a mono-nucleotide. Furthermore, a set of schemes is introduced in which short-range internucleotide distances in A-, B-, and Z-DNA, as well as RNA helices are indicated which are important for NMR structural studies of nucleic acids. For completeness sake we also list here short important proton–proton distances, that are not shown in *Figures A1* or *A2*: $d_i(2';2'') = 1.8$ Å, $d_i(5';5'') = 1.8$ Å, $d_i(A\text{-}H2;A\text{-}NH_26) = 4.4, 5.2$ Å, $d_i(A\text{-}NH6;A\text{-}NH6) = 1.7$ Å, $d_i(A\text{-}NH_26;A\text{-}H8) = 4.8, \underline{6.1}$ Å, $d_i(G\text{-}NH1;G\text{-}NH_2) = \underline{2.3}, 3.4$ Å, $d_i(C\text{-}NH_24;C\text{-}H5) = 2.4, \underline{3.6}$ Å, $d_i(C\text{-}NH_2H;C\text{-}H6) = 4.6, \underline{5.3}$ Å, $d_i(C\text{-}H5;C\text{-}H6) = 2.4$ Å, $d_i(T3;M) = 4.9$ Å, $d_i(TM;T\text{-}H6) = 2.9$ Å. The underlined distances are to the hydrogen-bonded proton in Watson–Crick base pairs.

Figure A1. (a–d) (overleaf) P-χ planes with contour lines indicating the intranucleotide distances of the deoxyribose ring protons H2', H2", H3', and H4' to purine H8 (a), to pyrimidine H6 (b), to cytosine H5 or thymine CH$_3$ (c), to adenine H2 (d). (e) (p. 286) P-γ planes with contour lines indicating the intranucleotide distances from H5' to the sugar ring protons H1', H2', H2" and H3'. The corresponding distances from H5" are obtained from these contour plots by $-120°$ shift of γ. (f) (p. 286) P-χ planes with contour lines indicating the intranucleotide distances to the pyrimidine H6 proton from H5' or H5" for the three staggered conformation of γ ($g^+ = 60'$, $g^- = -60°$ and $t = 180°$).

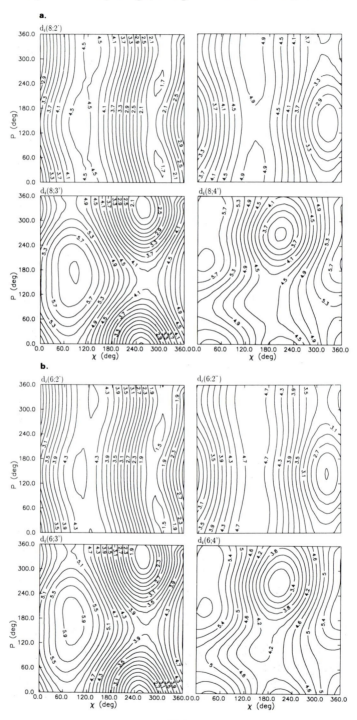

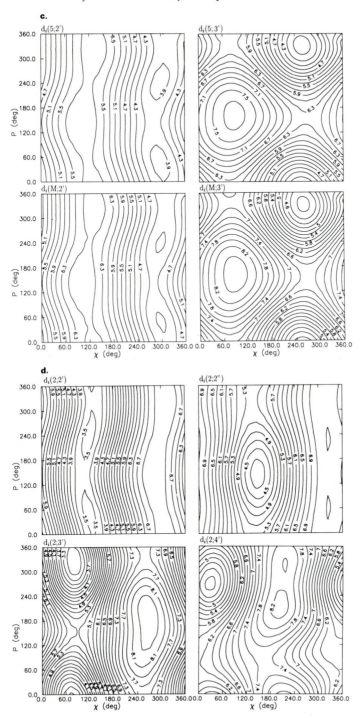

e.

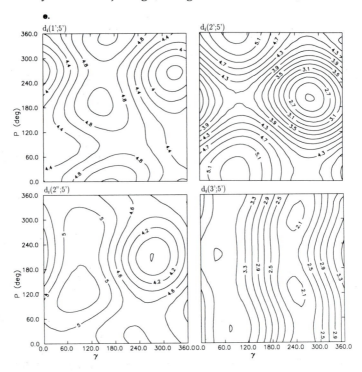

f.

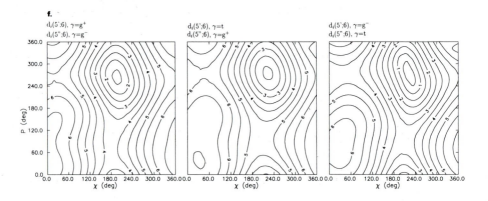

Figure A2. (following pages) Overview of short sequential and interstrand proton–proton distances for all possible combinations of base-base stacking in A-DNA (a), B-DNA (b), RNA (c), and Z-RNA (d). (a, b: page 287; c, d: page 288) The meaning of the symbols is: 0–2.5 Å (*thick solid line*), 2.5–3.0 Å (*solid line*), 3.0–4.0 Å (*dashed line*), 4.0–5.0 Å (*dotted line*).

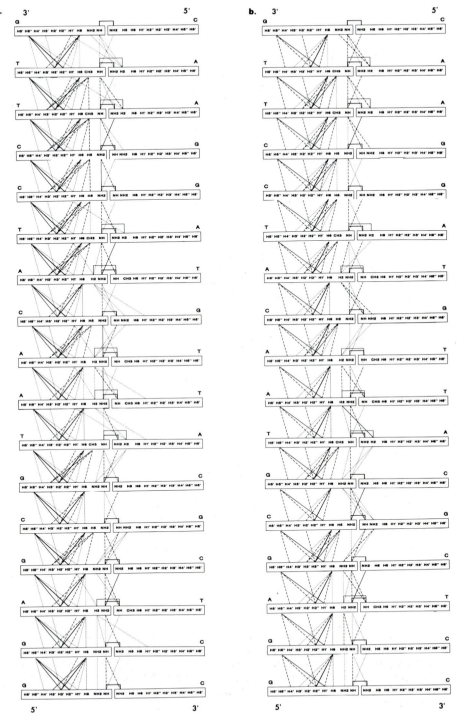

c.

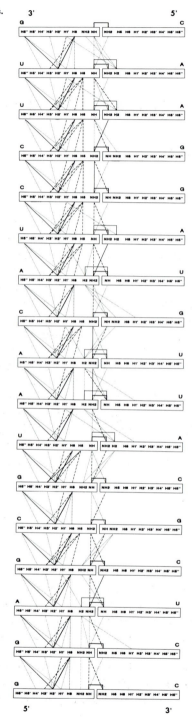

d.

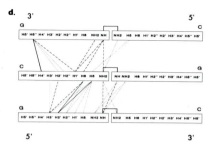

9

¹H NMR studies of oligosaccharides

S. W. HOMANS

1. Introduction

Proton NMR is a very valuable technique for the structural analysis of oligosaccharides since it is possible to determine primary sequence as well as three-dimensional structure. In view of their branched nature, there is at present no cyclical degradative chemical or enzymatic sequencing strategy for oligosaccharides which could potentially be automated as has been done for proteins and nucleic acids, and NMR is probably the most useful technique in this regard. Despite their relatively small size, oligosaccharides generally give remarkably complex proton NMR spectra. The primary reason for this is the poor spectral dispersion of ring protons, which presents problems due to severe resonance overlap and the increased probability of strong coupling effects. Resonance assignment strategies thus differ significantly from those used in studies of proteins and nucleic acids, although the experimental techniques are the same. Below we shall discuss the practical aspects of these techniques primarily by reference to a single compound, namely GlcNAcβ1–4GlcNAc. This disaccharide exhibits all of the idiosyncracies of oligosaccharide NMR spectra while at the same time generating spectra which are simple enough to allow casual interpretation by the non-specialist.

2. Sample preparation

2.1 Aqueous solvents

In most proton NMR studies to date, the oligosaccharide has been examined in aqueous solution (1). In general, the exchangeable protons are not of particular interest (see Section 4.2, however), and it is thus convenient to dissolve the sample in deuterium oxide. It is important to realize that the anomeric (C-1) proton resonances of certain monosaccharide residues resonate close to the residual ¹H²HO resonance, and hence it is important to reduce the intensity of this as far as possible. To this end, the sample should be repeatedly flash-evaporated from good quality (99.96%) deuterium oxide to exchange fully hydroxyl protons and bound water. For sample concentrations in the range 0.5–1 mM and above, it is sufficient to dissolve the sample in 99.96% ^2H_2O for analysis. If the sample concentration is in the range 200–500 μM, then dissolution in 99.996% ^2H_2O in a dry-box gives a significant reduction in the intensity of the $^1H^2HO$ resonance. If the sample concentration is lower than this, then it is necessary to flash-evaporate from 99.96% followed by 99.996% ^2H_2O, and then to dry the sample cryogenically, i.e. by suspension in an evacuated sealed chamber immersed in liquid nitrogen. The sample is then finally dissolved in 99.996% ^2H_2O. While it is possible to record proton NMR spectra of oligosaccharides at submillimolar concentrations without such detailed sample preparation, the additional effort yields a very worthwhile improvement in spectral quality.

In view of the fact that proton resonance linewidths of oligosaccharides are small in comparison with proteins and nucleic acids, it is possible to work with lower concentrations of material. In our laboratory, we routinely acquire COSY spectra at concentrations below 100 μM with workable signal-to-noise ratio. However, in order to achieve this it is often necessary to minimize the line-broadening effects of dissolved oxygen and paramagnetic impurities. The former can be removed by degassing the sample, or by dissolution in previously degassed ^2H_2O. The latter are most easily removed by passage through a small metal chelating column such as Chelex prior to flash-evaporation.

The use of buffer solutions is not generally necessary for neutral oligosaccharides. If the charge state is of no consequence, it is also unnecessary to buffer dilute solutions of oligosaccharides containing a charged group such as phosphate. However, certain glycosidic linkages such as α2–6 linked sialic acids are highly acid labile, and sialylated oligosaccharides should thus not be dissolved in unbuffered solutions.

2.2 Other solvents

In view of their hydrophilic nature, oligosaccharides are soluble in very few non-aqueous solvents, unless derivatized by, for example, permethylation. Indeed, the only non-aqueous solvent known to the author which has been in common use for NMR studies on underivatized oligosaccharides is dimethylsulfoxide (DMSO). This is a useful solvent since it is possible to observe protons (OH and NH) which otherwise exchange in deuterium oxide (2). The resonances from these protons can give useful information on intramolecular hydrogen bonding and moreover are valuable as an aid to primary sequence determination (Section 4.2). Unfortunately, DMSO is not a pleasant solvent, and is difficult to obtain free from impurities. For our studies we have found that deuterated DMSO commercially available at 99.96% purity is adequate. If the exchangeable protons are of interest, then again it is essential to dry the sample thoroughly prior to dissolution in DMSO, since the OH protons in particular will exchange rapidly with any traces of residual water.

2.3 Chemical shift references

In aqueous solutions, it appears to have become common practice to reference all chemical shifts indirectly to acetone, which has a shift of 2.225 p.p.m. relative to DSS at 25°C(1) With studies in DMSO solution, the residual DMSO resonance at 2.5 p.p.m. is generally used as a reference.

3. Assignment methods

3.1 Overall morphology of oligosaccharide NMR spectra

A typical oligosaccharide ¹H NMR spectrum is very complex, despite the fact that the number of proton resonances is generally much smaller than for protein or nucleic acid spectra (3). The primary reason for this complexity is the poor spectral dispersion of non-exchangeable protons. The anomeric (C-1) proton is an exception, and always resonates at low field due to the electron withdrawing properties of the ring oxygen. Due to mutarotation, the reducing termini of oligosaccharides in aqueous solution exhibit two low field resonances corresponding to the α and β anomers. These are easily distinguished from other anomeric proton resonances by virtue of their intensity, as shown in *Figure 1*.

3.2 Resonance assignment

The monosaccharide residues of an oligosaccharide in ²H₂O solution consist in general of seven non-exchangeable protons. The vast majority of these residues are in the pyranose form, and the non-exchangeable protons represent a seven-spin system (*Figure 2*). Proton resonance assignments invariably begin with the well resolved anomeric proton. In principle, it is possible to assign further each monosaccharide residue by stepwise J-correlation around the ring using experiments such as COSY (see below). However, the poor spectral dispersion in oligosaccharide proton NMR spectra invariably presents severe difficulties with this simple approach. This is due not only to the high probability of cross-peak overlap, but also to the presence of strong coupling. The assignment problem can thus only be tackled efficiently with a variety of multi-dimensional NMR methods.

Below we describe various techniques which have been effective in the simplification of the assignment problem. We shall not describe the theoretical details of these experiments, since several texts are now available which serve this purpose (4–6). Rather, we shall focus on those practical aspects which are important for the acquisition of high quality spectra. It should be mentioned at the outset that no single 'assignment protocol' can be defined for oligosaccharides, since the exact nature of the spin systems within each monosaccharide residue is highly dependent upon the structure. For

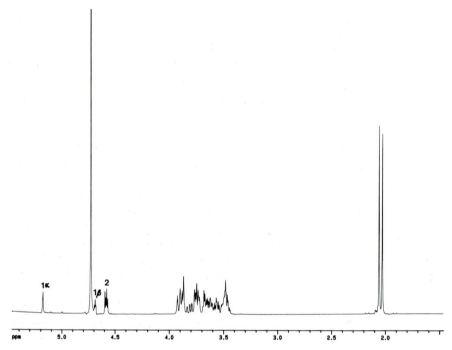

Figure 1. ^{1}H NMR spectrum of GlcNAcβ1–4GlcNAc. The H^1 resonances of –4GlcNAc (1) and GlcNAcβ1– (2) are labelled.

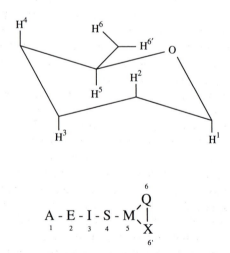

Figure 2. Diagrammatic representation of the non-exchangeable protons for a gluco-pyranoside. These form a spin system of the type AEISMQX, where the standard nomenclature implies that all protons are weakly coupled. However, various degrees of strong coupling are often found in the general case.

292

example, a monosaccharide residue which exhibits a weakly coupled spin system with good spectral dispersion in a given oligosaccharide may easily be assigned by use of COSY alone, whereas the same residue, when O-glycosidically substituted in a second oligosaccharide, may exhibit a much more complex strongly coupled spin system by virtue of the large chemical shift perturbations induced by the substitution, and may require an arsenal of techniques for full assignment.

3.2.1 Homonuclear ^{1}H-^{1}H correlated spectroscopy (COSY)

As in NMR studies on other macromolecules, the COSY experiment (7) is perhaps the most useful assignment aid. In the case of oligosaccharides, the value of COSY is that it is possible in principle to obtain sequential resonance assignments stepwise around the ring starting from the well resolved C-1 proton (H^1) of each residue (8). However, the effects of strong coupling and poor spectral dispersion are unfortunately still prevalent in oligosaccharide COSY spectra despite the second dimension. Complete resonance assignments are thus difficult to obtain and high quality spectra are vital in this regard.

The key to a successful result lies in the optimization of digital resolution. Obviously the block size within which data are acquired should be as large as practically possible, and it is important to maintain a spectral width which is as small as possible in each dimension. For this reason, resonances which are not coupled to ring protons, such as N-acetamido methyl groups should be aliased into a region of the spectrum which is clear of resonances. In the case of GlcNAcβ1–4GlcNAc, the spectral width can be approximately halved by this procedure, and the improvement in quality of the spectrum is seen in *Figure 3*. Those familiar with 2D NMR spectroscopy will probably have noted that the spectra in *Figure 3a* and *3b* were acquired in the 'absolute value' mode, whereas COSY spectra of proteins and nucleic acids are normally acquired in the pure phase (phase-sensitive) mode. There are good theoretical reasons for this such as improved resolution due to the absence of the 'phase twist' lineshape (9–12). However, resonance linewidths in oligosaccharide NMR spectra are much narrower than in proteins and nucleic acids, and good results can be obtained from 'magnitude mode' specta. Indeed, the author finds these in many respects superior to absorption mode spectra for assignment purposes, since it is easier to perform a mental 'cluster analysis' of cross-peaks from each monosaccharide residue. This preference is contrary to theoretical advantages (12), and is admittedly entirely subjective. For comparison, in *Figure 3c* we illustrate a pure absorption mode COSY spectrum of GlcNAcβ1–4GlcNAc which was recorded under otherwise identical conditions to *Figure 3b*. The reader can decide for himself which presentation most easily lends itself to interpretation.

Since resonance linewidths in oligosaccharide NMR spectra are relatively narrow as mentioned above, reasonable sensitivity is obtained even with the

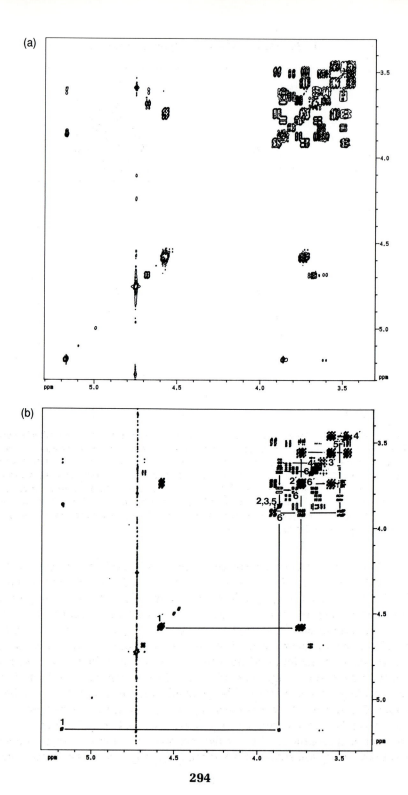

(c)

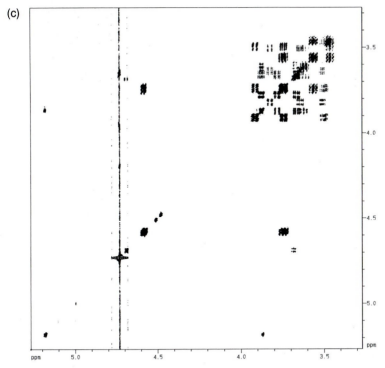

Figure 3. ^1H-^1H COSY spectra of GlcNAcβ1–4GlcNAc recorded (a) in magnitude mode with a sweep width of 2000 Hz in each dimension, and 1 K × 1 K real datapoints, (b) in magnitude mode with a sweep width of 1000 Hz in each dimension, and 1 K × 1 K real datapoints, (c) in 'pure absorption' mode with a sweep width of 1000 Hz in each dimension, and 1 K × 1 K real datapoints. Time domain data were apodized in each case by application of unshifted sine-bell functions in each dimension. In (b), assignment pathways are traced out for GlcNAcβ1– and –4GlcNAcα as described in the text, where the primed assignments refer to GlcNAcβ1–.

application of rather severe weighting functions in each dimension, such as the unshifted sine-bell. In fact both magnitude mode and absorption mode COSY experiments require weighting functions which ideally are of 'pseudo-echo' form to remove dispersive lineshape contributions from phase twisted lines in the former and from in-phase dispersive diagonal peaks in the latter. If sensitivity is a problem, then it is mandatory to utilize an experiment which has high inherent resolution without further resolution enhancement, such as absorption mode double quantum filtered COSY (Section 3.2.3). In deciding whether adequate sensitivity will be available without acquisition of a large number of t_1 increments, it is convenient to inspect a spectrum recorded for a single increment. In the case of magnitude mode COSY, the first ($t_1 \sim 0$) increment is appropriate, whereas the second t_1 increment is appropriate in the case of absorption mode COSY using time proportional phase incrementation (TPPI), since the first increment contains essentially no signals (sinusoidal modulation in t_1). As a rule of thumb, if any signals are observable, then sensitivity will be adequate. It should be remembered that the sensitivity of a 2D experiment is on the order of a conventional 1D experiment with the same total number of scans. Thus, the appearance of a few signals with poor signal-to-noise ratio in the first t_1 increment does not reflect the overall sensitivity of the 2D experiment, where a total of 512 or 1024 increments would generally be acquired.

In many circumstances, it may not be possible to obtain complete, unambiguous proton resonance assignments in oligosaccharides from COSY alone. Incomplete assignments arise from the apparent lack of cross-peak connectivities between neighbouring protons in the ring. The lack of cross-peak connectivities can in turn arise from two mechanisms. First and most common is the presence of two strongly coupled neighbouring protons, which gives rise to a cross-peak very close to the diagonal, that is easily lost. An example of this is the H^4 and H^5 of GlcNAcβ1– in GlcNAcβ1–4GlcNAc which are strongly coupled, resulting in the apparent lack of a cross-peak correlating the two (see Section 3.3). Second, the presence of a very small scalar coupling (J) between two protons can result in a cross-peak of very low intensity. Since cross-peak resonances are in antiphase with respect to the active coupling, these will mutually cancel if the magnitude of J approaches or is below the natural linewidth. Clearly this problem gets progressively worse with the size of the oligosaccharide. Particular 'trouble-spots' in this regard are the connectivities between H^1 and H^2 of β-mannose (J ~ 1 Hz) and between H^4 and H^5 of galactose (J < 1 Hz). A third difficulty, which may occur in addition to the above, is a situation where a cross-peak is either missed or can not be defined unambiguously due to accidental overlap. It is thus generally necessary to employ additional methods for resonance assignment as described below.

3.2.2 Homonuclear Hartmann–Hahn spectroscopy (HOHAHA)

Many difficulties due to overlap and strong coupling may be overcome by use of HOHAHA (13). The value of this experiment (14), and its closely related counterpart total correlation spectroscopy (TOCSY) (15), is that under the correct conditions correlations are obtained from a given ring proton not only to its neighbour, but to other spins within the ring, (i.e. within the same coupling network). In this regard the experiment gives information analogous to multistep relayed correlation spectra (16), but in a more efficient manner. The HOHAHA experiment differs from COSY in that the second pulse is replaced by an 'isotropic mixing' period, wherein the Zeeman terms of the free precession Hamiltonian are suppressed. Practically, this is achieved by use of a spin-lock sequence, or by an efficient decoupling sequence such as MLEV-17 (14). A simple physical interpretation of the effect of the isotropic mixing Hamiltonian is that the spin system becomes essentially strongly coupled under its influence. Under these conditions the spins do not behave independently, as is the case in weakly coupled systems, and if the mixing period is sufficiently long coherence transfer can take place between all pairs of nuclei within the same coupling network, even if there is no direct coupling between them.

The experimental criteria for good HOHAHA spectra are essentially identical to COSY. However, the inherent resolution is good since all types of resonance in HOHAHA are essentially in positive absorption mode for an extended mixing period. Resolution enhancement functions are thus unnecessary, but apodization with, e.g. a Gaussian function to ensure that the free induction decay is close to zero at the end of acquisition in each dimension will prevent the appearance of 'sinc wiggles'.

The value of the technique is best illustrated with an example, and a HOHAHA spectrum of GlcNAcβ1–4GlcNAc for a spin-lock time of 125 msec is shown in *Figure 4*. It is seen that coherence transfer is propagated between all spins within the same monosaccharide ring. Although in general terms the large increase in cross-peak numbers would appear to aggravate the assignment problem, ring proton assignments can be determined along the ω_2 coordinate of the resolved H^1 resonance of each monosaccharide residue, by use of the characteristic multiplicity of each proton resonance. This multiplicity is essentially constant in view of the fixed ring geometry of pyranoses.

Assignment of the terminal GlcNAc from *Figure 4*, in common with all β-gluco configuration monosaccharides, presents some difficulty since the coupling constants between neighbouring endocyclic protons are of essentially equal magnitude. This is not usually problematic if additional (but incomplete) coupling information is available from COSY, but in particularly difficult situations techniques derived from three-dimensional NMR can be utilized (see Section 3.2.5).

3.2.3 Double quantum filtered COSY (DQF-COSY)

Most macromolecular assignment strategies benefit from application of multiple quantum filtration methods (10–17, 19). The purpose of these is to exploit the spin coupling topology of the spin system, or a subset thereof, in order to reduce the complexity of the two-dimensional spectrum. In the case of multiple quantum filtered COSY methods, the multiple quantum filter can be thought of as preceding the COSY experiment proper, and hence the experimental criteria are exactly those for COSY. A double quantum filter is transparent to systems of two or more coupled spins, and is a trivial example since all resonances will pass a double quantum filter and appear in a DQF-COSY spectrum except singlets. At first sight this might appear to be a useful way of removing singlet solvent resonances such as $^1H^2HO$, but the suppression is not good enough to achieve this in an efficient manner. In fact a DQF-COSY spectrum is similar to COSY in information content, but the inherent resolution is higher, because the lineshapes of the resonances on the diagonal are different. Whereas these are in phase dispersive in COSY, they are

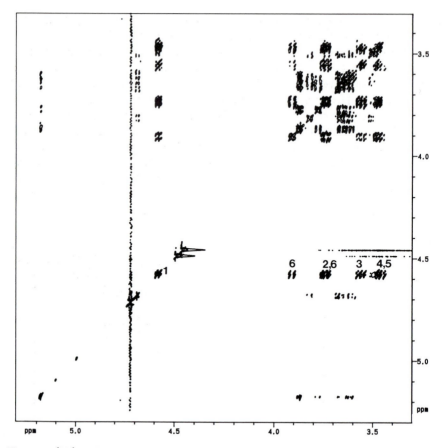

Figure 4. ¹H-¹H HOHAHA spectrum of GlcNAcβ1–4GlcNAc. This spectrum was recorded in 'pure absorption' mode using a sweep width of 1000 Hz in each dimension, and comprises 1 K × 1 K real datapoints. Time domain data were apodized by use of a Lorentzian to Gaussian transformation function. Cross-peak assignments for GlcNAcβ1– are indicated.

predominantly antiphase absorptive in DQF-COSY, and severe resolution enhancement functions are thus not necessary in the latter. This advantage of DQF-COSY is partially offset by the reduced sensitivity; cross-peak amplitudes are approximately halved in comparison with COSY. However, the effective sensitivity may be higher in view of the requirement for sensitivity limiting weighting functions in the latter.

3.2.4 Triple quantum filtered COSY (TQF-COSY)

A triple quantum filter is transparent to systems of three or more coupled spins with two resolvable couplings. Thus all singlets, together with the H¹ resonances should be purged from a TQF-COSY spectrum (20). This is illustrated in *Figure 5* for GlcNAcβ1–4GlcNAc. It is seen in practice that suppression of H^1 resonances is imperfect. This is due primarily to the fact that next-neighbour (> three bond) couplings are not zero, and are partially resolved, particularly in small molecules. The appearance of cross-peaks in TQF-COSY requires additionally that the three spins are coupled to each other. The only spins which in general satisfy this criterion are H^5 and H^6 (i.e. the MQX spin subsystem in *Figure 2*), and hence the COSY spectrum is essentially purged of all cross-peaks except those correlating the resonances of these protons. The suppression is again imperfect in *Figure 5* due to partially resolved next-neighbour couplings together with strong coupling effects. However, the three relevant coupling networks can easily be identi-

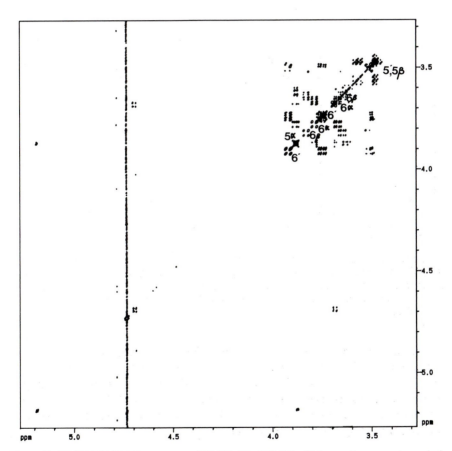

Figure 5. ¹H-¹H TQF COSY spectrum of GlcNAcβ1–4GlcNAc. This spectrum was recorded in magnitude mode using a sweep width of 1000 Hz in each dimension, and comprises 1 K × 1 K real datapoints. Time domain data were apodized in each dimension by use of an unshifted sine-bell function. The H^5 and H^6 resonance assignments are labelled in obvious notation, where the primed assignments refer to GlcNAcβ1–.

fied. The TQF-COSY technique is a valuable complement to HOHAHA, since it is not always possible to transfer coherence from H^1 to H^6, particularly when a small coupling constant is present, such as $^3J_{H^4-H^5}$ in galactose residues. In such cases, data from HOHAHA and TQF-COSY can be 'patched' together to yield a full assignment for the residue.

3.2.5 Three-dimensional (3D) NMR methods

At the time of writing, 3D NMR methods are becoming routine in the assignment and conformational analysis of proteins and nucleic acids (21–23). In contrast, only one 3D NMR experiment has been reported on an oligosaccharide to our knowledge (24). Due to the smaller number of proton resonances, it is unlikely that 3D NMR methods will be required routinely for oligosaccharide analysis. However, circumstances arise where 3D NMR techniques are of value. An example is in the unambiguous assignment of β-gluco configuration monosaccharide residues, which may present some difficulty as described above (Section 3.2.2). In these situations, application of 3D HOHAHA-COSY is of great value (25). The pulse scheme for this technique is illustrated in *Figure 6*, and in common with most 3D experiments, is essentially a combination of the pulse sequences of the corresponding two-dimensional experiments. In the resulting three-dimensional spectrum (*Figure 6*), the plane defined by $\omega_2 = \omega_3$ will comprise HOHAHA peaks, whereas the plane defined by $\omega_1 = \omega_2$ will comprise COSY peaks. A section through the three-dimensional spectrum at the ω_1 coordinate of H^1 of a given residue will therefore comprise HOHAHA peaks along the diagonal $\omega_2 = \omega_3$, with COSY-type correlations created by intersection of the plane $\omega_1 = \omega_2$ (*Figure 6*). For an extended HOHAHA spin-lock time, this section is essentially a COSY spectrum selectively extracted for a given monosaccharide residue. Since H^1 of each residue is well resolved, it is convenient to replace the preparation pulse by a selective pulse centred on this proton (*Figure 6*). This gives rise to a 2D analogue of 3D HOHAHA-COSY, and the resulting saving in acquisition time is analogous to that achieved by use of selective pulses to generate the one-dimensional analogues of two-dimensional spectra. The value of this technique is illustrated in *Figure 7* for GlcNAcβ1–4GlcNAc.

3.2.6 Nuclear Overhauser effect spectroscopy (NOESY)

The NOESY experiment (26) is a useful assignment aid when combined with one or more of the coherence transfer methods described above. In the general case, both intraresidue and interresidue NOEs are observed with reasonable sensitivity at 500 MHz (27, 28) where the rotational correlation times of oligosaccharides composed of six or more residues are comfortably in the spin diffusion limit. While interresidue NOE connectivities depend upon solution conformation, and hence are of limited value in assignment, intra-residue NOEs are of value since they provide an independent assessment of proposed assignments derived from coherence transfer methods. Clearly all

(a)

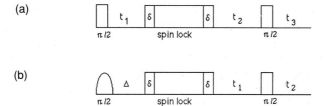

(b)

(c)

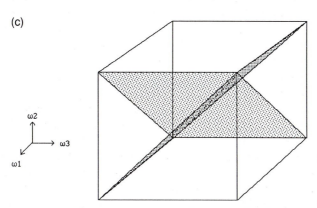

Figure 6. (a) Pulse scheme for three-dimensional HOHAHA-COSY. The first pulse is a non-selective $\pi/2$ pulse followed by the t_1 period. The spin-lock period comprises an MLEV-17 sequence with trim pulses (δ) before and after. The remainder of the sequence is identical to a conventional COSY experiment. (b) Pulse scheme for the acquisition of the two-dimensional analogue of three-dimensional HOHAHA-COSY. The initial pulse is a selective $\pi/2$ pulse centred on H^1 of the residue of interest followed by a short delay $\triangle$ which replaces the t_1 period in (a). (c) Schematic 3D NMR spectrum. In the case of HOHAHA-COSY, the plane $\omega_2 = \omega_3$ will comprise HOHAHA resonances, whereas the plane $\omega_1 = \omega_2$ will comprise COSY resonances. A cross-section perpendicular to ω_1 at the H^1 resonance of a given residue gives a 'COSY' spectrum of that residue alone.

cross-peaks in the NOESY spectrum of an oligosaccharide should be assignable in a manner which is self-consistent with other methods.

3.3 An assignment case study

As an example of resonance assignment, we shall illustrate the complete assignment of GlcNAcβ1–4GlcNAcα in the proton spectrum of GlcNAcβ1–4GlcNAc. The first point to bear in mind is that, since the disaccharide has a reducing terminus, there are effectively two compounds present in solution. The reducing terminal sugar has very different shifts in the α and β anomeric configuration, and hence we are looking for two coupling networks which we expect to be non-degenerate. In contrast, the non-reducing terminal GlcNAc residue, being quite distant from the anomeric centre of the reducing terminus,

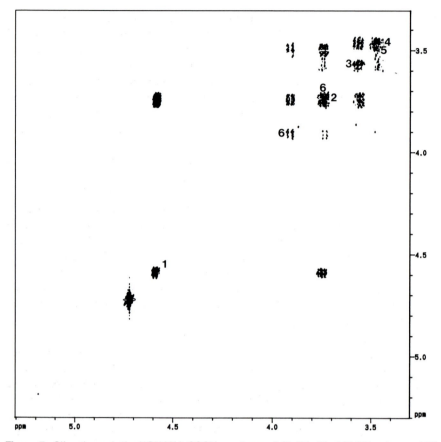

Figure 7. Slice through the HOHAHA-COSY spectrum of GlcNAcβ1–4GlcNAc at $\omega_1 = 4.59$ p.p.m., corresponding with the H^1 resonance of GlcNAcβ1–. The correct, complete assignments for GlcNAcβ1– are shown in obvious notation. Note that the H^4 and H^5 are strongly coupled. The spectrum was recorded with a delay $\triangle$ (see *Figure 6b*) of 1 msec, an MLEV-17 spin-lock of 120 msec duration, and trim pulses of 2.5 msec each. A selective Gaussian pulse of 80 msec duration was centred on the H^1 resonance of GlcNAcβ1–. The spectrum was acquired in the 'pure absorption' mode by use of time proportional phase incrementation on all pulses preceding t_1.

should be less sensitive to the influence of each anomer, and hence we are looking for two coupling networks which will probably be almost degenerate. In this trivial example, assignment of the non-reducing terminal GlcNAc is begun by the identification of two H^1 resonances of almost degenerate shift at around 4.57 p.p.m. (*Figure 3b*), with an intensity ratio which echoes the relative anomeric populations (*Figure 1*). This is a useful starting point which also holds for large oligosaccharides, where many anomeric proton resonances might occur in this region of the spectrum. The H^2 resonances at around 3.74 p.p.m., which are also almost degenerate, can readily be assigned from the COSY spectrum, from the pair of almost degenerate cross-peaks at $\omega_1 = 4.57$, $\omega_2 = 3.74$ p.p.m. The assignment of H^3 is more difficult, but a pair of almost overlapping cross-peaks at $\omega_1 = 3.55$, $\omega_2 = 3.74$ p.p.m. in *Figure 3b* assigns these at around 3.55 p.p.m. In the general case where digital resolution may be lower or linewidths greater, it may not be possible to observe cross-peak multiplicities so easily. As an example, the H^3 assignments could erroneously have been made under conditions illustrated in *Figure 3a* from the presence of a cross-peak apparently correlating H^2 at $\omega_2 = 3.74$ p.p.m. and H^3 at around $\omega_1 = 3.90$ p.p.m. This ambiguity is easily resolved by application of HOHAHA-COSY (*Figure 7*). The H^4 assignments (3.46 p.p.m.) are then easily made from H^3 as shown in *Figure 3b*. Note that now these protons are essentially degenerate due to their greater distance from the anomeric centre of the reducing terminus. At first sight there appears to be no cross-peak along either the ω_1 or ω_2 coordinate of H^4 with which to assign H^5. Again HOHAHA-COSY is of value since the degenerate pairs of H^6 protons at $\sim$ 3.74 p.p.m. and $\sim$ 3.90 p.p.m. are seen to be correlated with the degenerate H^5 protons at $\sim$ 3.5 p.p.m. In fact a weak cross-peak is seen very close to the diagonal at $\omega_1 = 3.50$, $\omega_2 = 3.46$ p.p.m. in *Figure 3b*, whose low intensity results from the fact that H^4 and H^5 are strongly coupled. With experience, such effects can be recognized and exploited as assignment aids in COSY spectra without resort to more complex methods.

The H^1 of GlcNAcα is easily distinguished from its characteristic resonance at low field (5.18 p.p.m.) together with a relatively small coupling constant. The H^2 can immediately be assigned at $\sim$ 3.87 p.p.m. from the presence of a cross-peak at $\omega_1 = 3.87$ p.p.m., $\omega_2 = 5.18$ p.p.m. in *Figure 3b*. If this cross-peak is carefully inspected, it will be seen to possess a 'skewed' appearance. Correspondingly, the components of the H^1 doublet in the 1D spectrum are not of equal intensity. This phenomenon is known as virtual coupling, and is diagnostic of strong coupling between neighbour and next-neighbour protons. We therefore anticipate that H^2 and H^3 will have very similar shifts. Since there is no cross-peak in the vicinity of 3.87 p.p.m. which could correlate H^2 with H^3, we must assume that these are very nearly degenerate, and search for a cross-peak correlating the H^2/H^3 complex multiplet with H^4. Since there are several such cross-peaks along the ω_1 (or ω_2) coordinate of this complex multiplet, and since we are unsure of the exact multiplicity with respect to H^2/H^3, it is virtually impossible to determine which is relevant for assignment of H^4. In such circumstances it is sometimes possible to derive the correct cross-peak by default following further assignments. However, it is always safer to obtain a firm assignment from an alternative method, and in this instance HOHAHA is of value (*Figure 4*). A cross-section through the HOHAHA spectrum at $\omega_1 = 5.18$ p.p.m. shows all of the proton resonances within the same spin system (*Figure 8*). The two H^6 are easily assigned from their large geminal coupling constant ($\sim$ −12 Hz), and the complex H^2/H^3 multiplet is seen at around 3.87 p.p.m. The H^5 resonance can not be determined

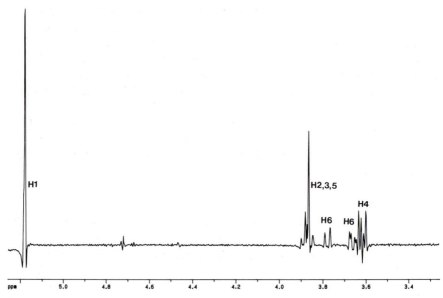

Figure 8. Section parallel to ω_1 at 5.18 p.p.m. from the HOHAHA spectrum of GlcNAcβ1–4GlcNAc. This shows an essentially complete subspectrum for –4GlcNAcα, which can be assigned as described in the text.

unambiguously from this cross-section. Indeed, there is apparently only one resonance which remains unassigned at 3.62 p.p.m. However, the H^5 resonance can be determined unambiguously from the COSY spectrum in *Figure 3b* via cross-peak connectivities with the two H^6, and is seen to overlap with H^2 and H^3. The H^4 must thus, by default, resonate at 3.62 p.p.m.

4. Determination of primary sequence

4.1 Fingerprinting

In view of the high sensitivity of proton chemical shifts to local environment, the proton NMR spectrum of an oligosaccharide with a particular sequence is essentially unique—the probability that an isomeric structure will have an identical spectrum is vanishingly small. The primary sequence of an unknown oligosaccharide can thus most easily be determined by comparison with a database of known structures. Due primarily to the work of Vliegenthart and co-workers (1), the number of *N*-linked oligosaccharides for which proton chemical shifts are available is well over 100. Although the number of proton resonance assignments in each of these structures is small, being restricted primarily to well-resolved resonances in one-dimensional spectra, these are nevertheless sufficient to give unambiguous sequences in many instances. In our own laboratory we routinely use the fingerprint method to determine initially if an unknown bears homology with a database of known structures. The key to the success of the method is that apparently small changes in primary sequence correspond to more significant modifications to three-dimensional structure, and hence to the NMR spectrum. As an example, the chemical shifts of the well resolved 'reporter resonances' of the biosynthetically related oligosaccharides GlcNAcβ1–2Manα1–6(GlcNAcβ1–2Manα1–3)Manβ1–4GlcNAc and GlcNAcβ1–2Manα1–6(GlcNAcβ1–4)(GlcNAcβ1–2Manα1–3)Manβ1–4GlcNAc are shown in *Table 1*. The presence of the additional GlcNAc residue in the second structure is seen to cause large chemical shift perturbations, illustrating the exquisite sensitivity of proton NMR to small structural changes.

4.2 *'Ab-initio'* method

The fingerprint method fails if the unknown bears little or no homology with the database of known structures. Under these circumstances, a more involved procedure, for which we have coined the term *'ab-initio'* is more appropriate (29). This relies upon knowledge of through-bond COSY-type connectivities and through-space NOESY-type connectivities between protons. The former determine the types of monosaccharide residue present in

---

Table 1. Comparison of 'reporter resonance' shifts for GlcNAcβ1–2Manβ1–6 (GlcNAcβ1–2Manα1–3)Manβ1–4GlcNAc and GlcNAcβ1–2Manα1–6(GlcNAcβ1–4) (GlcNAcβ1–2Manα1–3)Manβ1–4GlcNAc

```
      4'        3'                    GlcNAcβ1-2Manα1-6
   GlcNAcβ1-2Manα1-6                                    \
                     / 2    1         5 GlcNAcβ1-4Manβ1-4GlcNAcβ
                    Manβ1-4GlcNAcβ                      /
                     \                GlcNAcβ1-2Manα1-3
   GlcNAcβ1-2Manα1-3
      4        3
```

H-1 of	1	4.724	4.722
	2	4.770	4.699
	3	5.119	5.060
	4	4.558	4.555
	3'	4.919	5.003
	4'	4.558	4.549
	5		4.469
H-2 of	2	4.247	4.177
	3	4.189	4.247
	3'	4.108	4.148
NAc of	3	2.055	2.056
	4	2.054	2.059
	4'	2.052	2.050
	5		2.068

Data taken from (1).

---

the oligosaccharide, partly from characteristic resonance positions of ring protons, and from measurement of scalar spin couplings of neighbouring protons by analysis of cross-peak fine-structure. The latter can be used to determine how the monosaccharide residues are linked, since an interresidue NOE is usually measurable between H^1 of a given residue and the H^x of an adjacent residue, where X is the corresponding linkage position. However, this approach must be used with care since an NOE is occasionally measurable from H^1 to a proton adjacent to H^x, which may occur in addition to, or instead of, an NOE to H^x. It is thus prudent to determine the linkage positions by an alternative method such as GC-MS methylation analysis (30). In the case of small oligosaccharides, a simple alternative is to record a COSY spectrum of the sample in dimethylsulfoxide, so that the normally exchangeable hydroxyl protons are observable. It is a simple matter to assign these for a small oligosaccharide such as GlcNAcβ1–4GlcNAc (*Figure 9*). The absence

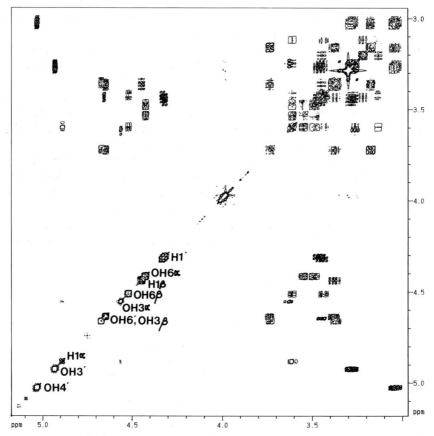

Figure 9. Partial ¹H-¹H COSY spectrum of GlcNAcβ1–4GlcNAc in dimethylsulfoxide. The assignments of exchangeable OH proton resonances are indicated in obvious notation.

of a resonance corresponding to the hydroxyl proton at C-4 of the reducing terminal GlcNAc in this spectrum confirms the sequence. Obviously, in this simple case NOE data are not required.

An alternative method relies upon measurement of long-range ^{13}C-^{1}H connectivities across glycosidic linkages by use of proton detected heteronuclear multiple bond correlation (HMBC) experiments (31). In principle, this method is superior to any other, since the presence of such connectivities defines the linkage position without ambiguity, and without recourse to other methods. However, the amount of material required is measured in milligrams, which is often an order of magnitude greater than that which is obtainable from biological sources.

5. Determination of three-dimensional structure

5.1 Choice of method—experimental or theoretical?

The primary determinants of oligosaccharide three-dimensional structure are the glycosidic torsion angles φ and ψ, together with ω in the case of 1–6 linkages (*Figure 10*). The preferred conformations about these linkages can in principle be measured from energy calculations together with experimental NOE and spin coupling constant measurements (20, 28, 31–38). It has been adequately demonstrated that NOE and spin coupling constant measurements in proteins and nucleic acids can give an average solution structure which compares very favourably with that derived by X-ray diffraction studies (3). The primary reason for the success of the NMR method lies in the very large number of available distance and angular constraints. These can be analysed using distance matrix algorithms in Cartesian or dihedral angle space to give an average solution structure which may then be refined by use of energy minimization and molecular dynamics algorithms together with an appropriate molecular mechanical force field. In principle, the solution conformations of oligosaccharides can be derived in a similar manner to that for

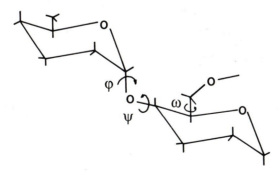

Figure 10. The definition of the torsion angles φ, ψ, and ω for oligosaccharides: φ = H1-C1-O1-CX, ψ = C1-O1-CX-HX, ω = C1-O1-C6-C5.

proteins and nucleic acids. However, the above approach has not been adopted for at least two reasons. First, the number of distance constraints available from NOE measurements is very much smaller than in proteins or nucleic acids. Often only one or two NOEs are available across glycosidic linkages, and since oligosaccharides generally exist as extended structures, no NOEs between distant regions of the molecule are generally observed. Second, full force fields for the conformational analysis of oligosaccharides have only recently become available (37–40). Since NOE data in oligosaccharides may not provide sufficient constraints when used alone, the availability of a full molecular mechanical force field with parameters for oligosaccharides is essential. In this way the complementarity between experimental and theoretical data can be exploited, and various approaches have been utilized, as described below.

5.2 Restrained energy minimization

Energy minimization of a model macromolecular system is a non-linear optimization problem (41). Given a set of independent variables $\mathbf{x} = (x_1, x_2, x_3, \ldots, x_n)$ and an objective function $V = V(\mathbf{x})$, the problem is to find a set of values for the independent variables $\mathbf{x}^*$, for which the function $V(\mathbf{x}^*) = \min(V(\mathbf{x}))$. In the case of a model macromolecule containing N atoms, the 3N components of $\mathbf{x}$ are the atomic coordinates and V is the potential energy. Given the availability of a suitable force field for oligosaccharides as described above, it is a straightforward matter to compute V from an initial set of Cartesian coordinates. Energy minimization algorithms utilize the fact that the potential energy function, (i.e. force field) is an explicit, differentiable function of the Cartesian coordinates, to calculate the potential energy and its gradient (which is the force on each atom) for any initial set of atomic positions $\mathbf{x}$. This information can be used recursively to generate a new set of coordinates in an attempt to reduce the total potential energy, and thus to optimize the molecular structure. The form of the force field varies according to its source, but these generally take the form (42, 43):

$$E_{tot} = \underset{\text{bond}}{\Sigma K_r(r - r_e)^2} + \underset{\text{valence angle}}{\Sigma K_\theta(\theta - \theta_e)^2} + \underset{\text{torsion angle}}{\Sigma K_n(1 + \cos(n\phi - \gamma))}$$

$$+ \underset{\text{Van der Waals}}{\Sigma \Sigma \epsilon[-(r^*/r_{ij})^{12} + 2(r^*/r_{ij})^6]} + \underset{\text{electrostatic}}{\Sigma (q_i q_j / \epsilon r_{ij})} + \underset{\text{hydrogen bond}}{\Sigma(C_{ij}/r_{ij}^{12} - D_{ij}/r_{ij}^{10})}$$

where the symbols have their usual meanings (42, 43). While it is straightforward to derive a local minimum energy structure using this procedure, it is extremely difficult to find the global minimum for all but the simplest structures. Even molecular systems with as few as ten atoms may have conformational spaces of such complexity that the global minimum is almost impossible to locate. However, by considering experimental data as additional criteria for energy minimization, the region of conformational space which is examined can be very much reduced. These experimental criteria are added to the theoretical model as additional terms in the force field, corresponding to distance constraints derived from NOE measurements or angular constraints derived from scalar spin coupling constant measurements. These can be implemented in the form of harmonic-type functions. The actual constraints included in the calculation are usually interresidue NOEs across the glycosidic linkage of contiguous residues, and angular constraints for exocyclic groups, e.g. about the C-5 C-6 bond of pyranoses, which in turn are derived from spin coupling constant measurements with the appropriate Karplus-type relation.

5.3 Adiabatic mapping

The global minimum energy structure of a model macromolecular system is by definition the structure at zero Kelvin. In many macromolecular systems under investigation, this may be a close approximation to the structure at physiological temperatures. However, in the case of oligosaccharides, it is appropriate to examine the dynamic properties of the structure at these temperatures since, as we have discussed above, there may exist regions of considerable flexibility within the molecule which will not be apparent from consideration of the minimum energy structure alone. The simplest method by which dynamics of the system can be investigated is by estimation of the energy required to stimulate transitions between two or more conformations of a given molecule (44–46). Given a suitable reaction coordinate such as rotation about one of the glycosidic angles ϕ or ψ (defined as ϕ = H1-C1-O1-CX and ψ = C1-O1-CX-HX, where CX and HX are aglyconic atoms), the dynamics of the system is assumed to follow directly from the dependence of conformational energy on the reaction coordinate. As an example, consider the potential energy surface for rotation about both ϕ and ψ in a disaccharide fragment such as Manα1–3Man (*Figure 11*). This surface was generated by deformation of the minimized geometry of the disaccharide in a series of small steps (10°) about each glycosidic angle, with energy minimization after each step. The implication is that the glycosidic linkage will exist with orientations defined by the low energy regions of *Figure 11*. This procedure is known as adiabatic mapping, and the name derives from the implicit assumption that the relaxation produced by energy minimization would occur adiabatically on the time scale of the transition during actual dynamics. Although adiabatic mapping is simple to implement, it suffers from several disadvantages. The most important of these is that it is difficult to incorporate experimental data into the analysis. This can be overcome to an extent by calculation of predicted parameters such as the NOE for a suitable number of discrete geometries over the potential surface with weighted populations calculated from the Boltzmann distribution (45, 46). However, this is of limited value in view of a second disadvantage, namely that the potential surface is a rough

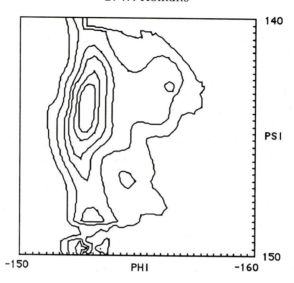

Figure 11. Molecular mechanical potential surface for the disaccharide Manα1–3Manα obtained by independent variation of ϕ and ψ in 10° steps with energy minimization. Contours are plotted at 0.5, 1.0, 2.0, 3.0, 5.0, and 8.0 kcal above the global minimum.

approximation to the enthalpy of the system, and contains no entropic contribution to the energy.

5.4 Restrained and free molecular dynamics simulations

A more accurate representation of the dynamics of a molecular system at a given temperature can be obtained by use of molecular dynamics simulations (47). Imagine the motion of a particular atom along a coordinate x. Given the position at time t, $x(t)$, the position after a short time interval, Δt, is given by the Taylor series:

$$x(t + \Delta t) = x(t) + x(t)\Delta t + x(t)\Delta t^2/2 + \ldots$$

Since the force on each atom is given by the negative gradient, (i.e. derivative) of the potential energy function with respect to position, then velocity $x(t)$ and acceleration $x(t)$ can be computed from Newtons laws. Given the position $x(t)$ of the atom and suitable approximations for contributions from higher derivatives, $x(t + \Delta t)$ can be calculated with good precision. The dynamics of the system can thus be determined with respect to time. The great advantage of the molecular dynamics approach is that both distance and angular constraints can be included in the potential energy function just as in restrained minimization, such that the system is constrained in conformational space during the simulation to those regions consistent with the experimental data. In addition, conformational space is explored in a fashion that directly reflects the density of states, so entropic and other thermal effects are naturally included.

5.5 Simulated annealing

Closely related to molecular dynamics methods is the technique termed simulated annealing (48). This is a useful approach to approximate the global minimum energy configuration of a macromolecular system. Essentially, the system is equilibrated at a high temperature such that the conformational space explored by the system is relatively large. The dynamics of the system is then simulated at slowly decreasing temperatures, until a very low final temperature is reached, at which point the system should theoretically exist in a configuration close to the global minimum. The success of the approach depends upon slow cooling—if this is too fast, then the system will become trapped in a local minimum. In practical terms it becomes almost impossible to cool the system slowly enough that the global minimum is obtained for each of a series of random starting structures. Nevertheless, it is usually possible to generate structures which have considerably lower energy than those derived from straightforward energy minimization.

Acknowledgements

The author gratefully acknowledges support from the Wellcome Trust and the Lister Institute for a Centenary Research Fellowship.

References

1. Vliegenthart, J. F. G., Dorland, L., and van Halbeek, H. (1983). *Adv. Carb. Chem. Biochem.*, **41**, 209.
2. Poppe, L., Dabrowski, J., Lieth, C.-L., Koike, K., and Ogawa, T. (1990). *Eur. J. Biochem.*, **189**, 313.
3. Wuthrich, K. (1986). *NMR of proteins and nucleic acids*. Wiley, New York.
4. Ernst, R. R., Bodenhausen, G., and Wokaun, A. A. (1987). *Principles of NMR in one and two dimensions*. Clarendon Press, Oxford.
5. Kessler, H., Gehrke, M., and Griesinger, C. (1988). *Angew. Chem. Int. Ed. Engl.* **27**, 490.
6. Homans, S. W. (1989). *A dictionary of concepts in NMR*. Oxford University Press, Oxford.
7. Aue, W. P., Bartholdi, E., and Ernst, R. R. (1976). *J. Chem. Phys.* **64**, 2229.
8. Homans, S. W., Dwek, R. A., Fernandes, D. L., and Rademacher, T. W. (1983). *Biochim. Biophys. Acta*, **760**, 256.
9. Bachman, P., Aue, W. P., Muller, L., and Ernst, R. R. (1977). *J. Magn. Reson.*, **28**, 29.
10. Rance, M., Sorensen, O. W., Bodenhausen, G., Wagner, G., Ernst, R. R., and Wuthrich, K. (1984). *Biochim. Biophys. Res. Commun.*, **117**, 479.

11. States, D. J., Haberkorn, R. A., and Ruben, D. J. (1982). *J. Magn. Reson.*, **48,** 286.
12. Keeler, J. and Neuhaus, D. (1985). *J. Magn. Reson.*, **63,** 454.
13. Homans, S. W., Dwek, R. A., Boyd, J., Soffe, N., and Rademacher, T. W. (1987). *Proc. Natl Acad. Sci. U.S.A.*, **84,** 1202.
14. Bax, A. and Davis, D. G. (1985). *J. Magn. Reson.*, **65,** 355.
15. Braunschweiler, L. and Ernst, R. R. (1983). *J. Magn. Reson.*, **53,** 521.
16. Homans, S. W., Dwek, R. A., Fernandes, D. L., and Rademacher, T. W. (1984). *Proc. Natl Acad. Sci. U.S.A.*, **81,** 6286.
17. Piantini, U., Sorensen, O. W., and Ernst, R. R. (1982). *J. Am. Chem. Soc.*, **104,** 6800.
18. Shaka, A. J. and Freeman, R. (1983). *J. Magn. Reson.*, **51,** 169.
19. Sorensen, O. W., Levitt, M. H., and Ernst, R. R. (1983). *J. Magn. Reson.*, **55,** 104.
20. Homans, S. W., Dwek, R. A., Boyd, J., Mahmoudian, M., Richards, W. G., and Rademacher, T. W. (1986). *Biochemistry*, **25,** 6342.
21. Vuister, G. W. and Boelens, R. (1987). *J. Magn. Reson.*, **73,** 328.
22. Griesinger, G. W., Sorensen, O. W., and Ernst, R. R. (1987). *J. Magn. Reson.*, **73,** 574.
23. Oschkinat, H., Griesinger, C., Kraulis, P., Sorensen, O. W., Ernst, R. R., Gronenborn, A. M., and Clore, G. M. (1988). *Nature (London)*, **332,** 374.
24. Vuister, G. W., Waard, P. D., Boelens, R., Vliegenthart, J. F. G., and Kaptein, R. (1989). *J. Am. Chem. Soc.*, **111,** 772.
25. Homans, S. W. (1990). *J. Magn. Reson.*, **90,** 557.
26. Macura, S. and Ernst, R. R. (1980). *Mol. Phys.*, **41,** 95.
27. Homans, S. W., Dwek, R. A., Fernandes, D. L., and Rademacher, T. W. (1983). *FEBS Lett.*, **164,** 231.
28. Homans, S. W., Dwek, R. A., and Rademacher, T. W. (1987). *Biochemistry*, **26,** 6553.
29. Homans, S. W. (1990). *Prog. NMR Spectrosc.*, **22,** 55.
30. Aspinall, G. O. (1982). In *The polysaccharides*. Vol. 1 (ed. G. O. Aspinall), p. 34. Academic Press, New York.
31. Lerner, L. and Bax, A. (1987). *Carbohydr. Res.,* **166,** 35.
32. Brisson, J.-R. and Carver, J. P. (1983). *Biochemistry*, **22,** 3671.
33. Brisson, J.-R. and Carver, J. P. (1983). *Biochemistry, * **22,** 3680.
34. Paulsen, H., Peters, T., Sinnwell, V., Heume, M., and Meyer, B. (1986). *Carbohydr. Res.*, **156,** 87.
35. Homans, S. W., Pastore, A., Dwek, R. A., and Rademacher, T. W. (1987). *Biochemistry*, **26,** 6649.
36. Homans, S. W., Dwek, R. A., and Rademacher, T. W. (1987). *Biochemistry*, **26,** 6571.
37. Scarsdale, J. N., Ram, P., Prestegard, J. H., and Yu, R. K. (1988). *J. Comput. Chem.*, **9,** 133.
38. Homans, S. W. (1990). *Biochemistry*, **29,** 9110.
39. Ha, S. N., Giammona, A., Field, M., and Brady, J. W. (1988). *Carbohydr. Res.*, **180,** 207.
40. Edge, C. J., Singh, U. C., Bazzo, R., Taylor, G. L., Dwek, R. A., and Rademacher, T. W. (1990). *Biochemistry*, **29,** 1971.

41. Fletcher, R. (1980). *Unconstrained optimization*. Practical methods of optimization series, Vol 1. Wiley and Sons, Chichester.
42. Weiner, S., Kollman, P. A., Case, D. A., Chandra Singh, U., Ghio, C., Alagona, G., Profeta, S. P., and Weiner, P. (1984). *J. Am. Chem. Soc.*, **106,** 765.
43. Weiner, S. J., Kollman, P. A., Nguyen, D. T., and Case, D. A. (1986). *J. Comput. Chem.*, **7,** 230.
44. Imberty, I., Tran, V., and Perez, S. (1989). *J. Comput. Chem.*, **11,** 205.
45. Cumming, D. A. and Carver, J. P. (1987). *Biochemistry*, **26,** 6664.
46. Cumming, D. A. and Carver, J. P. (1987). *Biochemistry*, **26,** 6676.
47. McCammon, J. and Harvey, S. (1987). *Dynamics of proteins and nucleic acids*. Cambridge University Press.
48. Clore, G. M., Brunger, A. T., Karplus, M., and Gronenborn, A. M. (1986). *J. Mol. Biol.,* **191,** 523.

10

Structure determination from NMR data I. Analysis of NMR data

IGOR L. BARSUKOV and LU-YUN LIAN

1. Introduction

This chapter deals with the strategies which are used to derive the experimental data required as inputs for calculation of the three-dimensional structure of a protein. The experiments used for spectral assignment described in Chapters 4 and 5 are also used for obtaining the data for structure calculations, although for the latter purpose, a more quantitative analysis is required. The principal structural data derived from NMR are scalar (J) couplings and NOEs, which are translated into restraints on dihedral angles and interproton distances, respectively. It is important to bear in mind that NMR leads to *constraints* on angles and distances, not to exact values of these parameters. The quality of the solution structure obtained has been shown to be proportional to the number of constraints used in the structure calculations.

We describe here methods for obtaining these constraints and cover the stage following complete sequential resonance assignment up to the point just before the computational stage for structure calculations which is the subject of the following chapter. Once the assignment of the NMR spectrum has been completed, the determination of the solution structure of a protein using NMR is usually carried out in two stages. The first stage involves the delineation of the secondary structure elements. This procedure is important as it can lead to the resolution of ambiguous NOEs and aid the complete analysis of the 'long-range' tertiary NOEs that are required for full structure calculations. Secondary structure elements can be identified on the basis of a number of readily measurable NMR parameters associated with backbone resonances: sequential short-, medium-, and long-range[a] NOEs, H^N-H^α coupling constants, and amide H^N solvent-exchange rates. The second phase of the structure determination process involves the use of computational methods such as distance geometry and molecular dynamics calculations, described in the following chapter. At this stage, additional NMR information, including

---

[a] Short-, medium-, and long-range NOEs are commonly defined as interproton distances of up to 2.8 Å, 3.4 Å, and 4.2 Å, respectively.

the stereospecific assignments of diastereotopic protons and non-sequential long-range NOEs, is used. A summary of the main experiments that need to be performed in order to obtain the input restraints for structural calculations is given in *Table 1*. Also included in the table for completeness are additional experiments which have been shown to provide useful scalar coupling information on small molecules but which have not been routinely used for proteins.

The main limitations to the methods outlined here arise from unfavourable transverse relaxation times (T_2), from resonance overlap in large proteins, and from internal mobility and multiple conformations. Particularly for proteins with M_r larger than 15 000, these problems may mean that it will not be possible to obtain complete *quantitative* data. In addition, the different types of secondary structure elements present in a protein make a significant differ-

Table 1. Experiments required to obtain input restraints

Experiments[a]	Conditions/parameters[b]	Information obtained
DQF-COSY	4096 × 1024; H_2O	$^3J_{HN-H\alpha}$ $^3J_{H\alpha-H\beta}$ (Ile,Val,Thr)
P.E.COSY	4096 × 1024; H_2O	$^3J_{HN-H\alpha}$ (Gly)
P.E.COSY	4096 × 1024; ^2H_2O	$^3J_{H\alpha-H\beta}$ (Tyr, Phe Trp, Asn, Asp, Ser, Cys, His)
^{15}N-^{1}H HMQC-J	4096 × 1024; H_2O	$^3J_{HN-H\alpha}$
J-modulated [^{15}N-^{1}H] COSY	4096 × 256; H_2O	$^3J_{HN-H\alpha}$
HOHAHA or COSY	2048 × 512; ^2H_2O	H^N exchange rates
^{15}N-^{1}H HMQC or HSQC	2048 × 512; ^2H_2O	H^N exchange rates
NOESY	2048 × 512; H_2O[c]	H-H distances from amide protons
NOESY	2048 × 512; ^2H_2O	H-H distances between aliphatic protons
HETERO-RELAY (^{15}N- and/or ^{13}C- labelled proteins)	4096 × 512; H_2O	$^3J_{HN-H\alpha}$
NOESY (^{15}N- and/or ^{13}C- labelled proteins)	4096 × 512; H_2O	$^3J_{HN-H\beta}$ $^3J_{HN(i+1)-H\alpha(i)}$ $^3J_{HN-H\beta}$ $^3J_{H\beta-HN}$ (side chain)

[a] For large proteins, multi-dimensional versions of these experiments, with one dimension being the ^{13}C or ^{15}N chemical shift, should be used.

[b] (i) The minimum number of t_1 increments are given. The number of increments will also depend on the size of the protein (see text). (ii) Data points are given in the following format: complex points (ω_2) × real points (ω_1).

[c] For water suppression, a non-excitation pulse such as a 1$\bar{1}$ pulse either on its own or in combination with a soft presaturation pulse should be used.

ence to how easy or difficult will be the tasks of sequential assignment and quantification of NOE data. These problems should not preclude the use of NMR to determine a three-dimensional structure; they simply mean that the structure obtained may not be as precise as one where a large number of unambiguous restraints is available.

The chapter is divided into three major sections:

- experimental procedures for obtaining NMR-derived torsional angle restraints (including the determination of scalar coupling constants and the stereospecific assignment of prochiral protons and methyl groups), and the measurement of amide hydrogen exchange rates
- secondary structure determination
- quantitative analysis of NOE data for conversion into distance constraints

Throughout this chapter, two proteins will be used as examples to illustrate some of the methods described; IgG binding domain II of protein G (61 amino acid residues) and dihydrofolate reductase (162 amino acid residues).

2. Scalar coupling constants and amide hydrogen exchange rates

Although the NMR methods for structure determination makes use primarily of distance constraints derived from NOEs, the determination of high quality structures requires in addition measurements of vicinal (three bond) scalar coupling constants. These coupling constants can provide local structure information which is complementary to that from the NOE data. Of particular interest are the H^N-H^α coupling constant, $^3J_{HN\text{-}H\alpha}$, which can be directly related to the polypeptide backbone dihedral angle ϕ, and the H^α-H^β coupling constant(s), $^3J_{H\alpha\text{-}H\beta}$, which gives the sidechain torsion angle χ_1. These torsion angles are shown in *Figure 1*.

2.1 Determination of $^3J_{HN\text{-}H\alpha}$ coupling constants

The coupling constant $^3J_{HN\text{-}H\alpha}$ is related to the torsion angle ϕ, as discussed in Chapter 11. An accurate measurement of this coupling constant is an advantage since it allows one to detect any distortion of regular secondary structures and allows additional constraints to be set for regions where no regular secondary structures are evident. Attempts should therefore be made to obtain as many accurate $^3J_{HN\text{-}H\alpha}$ values as possible.

2.1.1 The DQF-COSY experiment

For proteins with M_r up to 15000, a phase-sensitive DQF-COSY spectrum with extremely high resolution in F_2 (*Protocol 1*) is most frequently used to measure the H^N-H^α coupling constant (except for glycine residues).

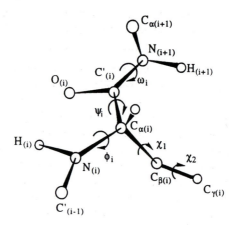

Figure 1. Diagram of a polypeptide segment showing the relevant torsional angles which can be obtained from NMR data for defining backbone and side chain conformations.

Protocol 1. Determination of $^3J_{HN\text{-}H\alpha}$ values

1. Obtain a DQF-COSY spectrum.
   (a) For proteins of $M_r < 10000$, use at least 1024 t_1 increments with 8192 or 4096 real data points in t_2. This will give a resolution of between 0.6 and 1.7 Hz per point when typical spectral widths of 5 to 7 kHz are used. The number of t_2 data points is limited by the amount of disk storage space available both for data acquisition and data processing.
   (b) For proteins of $M_r > 10000$, T_2 relaxation has to be taken into account when considering the optimum values t_2 acquisition time and $t_{1,max}$. Typically, 512 $t_1$increments of 4096 real points is sufficient.

2. Use as low a presaturation power as possible for water suppression in order to minimize the bleaching of cross-peaks close to the water chemical shift in the F_1 dimension. It may be necessary to obtain DQF-COSY spectra at more than one temperature.

3. Process the spectrum with zero-filling once in F_1 and up to 16 K complex points in F_2. It is necessary to try several window functions in order to obtain the best resolution without substantial loss in the signal-to-noise ratio and without causing severe distortion to the lineshape. Gaussian (with GB = 0.3 and LB = −10) and shifted sine-bell squared (shift by $\pi/30$) window functions have been found to be good starting functions.

4. Make sure that the spectrum is properly phased. It may be necessary to make minor adjustments to the phases after two-dimensional Fourier transformation.

5. Measure the H^N-H^α coupling constants for all non-glycine residues using the antiphase splittings in the cross-sections parallel to F_2.

6. The accuracy of the coupling constant measurements can be improved by using the spectral simulation procedure described in *Protocol 3*.

In practice, however, this approach has been shown to be complicated by the effects of finite linewidth and limited digital resolution (2). The minimum separation between the antiphase components of a COSY cross-peak is equal to approximately 0.576 times the linewidth at half-height. If the true separation is smaller than this, the only effect on the COSY cross-peak is a reduction in the overall intensity, the apparent separation remaining at about 0.576 times the linewidth at half-height. The result of this is that splittings observed in antiphase multiplet structures are unreliable measures of those coupling constants which are smaller than the observed linewidth. Therefore, only in very small proteins ($M_r < 5000$), where the linewidths are less than 10 Hz, can this method be reliably used to obtain quantitative $^3J_{HN-H\alpha}$ values. For proteins of M_r between 5000 and 15000, even with high digital resolution in F_2, the natural linewidths in proteins are usually in excess of 10 Hz, so that the measured antiphase doublet splittings of less than 5 Hz are overestimates of the corresponding couplings (although corrections can be made when the linewidth at half-height, Δ, is known). In practice, only a qualitative analysis of the DQF-COSY spectrum is possible in order to 'group' the coupling constant values into the categories of > 10 Hz, < 5 Hz, and 6–8 Hz. For even larger proteins ($M_r > 15000$), it is impossible to estimate the $^3J_{HN-H\alpha}$ values as these values will all appear to be greater than 8 Hz. In this case, several heteronuclear experiments may be useful when ^{15}N-labelled protein is available: the ^{15}N-^{1}H HMQC-J (3), the J-modulated [^{15}N-^{1}H]-COSY (4), and the HETERO-RELAY (5) experiments.

2.1.2 The ^{15}N-^{1}H HMQC-J experiment (3)

Figure 2 shows an example of an ^{15}N-^{1}H HMQC-J spectrum. The intrinsic linewidth of the cross-peaks in the ^{15}N direction (ω_1) in this type of experiment is less than those of the amide protons in the homonuclear ^{1}H experiments. This is because the cross-peaks in ^{15}N-^{1}H HMQC-J spectrum are generated via heteronuclear multiple quantum coherence and hence do not suffer from homonuclear dipolar broadening in ω_1. In addition, the format of the 2D spectrum means that there will be no loss of cross-peaks as a result of them being 'under' the water signal. For most proteins, an intrinsic digital resolution of less than 1 Hz per point in the raw data is easily achieved since the normal spectral width along the ^{15}N dimension is between 1.5 to 2 kHz. The $^3J_{HN-H\alpha}$ value is measured from the separation between the two in-phase components of the ^{15}N-^{1}H cross-peak. For quantitative purposes, this method of determination of $^3J_{HN-H\alpha}$ is favoured over the conventional homonuclear DQF-COSY experiment. *Protocol 2* describes the procedure that can be followed to obtain coupling constants (6).

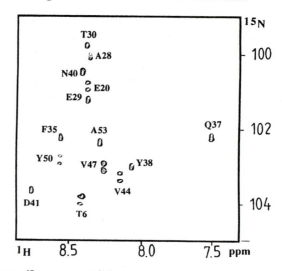

Figure 2. Part of the ^{15}N (ω_1-axis)-H^N ^{1}H (ω_2-axis) region of the HMQC-J spectrum of ^{15}N-labelled IgG-binding domain II of protein G, at 297 K, pH 4.2. Three types of cross-peaks are observed: doublets from residues in the β-sheet (T6, E20, V47, V44, Y50); singlets from residues in the α-helix (T30, F35, Y38, N40, D41), and doublets with reduced splittings from residues at the beginning of the helix (A28), in the helix (Q37), or in a turn (A53).

Protocol 2. Heteronuclear multiple quantum coherence experiment for obtaining $^3J_{H^N\text{-}H\alpha}$ values

1. Obtain an ^{15}N-^{1}H HMQC-J spectrum with at least 1024 t_1 increments of 2048 complex points.

2. Process the data with a Gaussian apodization function in F_1 using three or four different negative line-broadening parameters (-4, -6, -8, and -10 Hz) with a Gaussian function of between 0.3 and 0.5 to optimize digital resolution. Zero-fill the F_1 dimension to 8192 complex data points to give a final digital resolution of about 0.4 Hz per point.

3. Use the spectral simulation approach described in *Protocol 3* below to improve the accuracy of the measured $^3J_{H^N\text{-}H\alpha}$ values.

2.1.3 The J-modulated [^{15}N-^{1}H]-COSY experiment (4)

The pulse sequence used for this experiment is the same as the one used for the HMQC experiment except that an additional mixing period t_m is included at the end of the sequence. To simplify the data analysis, it is advantageous to perform this experiment without heteronuclear decoupling. The 2D spectrum shows cross-peaks composed of two antiphase components separated along ω_2 by the one bond heteronuclear ^{15}N-^{1}H coupling constant. The intensity of the heteronuclear cross-peak is modulated by $\cos(\pi \cdot {}^3J_{HN\text{-}H\alpha} \cdot t_m)$, as shown in *Figure 3(a)*; hence the ${}^3J_{HN\text{-}H\alpha}$ value can be estimated by varying the t_m period in the experiment. With a small series of three or four experiments with suitably selected t_m values, this experiment can be used to classify the ${}^3J_{HN\text{-}H\alpha}$ coupling constants into groups according to their magnitudes. In fact, as shown in *Figure 3(c)*, it is possible to choose t_m values so as to distinguish

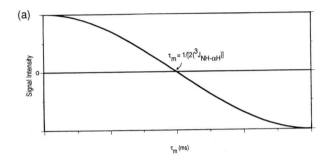

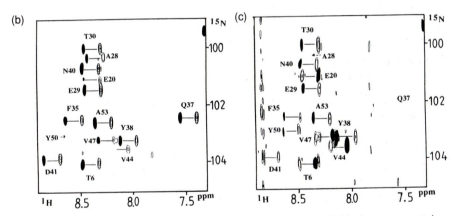

Figure 3. (a) Plot of signal intensity versus mixing times for individual components in a cross-peak in a J-modulated [^{15}N-^{1}H]-COSY experiment. Theoretically, a null in intensity occurs when t_m is set to $1/\{2({}^3J_{HN\text{-}H\alpha})\}$. (b) Part of the ^{15}N (ω_1-axis) ^{1}H (ω_2-axis) region of the HMQC-J spectrum of ^{15}N-labelled IgG-binding domain II of protein G, at 297 K, pH 4.2, t_m = 50 msec. Each cross-peak is identified by a straight line connecting the two antiphase components of the fine-structure which are separated by the one bond hetero-nuclear coupling constant $^1J_{(N\text{-}HN)}$. Positive peaks are shaded in black; negative peaks contain several contour levels. At this short mixing time, the cross-peaks for large and small ${}^3J_{HN\text{-}H\alpha}$ have similar phase characteristics. The cross-peak from Y50 is nulled at this mixing time. (c) Same as (b) expect that t_m = 100 msec. Note that cross-peaks from residues T6, E20, V47, and V44 are from the β-sheet region of the protein, have ${}^3J_{HN\text{-}H\alpha} > 9$ Hz and hence have phase characteristics opposite to the other cross-peaks which are from residues with ${}^3J_{HN\text{-}H\alpha}$ values < 7 Hz. The cross-peaks from A28 are nulled at this mixing time.

amide protons in a β-sheet from those in a helix; at appropriate t_m values the two sets of cross-peak have opposite phase characteristics. For more accurate coupling constant measurements it is necessary to consider effects such as relaxation and the presence of other coherences when plotting the time-course of the signal intensity. The main limitation of this method is that long t_m delays are necessary for small coupling constants, leading to decreased sensitivity due to relaxation.

2.1.4 The HETERO-RELAY experiment (5)

The third method is a triple resonance experiment involving the transfer process $^1H^\alpha \rightarrow {}^{13}C^\alpha(\omega_1) \rightarrow$ selective $^{15}N \rightarrow {}^1H^N(\omega_2)$. It relies on the displacement of the two components in a ^{13}C-$^1H^N$ correlation cross-peak, which gives a measurement of the passive coupling constant, in this case $^3J_{H^N\text{-}H^\alpha}$. A schematic representation of the ^{13}C-amide proton cross-peak is shown in *Figure 4(a)*. This method has not been used routinely for larger proteins and its general usefulness has yet to be proven.

2.1.5 Spectral simulation

With the use of heteronuclear experiments the molecular weight threshold below which scalar couplings can be determined with some accuracy is increased; it is possible to obtain fairly good estimates of the $^3J_{H^N\text{-}H^\alpha}$ values for proteins of M_r up to about 16000. To obtain quantitative values of $^3J_{H^N\text{-}H^\alpha}$ it is necessary to perform spectral simulations to account for the dispersive elements present in the cross-peak and for any overlap of the multiplet components. Multiplet overlap leads to overestimation of coupling constants in a DQF-COSY spectrum and their underestimation in the heteronuclear spectra. This problem can be severe enough to make the splittings in all the DQF-COSY cross-peaks appear to be similar and to lead to the disappearance of splittings in the HMQC-J cross-peaks when the coupling constant values are small. Contributions from the dispersive components to the lineshape which are present in HMQC-J spectrum lead to an overestimation of coupling constants. *Protocol 3* describes the spectrum simulation procedure that can be followed to correct for these effects (7). It can be applied to both homo- and heteronuclear spectra.

Protocol 3. Spectral simulation

1. Select the appropriate cross-sections of a H^N-H^α cross-peak in the 2D spectrum.

2. An initial guess of the value of $^3J_{H^N\text{-}H^\alpha}$ is estimated from the separation between the components of a doublet (antiphase for COSY spectra, in-phase for ^{15}N-^1H HMQC-J spectra).

3. Simulate the cross-sections using the initial guess values of $^3J_{H^N\text{-}H^\alpha}$ and the

linewidth, together with the appropriate apodization function, in the following basic equations:

$$f(real) = G \sin(2\pi\omega_1 t_1)\cos(\pi J_{H^N-H^\alpha} \omega_1)\exp(-\pi t_1/T_2)$$
$$f(imaginary) = G \cos(2\pi\omega_1 t_1)\cos(\pi J_{H^N-H^\alpha} \omega_1)\exp(-\pi t_1/T_2)$$

where G is the window function used in the processing of the original 2D data set—for example, Gaussian or sine-bell squared.

4. Fit the simulated and experimental spectra using a squares fitting procedure such as a downhill simplex method (the protocol can be taken directly from reference 8). The parameters fitted will be the coupling constant, linewidth, and chemical shift.

5. For cross-sections where the signal-to-noise ratio is less than 5:1, the fitting procedure will not work well. In this case, it is necessary to simulate spectra with several initial pairs of coupling constant and linewidth values that give the experimentally observed splitting. These simulated spectra are then compared with the experimental cross-sections and the best agreement with the experimental coupling constant and linewidth is selected.

To estimate the precision of the corrected values of $^3J_{H^N-H^\alpha}$ spectra can be simulated for different initial guesses of the coupling constant and linewidth values. Values of $^3J_{H^N-H^\alpha}$ within ranges of $\pm$ 0.4 Hz for large couplings and $\pm$ 1.0 Hz for small couplings should be obtained.

2.2 Determination of $^3J_{H^\alpha-H^\beta}$ coupling constants and stereospecific assignments

The values of the two coupling constant across the C_α—C_β bond, $^3J_{H^\alpha-H^{\beta2}}$ and $^3J_{H^\alpha-H^{\beta3}}$, give information about the χ_1 dihedral angle in amino acid side-chains, as discussed in Chapter 11. Furthermore, when used in conjunction with NOE data, measurement of these coupling constants allows stereo-specific assignment of the prochiral β-methylene protons to be made.

The conformation about the C_α—C_β bond is very likely either to correspond to one of the three low energy rotamers (see *Table 2*), or to be an average of two or more of these rotamers. Thus, it may be possible to identify the rotameric state for a given side chain, and hence the stereospecific assignment of its β-protons, even if precise values of $^3J_{H^\alpha-H^\beta}$ can not be determined. *Table 2* summarizes some of the relevant NMR parameters, and shows how stereo-specific assignments of β-methylene protons can be made on the basis of:

- $^3J_{H^\alpha-H^{\beta2}}$ and $^3J_{H^\alpha-H^{\beta3}}$
- the intraresidue NOEs $d_{\alpha\beta2}$, $d_{\alpha\beta3}$, $d_{N\beta2}$ and $d_{N\beta3}$

For each methylene proton a consistent set of at least three of these parameters must be obtained before any reliable stereospecific assignment can be

(a)

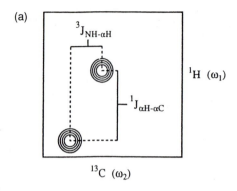

(b)

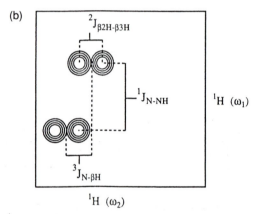

Figure 4. Schematic representation of cross-peaks from experiments that give heteronuclear coupling constant values for experiments carried out on ^{13}C-labelled (a) or ^{15}N-labelled (b) protein. (a) A ^{13}C (ω_1-axis)—H$^\alpha$ ^{1}H (ω_2-axis) cross-peak from a HETERO-RELAY experiment. The displacement in the ω_2 dimension gives the value of the homonuclear $^3J_{H^N-H\alpha}$ value. (b) A H^N-H$^\beta$ NOESY cross-peak; the displacement in the ω_2 dimension gives the value of the heteronuclear $^3J_{^{15}N-H\beta}$ value.

obtained. The heteronuclear long-range coupling constant can help to refine the assignments made in small to medium size proteins, particularly where there is ambiguity in the NOE data (see later).

For small proteins, the experiment most commonly used to obtain $^3J_{H^\alpha-H^\beta}$ values is one of the E.COSY-type (9), for example the P.E.COSY experiment (although DQF-COSY can be used for residues with single β-protons). To use this method, the two methylene protons must be sufficiently well separated in chemical shift. The experiment must be performed with as high a digital resolution as possible, bearing in mind that the optimum t_2 acquisition time and $t_{1,max}$ values that can be used depend on the T_2 relaxation times (see *Protocol 1*). It may be necessary to increase the temperature of the sample in

Table 2. NMR parameters used to define the three possible staggered conformations around the Cα—Cβ bond

χ$_1$	60°	180°	−60°
$^3J_{Hα\text{-}Hβ2}$ (Hz)	2.6–5.1	2.6–5.1	11.8–14.0
$^3J_{Hα\text{-}Hβ3}$ (Hz)	2.6–5.1	11.8–14.0	2.6–5.1
$^3J_{HN\text{-}Hβ2}$ (Hz)	−6	−1	−1
$^3J_{HN\text{-}Hβ3}$ (Hz)	−1	−1	−6
NOE(Hα-Hβ2)[a]	s	s	m–w [b]
NOE(Hα-Hβ3)	s	w–m [b]	s
NOE(HN-Hβ2)	w–m [b]	s–m [b]	s
r (HN-Hβ2)[c]	3.5–4.0	2.5–3.4	2.2–3.1
NOE(HN-Hβ3)	s–m [b]	s	w–m
r (HN-Hβ3)	2.5–3.4	2.2–3.1	3.5–4.0

[a] s = strong, m = medium, w = weak.
[b] For large proteins, when efficient spin diffusion is prevalent, the NOEs are biased towards stronger intensities. In this case a ROESY experiment is advisable.
[c] r (HN-Hβ) is the approximate interproton distance in Å.

order to reduce linewidths, although too high a temperature will increase internal mobility and hence cause averaging of the $^3J_{Hα\text{-}Hβ}$ values.

The P.E.COSY cross-peaks have characteristic multiplet patterns as shown in *Figure 5*. These patterns result from typical Hα-Hβ coupling constants for the different side chain conformations. The correct method of obtaining $^3J_{Hα\text{-}Hβ}$ values from the P.E.COSY spectrum is to measure the displacement caused by the passive scalar coupling in the well-resolved Hα-Hβ cross-peaks as shown in *Figure 5*. Accurate measurements of the coupling constants from this type of experiment require that the active coupling constant is significantly larger than the linewidth. For many proteins the intrinsic linewidths of the resonance can lead to cancellation of the Hα-Hβ cross-peak where the active $^3J_{Hα\text{-}Hβ}$ coupling is small (< 4 Hz). Consequently, in the staggered conformation, where χ$_1$ = −60° or 180°, only the lower cross-peak of *Figure 5(a)* is observed, as illustrated in *Figure 5(c)*. Nevertheless, it is still possible to estimate both values of $^3J_{Hα\text{-}Hβ}$, the small value being measured as the displacement of the large doublet and the large active coupling obtained directly from the large antiphase splitting. In this case, it may be necessary to use spectral simulations to obtain accurate coupling constant values. To establish the possible rotamer present a qualitative analysis of the E.COSY-type spectrum is sufficient in most cases.

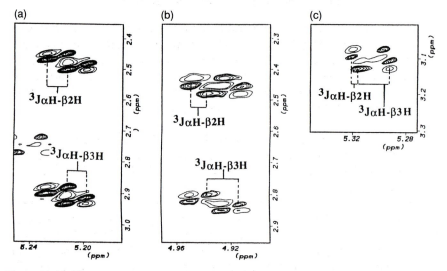

Figure 5. H$^\alpha$-H$^\beta$ cross-peaks from the P.E.COSY experiment on a sample of IgG-binding domain II of protein G in which all the amide protons had been exchanged for deuterium at 297 K, pH 4.2. (a) $^3J_{H^\alpha-H^\beta}$ values are averaged (residue N13); (b) one large and one small $^3J_{H^\alpha-H^\beta}$ value (residue Y50); (c) same as (b) except that the cross-peak with the small active coupling is cancelled (see text). When both $^3J_{H^\alpha-H^\beta}$ values are small, the multiplet pattern is similar to that in (a) except that the separation between the antiphase components of each H$^\alpha$-H$^\beta$ cross-peak is smaller than that shown in the figure.

Three combinations of $^3J_{H^\alpha-H^{\beta2}}$ and $^3J_{H^\alpha-H^{\beta3}}$ values are likely to be observed.

(a) If both $^3J_{H^\alpha-H^\beta}$ values are small (< 4 Hz) then χ_1 must be around 60°. Stereospecific assignment will depend on measurements of long-range heteronuclear coupling constants ($^3J_{15N^\alpha-H^\beta}$ or $^3J_{13C'-H^\beta}$; Section 2.3) or H^N-H$^\beta$ NOEs (see below).

(b) If one of the two $^3J_{H^\alpha-H^\beta}$ coupling constants is large (> 10 Hz) then the other must be small and χ_1 lies close to either −60° or 180°. In this case the homonuclear coupling constants alone can not distinguish between χ_1 = 180° and χ_1 = −60°. This distinction has most frequently been made from quantitative measurements of the NOEs between the amide and the β-protons (*Table 2*; see below). Heteronuclear coupling constants (Section 2.3) provide additional information that can assist with a stereospecific assignment.

(c) If the two $^3J_{H^\alpha-H^\beta}$ values are approximately equal, with values in the range from 6 to 8 Hz, and the relative intensities of the two H$^\alpha$-H$^\beta$ NOEs are also approximately equal, then the conformation about the C$_\alpha$—C$_\beta$ bond of the side chain is disordered. No stereospecific assignment can be made in this case. It must be pointed out that it may not be easy to distinguish a $\chi_1 = 60°$ rotamer ($^3J_{H\alpha\text{-}H\beta} \approx 4$ Hz) from one where conformational averaging exists ($^3J_{H\alpha\text{-}H\beta} = 6\text{–}8$ Hz), solely from the multiplet patterns in the P.E.COSY spectrum, due to limitations in the digital resolution and the finite linewidths. It is then necessary to examine the $d_{\alpha\beta}$ NOEs; in the $\chi_1 = 60°$ conformation, both $d_{\alpha\beta2}$ and $d_{\alpha\beta3}$ are strong.

It is clear from this discussion that the H^N-H^β NOE plays a crucial role in the stereospecific assignment of β-proton resonances. However, in many proteins the difference in magnitude between the H^N-$H^{\beta2}$ and H^N-$H^{\beta3}$ NOEs can become negligible as a result of efficient spin diffusion. In this case a ROESY experiment may be helpful. ROESY spectra contain cross-peaks resulting from both dipolar and scalar couplings; in proteins, these two cross-peaks have, respectively, the opposite and the same phase as the diagonal peak. If both scalar (either direct or relayed, as in TOCSY experiments) and dipolar couplings exist between a particular proton pair, such as an H^N-H^β cross-peak, the resultant phase of this cross-peak is the algebraic sum of the two contributions. Therefore, if the H^N-H^β distance is short, the corresponding ROESY cross-peak is of opposite phase to the diagonal; a large H^N-H^β distance will give an H^N-H^β ROESY cross-peak which is of the same phase as the diagonal peak.

For large proteins ($M_r > 15\,000$) it may be impossible to extract $^3J_{H\alpha\text{-}H\beta}$ values from the P.E.COSY spectra, due to increased linewidths and spectral overlap. Under these circumstances, qualitative estimates of the relative populations of the rotamers can be derived from the intensities of the resolved H^N-H^β cross-peaks in a 3D ^{15}N-separated HOHAHA-HMQC spectrum recorded at short mixing times (10). These intensities depend on the size of the relevant J-values (in this case $^3J_{HN\text{-}H\alpha}$ and $^3J_{H\alpha\text{-}H\beta}$) and the relaxation rates of all the protons involved in the magnetization transfer pathway (H^N, H^α, and H^β). If the two β-methylene protons are assumed to have similar relaxation times, good estimates of the relative intensities of the H^N-$H^{\beta2}$ and H^N-$H^{\beta3}$ cross-peaks for a particular amino acid residue can be obtained since $^3J_{HN\text{-}H\alpha}$ and the relaxation times of the amide and α-protons have the same effect on *both* the $H^N \rightarrow H^\beta$ magnetization transfers.

As many stereospecific assignments as possible should be made before embarking on the structural determination process, since it has been shown that this leads to a marked improvement in the quality of the structures obtained (11). Stereospecific assignments can be carried out either by a manual screening of the spin-spin coupling constants ($^3J_{H\alpha\text{-}H\beta2}$, $^3J_{H\alpha\text{-}H\beta3}$, $^3J_{HN\text{-}H\alpha}$) and the intraresidue and sequential NOEs, or by using programs such as HABAS (12) and STEREOSEARCH (13), which require the coupling constant and NOE data listed above as input data. Additional stereospecific assignments, using the program GLOMSA (14), can be made when preliminary structures become available.

2.2.1 Stereospecific assignments of valine and leucine methyl groups

The rotamers around the α-β bond of the valine residue are shown in *Table 3*, together with the relevant NMR parameters. The *trans* conformation usually predominates, with $^3J_{Hα-Hβ}$ being about 12 Hz. The first step in the assignment process is to obtain an accurate $^3J_{Hα-Hβ}$ value, using the DQF-COSY experiment together with spectral simulations. The stereospecific assignment can be made manually by examining the $d_{HN-Hγ}(i,i)$ (one methyl group shows a much larger NOE to the amide proton than the other) and the $d_{Hγ-HN}(i, i + 1)$ values. The other two rotamers can in principle be distinguished from each other on the basis of the $d_{HN-Hβ}$ NOE; this should be large for the g^- conformation and small for the g^+ conformation. The stereospecific assignment of the δ protons of leucines should be possible on the basis of long-range NOEs once its β-protons have been stereospecifically assigned. This is, however, rarely done.

2.3 Heteronuclear scalar coupling constants

The measurement of long-range heteronuclear scalar coupling constants, in particular $^{15}N^α$-$H^β$ vicinal coupling constants, can give additional dihedral angle restraints that are otherwise not available from homonuclear experiments. These restraints are especially useful for defining the local structures in a polypeptide chain (*Table 4*). Heteronuclear coupling constants can in principle be obtained provided that ^{13}C- or ^{15}N-labelled proteins are available, and secondly that the cross-peaks from which these coupling constants are to be extracted are sufficiently well resolved from other cross-peaks.

Table 3. NMR parameters used to define the three possible rotamers around the Cα—Cβ bond

Conformation	g^+	g^+	t
$^3J_{Hα-Hβ}$	2.6–5.1	2.6–5.1	11.8–14.0
NOE($H^α$-$H^{γ1}$) [a]	s	w–m [b]	s
NOE($H^α$-$H^{γ2}$)	w–m [b]	s	s
NOE(H^N-$H^{γ1}$)	s	s	w–m [b]
NOE(H^N-$H^{γ2}$)	s	w–m [b]	s

[a] s = strong, m = medium, w = weak.
[b] For large proteins, when efficient spin diffusion is prevalent, the NOEs are biased towards stronger intensities.

Table 4. Summary of torsion angles that can be defined by heteronuclear long-range coupling constants and typical values for these coupling constants

Torsion angle	Heteronuclear coupling [a]	J-values (Hz)
ψ	$^3J_{H\alpha(i)\text{-}^{15}N(i+1)}$	0–(−9)
ϕ	$^3J_{HN(i)\text{-}^{13}C\beta(i)}$	0–(−6)
χ_1	$^3J_{^{15}N\alpha(i)\text{-}H\beta(i)}$ [b]	(−1)–(−6)
$\chi_{2,3,4}$ [c]	$^3J_{H\beta\text{-}^{15}N}$ (side chain)	(−3)–(−4.5) [d]

[a] Only heteronuclear coupling constants that can be deduced from 2D homonuclear ^1H NMR experiments using ^{13}C or ^{15}N isotopically-enriched proteins are given. Heteronuclear long-range coupling constants such as $^3J_{C'\text{-}HN}$ and $^3J_{C'\text{-}H\alpha(i+1)}$ are also useful for defining torsional angles and these must be obtained using heteronuclear 2D methods.

[b] $^3J_{^{15}N(i)\text{-}H\beta(i)}$ approximately −1 Hz for nitrogen *gauche* to β-proton; $^3J_{^{15}N(i)\text{-}H\beta(i)}$ approximately −6 Hz for nitrogen trans to β-proton.

[c] χ_2 for Asn, χ_3 for Gln, χ_4 for Arg.

[d] Averaged values.

One way that has been suggested for obtaining some of these long-range $^3J_{^{15}N\alpha\text{-}H\beta}$ (i,i) and $^3J_{H\alpha\text{-}^{15}N\alpha}(i,i+1)$ coupling constants is to perform experiments such as NOESY, HOHAHA, ROESY, on ^{15}N- or ^{13}C-labelled proteins (15). The experiments chosen must give spectra that contain cross-peaks between a proton directly bonded to ^{15}N or ^{13}C and a proton which has a long-range scalar coupling to the *same* nitrogen or carbon. The cross-peak will have an E.COSY-type format (see above) and the displacement of the multiplet components will give the passive long-range coupling. A schematic representation of the H^N-H^α cross-peak patterns obtained with the NOESY pulse sequence is shown in *Figure 4(b)*. The region below the diagonal is normally used for the analysis of these spectra to take advantage of the inherently better digital resolution in the ω_2 dimension. The splitting in ω_1 is the large one bond heteronuclear coupling (active coupling). In practice, it is difficult to obtain these coupling constants reliably for large proteins since first these long-range heteronuclear coupling constants are in the range of 0 to −9 Hz, less than the proton linewidths, and secondly the H^N-H^β and the H^N-H^α regions of the 2D spectra are very crowded. The latter problem can be overcome by using proteins where only selected amino acid residues are labelled (see Chapter 5).

2.4 Amide hydrogen exchange rates

Slow exchange of an amide proton usually indicates that it is involved in an intramolecular hydrogen bond, although exchange can also be slowed by simple inaccessibility to solvent. If additional information from NOEs and coupling constants suggest that the residue concerned is part of a regular secondary structure, then it is safe to assume that the amide proton is part of the hydrogen bonding pattern characteristic of this structure and that the hydrogen bond acceptor can be identified. Amide proton exchange rates are usually estimated by following the intensity of the amide proton cross-peak as a function of time after dissolving a fully protonated sample of the protein in ^2H_2O. The 2D experiments that are commonly used to follow this time-course are TOCSY, COSY or, if an ^{15}N-labelled protein is available, a ^{15}N-^1H HSQC experiment (see also Chapter 5). Typical time intervals used are 4 hours, 8 hours, 24 hours, 48 hours, and several days after dissolution of the protein sample. When carrying out these experiments, it is essential to keep the total data acquisition time to a minimum for the early time intervals. This is normally achieved by using acquisition techniques that minimize any artefacts which may appear as a result of a reduction in the number of phase-cycling steps in the pulse sequence (16). Amide protons that are still observed after 24 hours in ^2H_2O are considered 'slowly exchanging'; since the purpose of this experiment is to support the NOE and scalar coupling data in delineating the secondary structure elements of the protein, it is not necessary to obtain a very precise measurement of the amide hydrogen exchange rates.

3. Secondary structure identification

Secondary structures in proteins have characteristic NOE patterns and scalar coupling constants, as summarized in *Table 5* and illustrated in *Figure 6*. Knowledge of the secondary structure elements in a protein helps with the structural determination process in two main ways. First, it allows the approximation of the dihedral angles that are standard for regular secondary structures. This approach adds to the number of input restraints without needing to measure every resolvable coupling constant in the NMR spectrum. Secondly, hydrogen bonding restraints may be added in regions assigned to a regular secondary structure, although care must be taken to ensure that the amide exchange rate data concurs with the secondary structure already deduced from the NOEs and coupling constants. Any discrepancy in the latter case implies a distortion of the regular structure or the presence of flexible regions.

The proton–proton distance characteristics of α-helices and β-sheets are illustrated for a polypeptide backbone in Chapter 11. Evidence of regular secondary structures from the NOE patterns and H^N-H^α coupling constants can be corroborated by analysis of the amide exchange rates. The presence of turns are implied when isolated strong d_{NN} and $d_{\alpha N}$ NOEs are observed; however, the type of turn is more difficult to identify. *Protocol 4* describes the procedure for determining secondary structure elements in a protein using NMR data.

Table 5. Parameters that are useful for secondary structure identification: short sequential and medium-range ^{1}H-^{1}H distances,[a] vicinal coupling constants, and amide hydrogen exchange rates

Parameter	α-Helix	3$_{10}$-Helix	β-Anti-parallel[b]	β-Parallel[b]
$d_{\alpha N}(i, i)$	2.6	2.6	2.8	2.8
$d_{\alpha N}(i, i + 1)$	3.5	3.4	2.2	2.2
$d_{\alpha N}(i, i + 2)$	4.4	**3.8**		
$d_{\alpha N}(i, i + 3)$	**3.4**	3.3		
$d_{\alpha N}(i, i + 4)$	4.2	(> 4.5)		
$d_{NN}(i, i + 1)$	**2.8**	2.6	4.3	4.2
$d_{NN}(i, i + 2)$	4.2	4.1		
$d_{\beta N}(i, i + 1)$	2.5–4.1	2.9–4.4	3.2–4.5	3.7–4.7
$d_{\alpha \beta}(i, i + 3)$	2.5–4.4	3.1–5.1		
$d_{\alpha \alpha}(i, j)$			2.3	4.8
$d_{\alpha N}(i, j)$			3.2	3.0
$d_{NN}(i, j)$			3.3	4.0
$^3J_{HN\text{-}H\alpha}$	(< 4)	(< 4)	(> 9)	(> 9)
NH exchange rate	**slow**[c]	**slow**[c]	**slow**[d]	**slow**[d]

[a] The most readily available parameters are highlighted in **bold**.
[b] $d_{\alpha\alpha}(i, j)$, $d_{\alpha N}(i, j)$, and $d_{NN}(i, j)$ refer to interstrand distances.
[c] The first four residues in the α-helix and the first three residues in the 3$_{10}$-helix will have fast amide proton exchange rates.
[d] Every second residue in the flanking strand will have slow amide proton exchange rates.

Protocol 4. Procedure for delineating secondary structure elements

1. Examine the NOESY spectrum in H_2O at mixing times ranging from 30 msec to 100 msec, concentrating first on the H^N-H^N and H^N-H^α regions of the spectrum.

2. (a) In the spectrum at short mixing times ($\tau_m < 50$ msec), the presence of a stretch of strong sequential $d_{NN}(i,i + 1)$ NOEs indicates that a helical secondary structure exists.

   (b) If (a) is true, examine the H^N-H^α region for the same mixing times. The sequential $d_{\alpha N}(i,i + 1)$ should be of lower intensity than both the $d_{NN}(i,i + 1)$ and the intraresidue $d_{\alpha N}(i,i)$ cross-peaks if a helix is present.

   (c) If (b) is true, analyse the H^N-H^α region at longer mixing times (> 80 msec) and look for $d_{\alpha N}(i,i + 2)$ and $d_{\alpha N}(i,i + 3)$ cross-peaks. These should be weak but observable in a regular α-helix.

   (d) Examine the $^3J_{HN\text{-}H\alpha}$ values for the residues that appear to be involved in the helix; these should be < 5 Hz.

   (e) Check the amide exchange rates for the same residues as (d); except

Protocol 4. *Continued*

for the first three residues in the deduced stretch of helix, the other amide protons may well be slowly exchanging.

3. (a) Examine the H^N-H^α region at short mixing times (< 50 msec). Stretches of very strong sequential $d_{\alpha N}(i, i + 1)$ cross-peaks are indicative of the presence of an extended polypeptide chain. These cross-peaks are the most intense ones in this region of the spectrum.

(b) It is necessary to establish whether a regular β-sheet is present or simply an extended segment of polypeptide chain. A β-sheet exists if all the following three criteria are satisfied: large $^3J_{H^N-H^\alpha}$ (> 10 Hz), very weak (almost unobservable) $d_{NN}(i, i + 1)$ cross-peaks, and slowly exchanging amide protons (for every second amino acid residue in strands at the edge of the sheet). In addition, the α-proton chemical shifts in a β-sheet are generally further downfield than those in a helix.

(c) If a β-sheet is present, is it parallel or anti-parallel? This can be deduced from the interstrand $d_{\alpha\alpha}(i, j)$ characteristics (see *Table 5*).

4. Isolated $d_{NN}(i, i + 2)$ or $d_{\alpha N}(i, i + 2)$ connectivities or two consecutive $d_{NN}(i, i + 1)$ connectivities are indicative of a tight turn. The other characteristics for the commonly found type I, I', II, and II' turns are $^3J_{H^N-H^\alpha}$ of between 4 to 9 Hz, and a hydrogen bond between the carbonyl of residue one and the amide proton of residue four in the turn, although this slowly exchanging amide proton can be difficult to identify.

In addition to the steps outlined in *Protocol 4* it is informative, where possible, to analyse the ^{15}N-^{1}H HMQC-J experiment qualitatively. The cross-peaks can be conveniently grouped into three distinct classed based on the splitting in the ^{15}N dimension: well resolved, small, and no measurable splittings (*Figure 2*). There is a good correlation between the secondary structure in which the residues occur and these qualitative classes of H^N-H^α splitting. The well resolved 'doublet' resonances are all from residues located in the β-sheets of the protein; signals with reduced splittings generally arise from residues located at the end of the β-sheet or the α-helix secondary structure, in the turns, or at the termini of the protein where conformational averaging is apparent. Residues located well within the α-helix have H^N-H^α cross-peaks that contain no discernible splittings. Therefore, using a qualitative analysis of the spectrum, it is possible to establish clearly the secondary structure elements which are present in a protein.

The points at which these secondary structure elements, particularly the α-helices, begin and end may not be obvious when using the above procedures for secondary structure identification. Accurate measurements of $^3J_{H^N-H^\alpha}$ help to improve the definitions of these termini. For the α-helices, these coupling constants have to be used in conjunction with the $d_{NN}(i, i +$

1,2,3), $d_{\alpha N}(i, i + 1,2,3)$, and $d_{\alpha\beta}(i, i + 3)$ NOEs. Accurate $^3J_{H^N\text{-}H^\alpha}$ values can also help with the assessment of the degree of regularity of the secondary structures. For a β-sheet, deviation from ideality is indicated by a reduction in the $^3J_{H^N\text{-}H^\alpha}$ values (from 8–10 Hz down to 5–7 Hz) together with the appearance of $d_{NN}(i, i + 1)$ NOEs. Distortions in α-helices are indicated by $^3J_{H^N\text{-}H^\alpha}$ values that are larger than those (< 6.0 Hz) expected within a perfect α-helical stretch.

4. Nuclear Overhauser enhancement and distance restraints

The principal structural information from NMR comes from NOE cross-peaks in 2D (nD) spectra. Before using NOEs for structure calculations one has to address three main problems:

- NOE cross-peak intensity measurement
- NOE cross-peak assignment
- converting NOEs into distance information

The method for dealing with these problems depend to a large extent on the complexity of the molecule under investigation and the aim of the study.

Prior to measuring cross-peak intensities, it is important to ensure that the spectrum has no severe base-plane distortions and that the window functions used during processing do not severely modify the intensities. The intensity of strong well resolved peaks can then be measured by simple numerical integration. Overlapping peaks or peaks close to distorted base-plane regions require more sophisticated approaches; these latter methods can also improve the accuracy of the measurement of the resolved peaks when signal-to-noise ratio is low.

The assignment of NOE cross-peaks for small molecules is straightforward and can usually be done on the basis of chemical shifts alone. Spectra of large proteins exhibit a great deal of resonance overlap and hence ambiguity in the assignment. Additional information has to be used to resolve this ambiguity.

When the overall motion of a molecule is fast the surrounding protons (next-nearest neighbours) have negligible influence on the intensity of the NOE between a given proton pair, and the intensity is directly related to the interproton distance. The main problem in such a case is internal mobility, which can greatly change the observed intensity. For slow overall motion, corresponding to large molecules such as proteins, there is an additional complication in that the arrangement of the next-nearest neighbour protons around a given pair has to be considered in order to derive the interproton distance—the problem of spin diffusion.

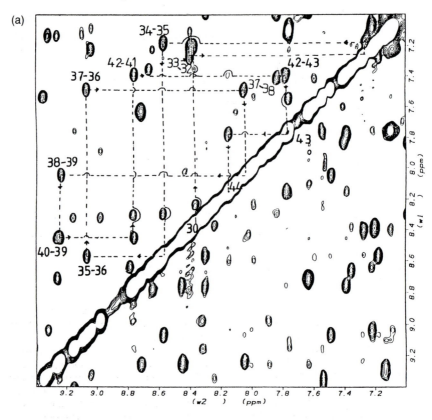

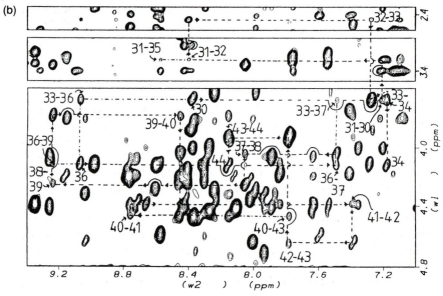

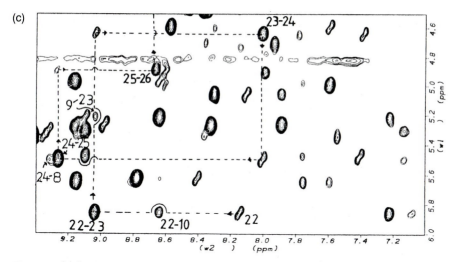

Figure 6. ^{1}H-^{1}H NOESY spectrum of the IgG-binding domain (II) of protein G, at 297 K, pH 4.2, t_m = 100 msec. 'Walks' along segments of the polypeptide backbone are indicated in each of the figures. (a) Part of the H^N-H^N region showing the $d_{NN}(i,i + 1)$ connectivities from residues T30 to V44, which are all within the α-helix of the protein. Although the α-helix starts from residue 28, the $d_{NN}(i,i + 1)$ connectivities from residues 28 to 30 are difficult to determine because their amide proton chemical shifts are very close together. Each cross-peak is denoted by two residue numbers: the first number corresponds to the residues whose H^N chemical shifts is in the ω_1 dimension; the second number, the amide proton chemical shifts in the ω_2 dimension. (b) H^N-H$^\alpha$ region showing resolved sequential H^N-H$^\alpha$ connectivities for residues in the α-helix. Interresidue $d_{\alpha N}(i,i + 1)$, $d_{\alpha N}(i,i + 3)$, and $d_{\alpha N}(i,i + 4)$ connectivities are labelled. Each cross-peak is denoted by two residue numbers in the following manner: i(H$^\alpha$, ω_1-axis)—j(H^N, ω_2-axis). (c) H^N-H$^\alpha$ region showing the resolved sequential and interstrand $d_{N\alpha}$ connectivities for residues in the β-sheet, which is made up of residues 6 to 25, with a type II turn from residues 14 to 17. Only the interresidue cross-peaks are labelled. The cross-peaks are denoted in the same manner as in (b).

4.1 Measurement of NOE cross-peak intensities

For an N-dimensional spectrum the intensity of the NOE cross-peak is represented by its integral value and does not depend upon the lineshape and multiplet structure. In the case of 2D spectra, the integral corresponds to the *volume* of the cross-peak. The factors that influence the accuracy of measurements of NOE cross-peak intensities are: lineshape, window function, peak overlap, and method of measurement. The routines useful for volume measurement and their characteristics are summarized in *Table 6*, and a flowchart describing how to choose which routine to use in different cases is shown in *Figure 7*. The simplest way to estimate cross-peak volume is to use the *amplitude of the maximum*, which is either measured directly or by counting the number of contour levels on the contour plot (algorithm 1, *Table 6*). This is strictly valid only if all cross-peaks have the same lineshape. This is

Table 6. Characteristics of algorithms used for NOE cross-peak volume measurements

Algorithm	Advantage	Disadvantage	Application
1. Estimation of the volumes from the maximum intensity of the cross-peak $V = kI_{max}$.	Easy to use. High sensitivity.	Accurate values only if lineshapes and linewidth are the same for all peaks or initial digital resolution is less than 30 Hz/point. Sensitive to peak overlap.	Fast way to create crude constraints.
2. Summation of spectrum intensities within a certain region around cross-peak.	Easy to use. No assumption on the lineshapes or digital resolution	Strong dependence on the region of integration, peak overlap, and signal-to-noise ratio.	Measurement of the volumes for strong well resolved peaks.
3. Evaluation of the volumes from the linewidths and amplitude in maximum $V = k\Delta\omega_1\Delta\omega_2 I_{max}$.	Easy to use. High sensitivity. Does not depend on the linewidth	Depends on the lineshape and peak overlap	Measurement of the volumes of sufficiently resolved peaks with similar lineshapes.
4. Fit rows and columns in the spectrum to the set of reference lines.	Fully automatic measurement of volumes once the set of reference lines created. Allows resolution of overlapping peaks.	Computationally extensive. Requires a complete set of reference lines which is time-consuming to create.	Measurement of volumes of poorly resolved peaks and peaks with different lineshapes.
5. Curve fitting in one row and one column for each peak to a theoretical lineshape	High sensitivity. Allows to resolve overlapped peaks.	Time-consuming due to required manual interaction.	Measurement of volumes of poorly resolved peaks and peaks with different lineshapes.

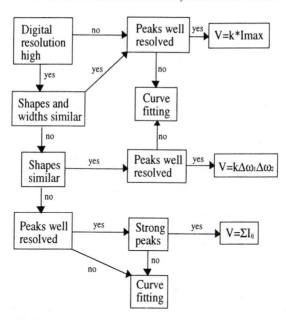

Figure 7. Schematic diagram of the protocol to use for cross-peak volume measurements.

the case when the digital resolution during acquisition is ~ 30 Hz per point or more—much larger than the intrinsic linewidth and coupling constants—so that the lineshape is mostly determined by the degree of truncation. This is particularly true in 3D spectra where the maximum amplitude usually represents cross-peak intensity to a sufficient degree of accuracy. The cross-peak volume is also influenced by the window function applied during data processing. The absolute value of the integral is proportional to the amplitude of the window function at the first data point. Thus integral intensities obtained with different window functions should be scaled accordingly.

When the digital resolution is sufficiently high the lineshape must be considered during the volume estimation. The volume of the cross-peak is roughly proportional to the linewidth which allows one to estimate the error introduced by neglecting the lineshape difference. If counting of contour levels is used to determined volumes, the approximate correction factor can be estimated as proportional to the area of the contour drawn at the peak half-height.

A simple method for the measurement of volumes is the numerical integration method, realized as the summation of spectrum amplitudes within a certain region around the cross-peak maximum (17, 18) (algorithm 2, *Table 6*). Unfortunately this method is extremely sensitive to the selection of the region, and to spectral artefacts and cross-peak overlap. The method can thus be used only for measuring the volumes of intense well resolved cross-peaks, and for rough estimation of the volumes in other cases. Several methods have been proposed to make volume measurements more accurate and robust. Under the assumption that the lineshape is the same for all cross-peaks, the only difference being in the linewidth, the volumes can be obtained from the linewidths and maximum amplitudes as represented by the following equation (19):

$$V = K.\Delta_1\Delta_2.I \qquad [1]$$

where I is the amplitude of the peak maximum, K the geometry factor, and Δ_1 and Δ_2 are the linewidths along ω_1 and ω_1 respectively (algorithm 3, *Table 6*). This method is liable to errors when the lineshapes of different signals differ due to multiplet structure but it seems to be less sensitive to the peak overlap than numerical integration. When applying this method, lineshapes are usually made more similar by introducing additional line-broadening.

Overlapped cross-peaks are difficult to treat by the above methods. In such cases integration can be achieved by fitting all rows and columns of the 2D NOESY spectrum to a set of reference lineshapes (20, 21) (algorithm 4, *Table 6*). The reference lineshapes are defined along ω_1 and ω_2 for each resonance by using a part of the row or column containing a well resolved cross-peak unambiguously assigned to this resonance. The procedure relies heavily on the completeness of the set of reference lines, and can be rather time-consuming.

Alternatively the volume of the peak can be evaluated from the 1D integrals in one row (I_1) and one column (I_2) drawn through the same point (*ij*) within the cross-peak and the signal amplitude $A(i,j)$ at this point (algorithm 5, *Table 6*) (22):

$$V = I_1I_2/A(i,j) \qquad [2]$$

The use of the row and column drawn through the maximum of the peak (or through another point at which the cross-peak is better resolved) increases the accuracy of the integration. If there is strong overlap of the cross-peak, the row and column integrals can be evaluated by curve fitting, which is well developed for 1D spectra. The fitting procedure can be made completely independent of reference lines; instead one uses the theoretical lines expected for NMR resonances, such as a mixture of Lorentzian and Gaussian lineshapes. The main disadvantage of the procedure is that a large amount of manual interaction is required.

The error ΔV in the measured cross-peak intensity can be evaluated from the equation:

$$\Delta V = N + \varepsilon V \qquad [3]$$

where N is the noise level in the spectrum, and ε the relative error associated with the procedure applied. The noise level is usually of the order of the volume of smallest peak which can be measured, and the relative error is about 20% if the appropriate procedure is applied for measurement.

As none of the methods for volume measurement is without its drawbacks, it is usually necessary to use a combination of different methods. In particular, resolved and unresolved peaks are usually treated differently. The flowchart shown in *Figure 7* helps one to choose the optimum protocol.

In addition to the errors associated with the measurement methods, several experimental factors which can influence the intensities should also be considered. Zero-quantum coherence present during the mixing time can not be removed by phase-cycling and therefore introduces an additional intensity component into intraresidue NOE cross-peaks. This contribution can be rather large for small molecules and at short mixing times. Random variation of the mixing time ($\pm$ 10%) or alternatively the introduction of a 180° pulse during the mixing time, its exact position within this mixing period being subject to random variation (23, 24) can help to solve the problem. For molecules of $M_r > 10\,000$ and for $\tau_m \sim 50$ msec or longer, the zero-quantum coherence decays to such extent that its contribution to the NOESY cross-peak intensity is negligible. In addition, the antiphase multiplet structure of zero-quantum peaks makes their integral intensity equal to zero.

Another factor which can influence NOE cross-peak intensity is the incomplete relaxation of z-magnetization during the relaxation time of the pulse sequence. Since this magnetization component is zero after the last pulse, the cross-peak amplitude is proportional to

$$1 - e^{-\frac{(AQ + D1)}{T_{1j}}}$$

where AQ is the acquisition time, D1 the relaxation delay, and T_{1j} the relaxation time of the proton whose chemical shift lies along ω_1. This correction factor should be applied when T_1 differs significantly for the different protons in the molecule. Usually it is sufficient to estimate the relaxation time for each type of proton. If selective excitation is used for water suppression another correction factor, representing the excitation profile, must be introduced.

In summary, when measuring NOE intensities in 2D spectra it is advantageous to start with a quick but crude evaluation using numerical integration, linewidth, and intensity estimation, or even just the intensity at the maximum. This can be combined with a peak picking and peak analysis routine and is sufficiently accurate to generate semi-quantitative restraints. The analysis of overlapped regions and the more accurate integration required for quantitative use of the NOE intensities can best be done with the spectral simulation techniques. A combination of several methods is also possible but attention should be paid to deriving the proper scaling factors.

4.2 Assignment of NOEs

The quality of the final structure obtained depends strongly on the number of NOEs which have been unambiguously assigned. Ambiguously assigned

NOEs can also be used by setting constraints averaged over all possible proton pairs, but they can not play a major role in the structure determination. For proteins with $M_r \leq 6000$, most NOEs can usually be assigned unambiguously just on the basis of 2D spectra. For larger proteins unambiguous assignment of the NOE cross-peaks can be difficult and much of the discussion in this section is focused on this class of proteins. Some of the ambiguity due to overlapping cross-peaks can be resolved by varying experimental conditions such as temperature or pH which, within certain limits, lead only to changes in chemical shifts without gross structural changes. More powerful and general ways to resolve the ambiguities are offered by heteronuclear 3D and 4D spectroscopy. It has been claimed that 3D spectroscopy of proteins uniformly labelled with ^{13}C and ^{15}N allows unambiguous assignments of all NOE cross-peaks for proteins of molecular weight up to 20 000 (25). The main disadvantage of the approach is the difficulty and cost of introducing isotope labels, particularly ^{13}C, into proteins. These methods are discussed in detail in Chapter 5. Homonuclear 3D spectroscopy has more limited application because of the marked decrease in sensitivity when going from two to three dimensions.

In cases where labelled protein is not available, additional NOE assignments can be made by using the intermediate structures obtained in the course of a structure calculation to resolve ambiguities (26). This is presented in *Protocol 5* and as a flowchart in *Figure 8* (assuming that sequence-specific proton assignments have already been done at this stage). First, NOE cross-peaks are quantified, if possible in spectra obtained under different condi-

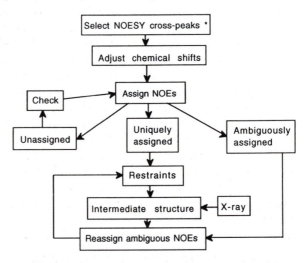

* Selection of cross peaks can be done semi-automatically and combined with volume measurement

Figure 8. Schematic diagram of the protocol used for NOE cross-peak assignment.

tions. This 'quantification' may involve just the determination of peak positions although information on the lineshape and multiplet structure (if resolved) can be very useful. This procedure can be automated for the well resolved peaks (see for example reference 27) but overlapping cross-peaks will require manual interaction. It is advantageous to measure volumes at this stage as described in previous section since it can save time in the long run.

In practice, there can be slight variations in chemical shift between different spectra even if the experimental conditions are the same. Thus, to increase the reliability of the assignment, chemical shift values have to be adjusted for each NOESY spectrum. This can be done semi-automatically by first finding the resolved intraresidue or sequential cross-peaks which can be easily assigned in the NOESY spectrum. NOE cross-peaks are then assigned on the basis of chemical shifts and lineshape information. Careful adjustment of chemical shift values allows the limits for shift variation to be set at ± 0.01 p.p.m. All protons having chemical shifts within this range of the shift of the observed cross-peak are considered as possible candidates for the assignment. Lineshape information can sometimes be applied qualitatively and from the various candidates for a particular cross-peak only those which are consistent with the proton-type derived from the lineshape are chosen as possible assignments, e.g. sharp singlets could correspond to carbon-bound protons of imidazole rings but not to amide protons. In principle, this process could be improved by using more sophisticated lineshape analysis at the expense of additional computer time; appropriate software has yet to be developed. (The assignment can be improved further by considering corresponding cross-peaks in spectra obtained under different conditions.)

After this first step, the cross-peaks are subdivided into three categories: uniquely-assigned, ambiguously assigned, and unassigned. Unassigned peaks are checked for possible errors and spectral artefacts. The uniquely assigned peaks are used to generate a set of crude distance constraints for the structure calculation (see Chapter 11). Once the constraints are reasonably well satisfied, the structures obtained are used to resolve the ambiguities in as many of NOE assignments as possible. There are no universal rules for this since the strategy can vary from molecule to molecule. Some general recommendations are presented in *Protocol 5*. Since the NOE intensity is approximately proportional to r^{-6}, an increase in the distance by a factor of 1.5 leads to decrease in the intensity by a factor of ~ 10. Thus, proton pairs with an internuclear distance 1.5 times larger than the smallest distance among the candidate proton pairs for the assignment can be eliminated.

Protocol 5. Assignment of NOE cross-peaks using structure calculation

1. Calculate the ensemble of structures satisfying unique NOE constraints using distance geometry or molecular dynamics procedures, as described in Chapter 11.

Protocol 5. *Continued*

2. For each ambiguously assigned cross-peak find the proton pair with the smallest internuclear distance, averaged over the ensemble, $<r_{ij}>_{min}$. If this distance is greater than a threshold distance r_{upper}, consider the cross-peak as unassignable at this stage. The value of r_{upper} can vary from molecule to molecule, but 8 Å is a good first guess.

3. Establish whether, among the possible candidates for assignment of the cross-peak, there are any other proton pairs with an interproton distance (averaged over the ensemble) smaller than 1.5 $<r_{ij}>_{min}$. If none, assign the cross-peak to the pair i,j.

4. If new unambiguous assignments are found in this way repeat the structure calculations with the modified set of constraints, followed by iteration through steps 1–3.

5. When there is no improvement to the assignments, increase the value of r_{upper}, and repeat steps 1–4.

6. Check cross-peaks for which there are no possible proton pairs with a distance less than 5 Å. The assignment could be wrong or missing.

For large proteins it may be difficult to calculate an initial structure solely on the basis of NOE information because of the small number of unambiguously assigned cross-peaks. For such proteins the starting structure can come from an X-ray structure of the protein itself or of a homologous protein. This structure can be used for the initial resolution of ambiguities followed by the procedure described above; in large proteins most of the cross-peaks will be ambiguously assigned and should be reassigned after each step of structure calculation.

4.3 From NOEs to distances

4.3.1 General theory

For isotropically tumbling molecules the intensities in a 2D NOESY spectrum obtained with the mixing time t_m, represented by the matrix $V(t_m)$, can be calculated using the equation (28):

$$V(t_m) = \exp(-Rt_m)V(0) \qquad [4]$$

where $V(0)$ represents the intensities of the diagonal peaks at $t_m = O$ (corresponding to the 1D spectrum) and R the relaxation matrix with elements:

$$R_{ii} = Q \sum_{i,j} (J_{0,ij}(\omega) + 3J_{1,ij}(\omega) + 6J_{2,ij}(\omega)) + R_{1,i}) \qquad [5]$$

$$R_{ij} = Q[6J_{2,ij}(\omega) - J_{0,ij}(\omega)]$$

with: $\qquad Q = 0.1\gamma^4\hbar^2$

and spectral densities:

$$J_{n,ij}(\omega) = \frac{1}{r_{ij}^6} \frac{\tau_{ij}}{(1 + (n\omega_0\tau_{ij})^2)} \qquad [6]$$

where τ_{ij} is the correlation time of the interproton vector between i and j and $R_{1,i}$ the 'leakage' rate due to relaxation mechanisms other than dipolar relaxation. In the absence of paramagnetic nuclei, dipolar interproton relaxation is usually predominant and $R_{1,i} \sim 0$.

It can be seen from eqn 4 that even though the off-diagonal elements of the relaxation matrix are directly related to the corresponding distance between two protons, the measured cross-peak volume is in general a function of the mutual arrangement of all protons in the molecule. Another complication is caused by the correlation time, which is in general an independent parameter for each proton pair. Thus the first approximation applied is to assume a single correlation time for the whole molecule, i.e. to consider it as a rotating rigid body (see *Figure 9*). Two different approaches are used to extract interproton distances from cross-peak intensities:

- the isolated spin pair approximation
- complete relaxation matrix analysis

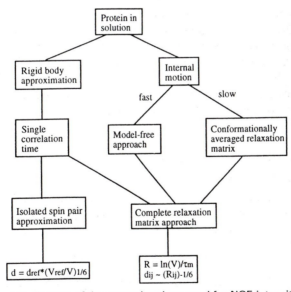

Figure 9. Schematic diagram of the approximations used for NOE intensity calculations.

4.3.2 Isolated spin pair approximation

The expansion of the exponential term of eqn 4 into a Taylor series gives the basis for the approximations used:

$$\exp(-\mathbf{R}t_m) = 1 - \mathbf{R}t_m + 0.5\mathbf{R}^2 t_m{}^2 - \ldots \qquad [7]$$

so that peak intensities are given by:

$$A_{ij} = \delta_{ij} - R_{ij}t_m + 0.5 \, \Sigma_k R_{ik}R_{kj}t_m{}^2 - \ldots \qquad [8]$$

Truncation of eqn 7 after the linear term represents the so-called isolated spin pair approach or ISPA (29). The cross-peak intensity depends only on the corresponding off-diagonal element of the relaxation matrix. The interproton distance can then be easily estimated from a comparison of the given cross-peak intensity V_{ij} with the intensity V_{ref} of a reference cross-peak corresponding to a proton pair at a known fixed distance r_{ref}

$$r_{ij} = r_{ref}(V_{ref}/V_{ij})^{1/6} \qquad [9]$$

For the reference distance, one can use either chemically fixed distances, such as the distance between geminal or aromatic ring protons, or conformationally fixed distances, such as backbone interproton distances in a β-sheet or an α-helix.

The neglect of the high-order terms in eqn 7 is valid for short mixing times. Under these conditions the dependence of cross-peak intensities on t_m is close to linear as can be seen in *Figure 10* for the typical time-dependence of the intensity of a cross-peak between tyrosine ring protons at a distance of 2.4 Å. With increasing correlation time, the region of linear-dependence becomes smaller and the maximum t_m value for which the ISPA approximation is valid decreases. The smaller the mixing time the better ISPA will work but the poor signal-to-noise ratio at short t_m makes intensity measurements difficult.

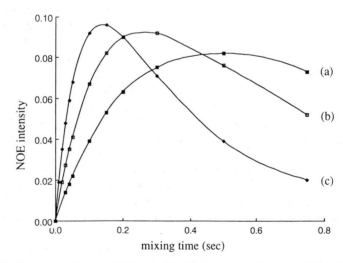

Figure 10. Time-dependence of the intensity of the cross-peak between Hδ and Hε protons of Tyr68 aromatic ring in X-ray structure of dihydrofolate reductase calculated for the correlation times (a) 2 nsec, (b) 4 nsec, and (c) 8 nsec.

The accuracy of ISPA is illustrated in *Figure 11*. The NOE intensities were calculated from the coordinates of a known structure (a protein G domain) using the full relaxation matrix; the calculations were performed for several different correlation times.[a] The distances were then evaluated from the calculated intensities using ISPA (eqn 9) with the methylene proton pair as a reference ($r_{ref} = 1.7$ Å) and plotted against the real distances. One can see that the ISPA distances are systematically shorter than the real distances. This arises because the reference distance is very short. In general, distances smaller than the reference are overestimated and large distances are underestimated. The error increases as the correlation time and the mixing time increase; for $t_m = 100$ msec, it can be as large as 60% for $\tau_c = 2$ nsec, 100% for $\tau_c = 4$ nsec, and 150% for $\tau_c = 8$ nsec. Allowance for such deviations should be made if the distances derived with ISPA are used for structure calculations.

The experimental error in distance Δr can be estimated from the error in cross-peak intensity measurement ΔV using the following equation (30):

$$\Delta r_{ij}/r_{ij} \approx \Delta V_{ij}/6V_{ij} \qquad [10]$$

For a typical case where the intensity error is $\pm 20\%$ for the large peaks and $\pm 100\%$ for the small ones, the error in distance is between $\pm 3\%$ and $\pm 20\%$. The error arising from the multiple spin effect is usually larger than the error introduced by the inaccuracy of intensity measurement. The limited experimental signal-to-noise ratio will prevent observation of small cross-peaks, thus introducing an *upper* limit on the interproton distance that can be detected by NOE. A noise level estimated for NOESY spectra of a protein at concentration of ~ 3 mM is used as the threshold in the plots of *Figure 11*.

One of the ways to deal with the inaccuracy of the distance determination by ISPA is to subdivide the intensities into several groups (see for example (31) and references therein). The distance limits corresponding to each group are then set to be sufficiently broad to take account of all the errors (including neglected internal mobility). The crudest—and safest—approach of this type is to assign upper bound restraints of 5–6 Å to any interproton distance corresponding to a detectable NOE cross-peak. This is a reasonable approach if one uses long mixing times to acquire a NOESY spectrum of a large molecule. With spectra obtained at shorter t_m it is possible to use more groups with the more popular subdivision of cross-peaks into strong, medium and weak, corresponding to absolute NOE intensity in the regions 6.1–14%, 4.1–6%, and 1–4% with the distance restraints 1.8–2.8 Å, and 1.8–3.3 Å, and 1.8–5.0 Å, respectively. (The estimation of the absolute intensity of an NOE cross-

---

[a] From eqn 5–7, the cross-peak intensity has the same functional dependence on τ_c as on t_m when $\omega_o\tau_c \gg 1$; hence the proportional changes in these parameters are interchangeable. This means that the intensity distribution will be the same for $\tau_c = 2$ nsec and $t_m = 100$ msec as for $\tau_c = 4$ nsec and $t_m = 50$ msec. Thus the plots in *Figure 11* can be used to evaluate the expected error for different combinations of t_m and τ_c.

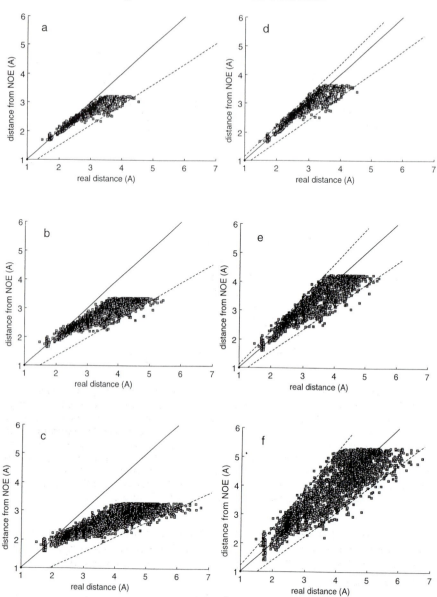

peak is rather tedious and requires its comparison with the diagonal peak or calculated intensities.) The apparent disadvantage of such an approach is the loss of a fair amount of the structural information. Nevertheless, the approach works well for obtaining the overall fold of a protein when a large number of NOEs can be identified, and its very crudity means that it is less sensitive to errors from multiple spin effects or differing local correlation times than a more quantitative approach would be.

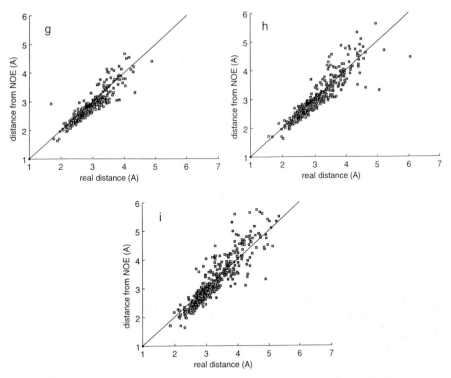

Figure 11. The correlation between interproton distances in the protein G structure and distances evaluated from NOE intensities at $t_m = 100$ msec and $\tau_c = 2$ nsec (a, d, and g), $\tau_c = 4$ nsec (b, e, and h) and $\tau_c = 8$ nsec (c, f, and i) using different approximations. (a–c) ISPA with methylene protons at 1.7 Å as reference; the *dashed line* corresponds to distance deviation of (a) 60%, (b) 100%, and (c) 150%. (d–f) ISPA with distances evaluated from eqn 14 using two references, methylene protons at 1.7 Å and aromatic ring protons at 2.4 Å; the *dashed lines* correspond to distance deviations of (d) −10% and +30%, (e) − 20% and +40%, and (f) −30% and +40%. (g–i) The results of MARDIGRAS calculations using NOE intensities corresponding to assigned NOE cross-peaks.

The distance evaluation can still be improved even within the framework of ISPA with the use of two different reference distances $r_{ref,1}$ and $r_{ref,2}$ with the corresponding intensities $V_{ref,1}$ and $V_{ref,2}$. Then a distance can be calculated from the equation:

$$r_{ij} = A + B(V_{ij})^{-1/6} \qquad [11]$$

with:

$$A = \frac{r_{ref,1} \, V_{ref,1}^6 - r_{ref,2} \, V_{ref,2}^6}{V_{ref,1}^6 - V_{ref,2}^6}$$

$$B = \frac{r_{ref,1} - r_{ref,2}}{V_{ref,1}^{-6} - V_{ref,2}^{-6}}$$

The results of using eqn 11 for distance evaluation with methylene protons at 1.7 Å and aromatic ring protons at 2.5 Å as reference distances are shown in *Figure 11 (d–f)*. The distances obtained from the cross-peak intensities are now distributed rather symmetrically about the real distances. Most of the evaluated distances for $\tau_m = 100$ msec are within the limits of $0.9r_{real} - 1.2r_{real}$ for $\tau_c = 2$ nsec, $0.8r_{real} - 1.4r_{real}$ for $\tau_c = 4$ nsec and $0.7r_{real} - 1.4r_{real}$ for $\tau_c = 8$ nsec. These limits are narrower than those obtained with just one reference distance. They can be used to set distance restraints when distances are obtained from eqn 11.

The accuracy of the distances obtained using ISPA for $\tau_c = 2$ nsec and $t_m = 100$ msec is reasonable for structural calculations, although the noise in the spectrum may prevent the observation of NOEs for distances larger than ~ 3.5 Å. Hence, large distances detected at longer mixing times can only be evaluated at much lower accuracy using ISPA. It is reasonable therefore to use a combined approach:

(a) Evaluate distances from NOE intensities at short mixing times using eqn 11. For $\tau_c = 2$ nsec use data obtained at $t_m = 100$ msec; similarly, for $\tau_c = 4$ nsec, $t_m = 50$ msec and for $\tau_c = 8$ nsec, $t_m = 25$ msec. The restraints from this set of NOEs can be set to 0.9r (lower bound) and 1.3r (upper bound), where r is the evaluated distance.

(b) Measure the spectra at longer t_m; use the new NOEs which appear only at longer mixing times to set distance restraints of 3.0–5.0 Å. For $\tau_c > 8$ nsec the upper limit can be increased up to 6 Å.

4.3.3 The complete relaxation matrix approach

The multispin relaxation effects can be fully accounted for if one manages to measure the complete intensity matrix **V**. The relaxation matrix can then be obtained from:

$$\mathbf{R} = \{- \ln[\mathbf{V}(t_m)/\mathbf{V}(0)]\}/t_m \qquad [12]$$

and the interproton distances can be derived from the corresponding off-diagonal elements of the relaxation matrix **R**. The measurement of *all* the intensities including diagonal peaks is an unattainable goal in real cases. Consequently, several iterative approaches have been developed to cope with incomplete intensity information. All the approaches are designed to minimize the deviation between measured and calculated intensities; they can be subdivided into two classes, shown schematically in *Figure 12*. The first type (*right-hand side* of *Figure 12*) utilizes the self-consistency of the relaxation matrix, modifying the matrix until there is agreement between *calculated* and *measured* intensities. In the second approach (*left-hand side* of *Figure 12*) the *structure* is iteratively modified.

The most popular program utilizing the approach of the first type is MAR-DIGRAS (30) although other similar programs have also been developed.

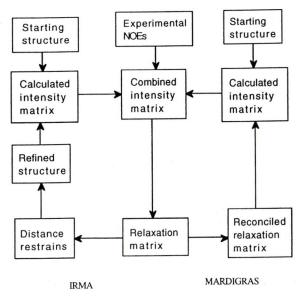

IRMA   MARDIGRAS

Figure 12. Schematic diagram of the approaches which make use of the full relaxation matrix analysis for obtaining distances from NOEs. *Left-hand side*: algorithms based on the self-consistency of the relaxation matrix. *Right-hand side*: algorithms based on the structure refinement.

The algorithm is shown diagrammatically on the *right-hand side* of *Figure 12*; it includes the following main steps.

(a) The theoretical intensity matrix is calculated for the initial model structure.

(b) The measured NOE intensities are used to substitute for corresponding calculated ones in the intensity matrix.

(c) From the 'combined' intensity matrix a new relaxation matrix is calculated using eqn 12.

(d) This relaxation matrix is checked and modified for consistency; this includes the correspondence between diagonal and off-diagonal elements, known values of off-diagonal elements for protons at fixed distances, and upper limits for the off-diagonal elements coming from the shortest possible interproton distance.

(e) From the modified relaxation matrix the intensity matrix is calculated again.

(f) The cycle is repeated until the predetermined convergence criterion is reached or the changes in the calculated intensities are below the set limit.

(g) From the final relaxation matrix, a set of distances corresponding to measured cross-peaks is obtained. These distances are then used in the structure calculation.

The second class of methods uses a comparison between the theoretical and experimental intensities to define how close the structure is to the experimental data while refining the structure with the help of distance geometry or molecular dynamics methods (see Chapter 11). In the 'direct' approach (32) the set of constraints for distance geometry is modified in a simple manner: if the theoretical NOE is larger than the experimental one, the corresponding target distance is decreased and *vice versa*. The structure is then refined with the corrected set of constraints and the process is repeated until convergency is achieved.

A more complex approach is utilized in the algorithm IRMA (33) (left-hand side of *Figure 12*). The combined intensity and relaxation matrices are obtained in the same way as in the MARDIGRAS approach. From the relaxation matrix a set of constraints is obtained and used to refine the structure with the help of molecular dynamics. The refined structure is then used to calculate the intensity matrix and the cycle is repeated until agreement between calculated and measured NOE intensities is achieved.

The difference between experimental and calculated intensities can be incorporated directly into the penalty function used for distance geometry or molecular dynamics calculations. The corresponding force field can be calculated by using the analytical expressions for the derivatives of the NOE intensities as obtained for Cartesian coordinates in (34) or dihedral angle space in (35). The derivatives can also be evaluated numerically for each proton (36).

Multispin effects can be better accounted for by using the NOE intensities measured at several mixing times for the distance calculations. Though it is not a standard option, it can easily be introduced into some programs, such as MARDIGRAS, by averaging the relaxation matrix over several mixing times.

The approaches which employ the full relaxation matrix produce structures which are in better agreement with the experimental data, and appear to be robust with the examples supplied. Since most of the calculations have been done for relatively small molecules with $\tau_c < 3$ nsec and $t_m \sim 100$ msec, one should not be surprised if the convergence is not as good for larger proteins. The results of test calculations using MARDIGRAS are shown in *Figure 11* *(g–i)*. The theoretically calculated NOE intensities corresponding to the assigned NOE cross-peaks were calculated for the known structure of protein G using the full relaxation matrix. Random noise was added at the level of experimental noise in a NOESY spectrum. An extended protein structure was used as the starting structure. The MARDIGRAS calculations were done for different correlation times. The agreement between 'real' and calculated distances is better than for ISPA, especially for correlation times greater than

4 nsec. It can be further improved if intensities at different mixing times are used in the calculations.

The computation time required for these approaches depends strongly on the number of atoms considered; it is usually longer for the programs using molecular dynamics and back-calculations. MARDIGRAS and similar programs lead quite rapidly to a set of restraints, but this must be followed by a structure calculation, and the total time for obtaining a structure can be quite comparable.

4.3.4 Molecular motion

The incorporation of intramolecular motion into structure calculations is described in Chapter 11; in this chapter we concentrate on its effects on the derivation of structural information from the NMR parameters.

The effect of intramolecular motion on relaxation depends on the time scale of the motion. In this section we consider only the effects of internal motions which are faster than relaxation rates. The approximations used when deriving distances from NOEs in the presence of internal motions further depend on the relationship between the rate of internal motion and the rate of overall tumbling, as shown schematically in *Figure 9*. When the internal motion is *slower* than the overall tumbling of the molecule the resultant relaxation matrix is the average over all possible arrangements of the protons (37). It can be calculated by using $<r^{-6}>_{ij}$, averaged over all possible conformations, instead of r^{-6}_{ij} for each proton pair in eqn 6. For example, the relaxation matrix should be averaged for each pair of chemically equivalent protons of the phenylalanine and tyrosine rings to allow for the 'flip' of these rings about their symmetry axes, since this flip can easily cause the appearance of NOE cross-peaks between protons which are 8 Å apart (37). The $<r^{-6}>$ averaging overestimates short distances, so that all the distances obtained from the NOE intensities appear to be somewhat shorter than the real ones; thus, they may not be compatible with those in a single structure or the structure averaged over the ensemble. Another consequence of motional flexibility is indirect magnetization transfer. This happens when in one conformation proton A is close to proton B and in the other conformation proton B is close to proton C. Then cross-peak between protons A and C appears in NOESY spectrum due to magnetization transfer from A to B in one conformation and then from B to C in the other. Indirect and direct magnetization transfer can in principle be distinguished by differences in the shapes of the corresponding NOE build-up curves (see also Chapter 6). To assess this difference in shape one needs to measure NOE intensities at short mixing times, where the signal-to-noise ratio is rather poor. The difference between direct and indirect cross-peaks can be seen more clearly in ROESY spectra where they have opposite phases (38).

When internal motion becomes *faster* than the molecular tumbling the spectral density function (eqn 9) must be modified. One common approximation is that of Lipari and Szabo (39) which leads to spectral density functions of the form:

$$J_{n,ij} = \langle r_{ij}^{-6} \rangle \left(\frac{S^2 \tau_c}{1 + (n\omega_0 \tau_c)^2} + \frac{(1 - S^2)\tau_e}{1 + (n\omega_0 \tau_e)^2} \right)$$ [13]

where the order parameter S^2 is the measure of spatial restriction of the motion and τ_e is the 'effective' correlation time of the internal motion. Both parameters can be estimated from ^{13}C and ^{15}N relaxation measurements, where the relaxation of the heteronucleus is governed mainly by the directly attached proton. The parameters obtained in this way are related to the motion of the vector between proton and heteronucleus, and can be significantly different from the parameters for the interproton vectors. Since the proton relaxation is influenced by several surrounding protons, it is impossible to derive motion parameters from proton relaxation measurements. When $\tau_e \ll \tau_c$ the correlation functions become:

$$J_{n,ij} = \langle r_{ij}^{-6} \rangle \frac{S^2 \tau_c}{1 + (n\omega_0 \tau_c)^2}$$ [14]

Molecular dynamics calculations (40) have shown that correlation functions for most of the interproton vectors decay over a short time period, after which their values do not change significantly. Using this plateau value of the correlation function as S^2 in eqn 14 improved the agreement between experimental and calculated NOE intensities (41). When S^2 was used as a parameter it was also possible to obtain a better theoretical fit of the experimental build-up curves for the protons at fixed distances such as methylene or aromatic ring protons (42). The S^2 varied between 1 and 0.6 for different proton pairs; this corresponds to not more than 10% error in terms of distance. The best one can do is either to evaluate the order parameters S^2 for protons at fixed distances or for heteronuclei and use them for the corresponding groups, or to combine NOE calculations with simulation of the internal motions (42). The latter is very time-consuming for large molecules (see Chapter 11).

For the extremely fast rotation of methyl groups, $\langle r^{-6} \rangle$ averaging must be replaced with $\langle r^{-3} \rangle$ averaging over three proton positions in the methyl group (43). This $\langle r^{-3} \rangle$ averaging gives results closer to the mean distance than does $\langle r^{-6} \rangle$ averaging. Thus when the methyl group is considered as a pseudo-atom the correction required is 0.3 Å (37; see Chapter 11).

4.3.5 Evaluation of the agreement with experimental data

The analysis of the quality of the structure obtained from the NMR data is discussed in Chapter 11, but it is worth including here some brief comments on the evaluation of the agreement between theoretical and experimental intensities. This is commonly done in terms of an 'R factor' analogous to that used in crystallography. Several different expressions for the R factor give similar results (44); one widely used form is:

$$R = \sqrt{\frac{\sum\limits_{i,j,t_m} W(t_m)\,[SV_{ij,c}(t_m) - V_{ij,e}(t_m)]^2}{\left[\sum\limits_{i,j,t_m} W(t_m)\,V_{ij,e}(t_m)\right]^2}} \qquad [15]$$

where $V_{ij,c}$ and $V_{ij,e}$ are, respectively, the calculated and experimental cross-peak volumes. The weighting factor, $W(t_m)$, is set either to 1, or to t_m in order to take account of decreasing intensities at short mixing times. S, the scaling factor, is given by $S = \Sigma_{ij}V_{ij,e}/\Sigma_{ij}V_{ij,c}$, with the summation done either over protons at fixed distances, such as geminal and ring protons, or over all proton pairs considered. Both approaches seem to lead to the same result.

A disadvantage of the expression for the R factor in eqn 15 is that it over-emphasizes the intensities corresponding to short interproton distances. To overcome this bias, a weighted R factor can be expressed in terms of distances, which are approximately proportional to $V^{-1/6}$, giving:

$$R_r = \sqrt{\frac{\sum\limits_{i,j,t_m} (SV_{ij,c}(t_m)^{-1/6} - V_{ij,e}(t_m)^{-1/6})^2}{\sum\limits_{i,j,t_m} V_{ij,e}(t_m)^{-1/3}}} \qquad [16]$$

In structure calculations with experimental NOEs, the R factor was found to decrease significantly over the first few iterations to a value of about 0.4 which did not change significantly with further iterations (44). This failure to improve the R factor much further is due to the low accuracy of measuring the volume of small (long-range) cross-peaks and to the disregard of intramolecular motion. Calculations performed with simulated experimental data produced much lower values for R (45).

4.4 Summary: the use of NOEs for the structure calculation

The particular approach to use depends on the type of molecule under investigation, data available, and information required. The strategy to follow when reasonably good NOESY spectra can be measured and a high quality protein structure is to be produced is presented in *Protocol 6*. In general the protocol can be divided into the following steps:

- rough estimation of distances from NOE intensities using ISPA
- structure calculation with generous constraints to take account of spin diffusion
- structure refinement using back-calculations and quantitative analysis of NOEs

Protocol 6. Using NOEs for structure calculations

1. Estimate the correlation time of the protein from its molecular weight.

2. Measure NOESY spectra at different mixing times; 100, 200, and 300 msec for τ_c = 2 nsec, 50, 100, and 200 msec for τ_c = 4 nsec, and 25, 50, and 100 msec for τ_c = 8 nsec are reasonable starting values.

3. Measure cross-peak volumes using the guide-lines given in *Table 6* and *Figure 7*.

4. Evaluate distances from the NOE cross-peak intensities in the spectrum at short mixing times (t_m = 100 msec for τ_c = 2 nsec, t_m = 50 msec for τ_c = 4 nsec, and t_m = 25 msec for τ_c = 8 nsec) using the isolated spin pair approximation and two reference distances (eqn 11). Set the upper and lower limits to 0.9r and 1.3r to allow for most of the spin diffusion effects under these conditions.

5. Constraints for distances involving methyl groups can be obtained from the same calibration curve if the peak intensity is divided by three for methyl to non-methyl cross-peaks, and by nine for cross-peaks between two methyl groups. The upper limit should be increased and the lower decreased by 0.3 Å for each methyl group involved in the constraint to allow for changes of methyl group orientation.

6. Use the additional NOEs appearing only in the spectra at longer mixing times as additional distance constraints: 3–4.5 Å for τ_c = 2 nsec, 3–5 Å for τ_c = 4 nsec and 3–6 Å for τ_c = 8 nsec. Add 1 Å to the upper limit for each methyl group involved.

7. Use the resultant constraints in structure calculations (see Chapter 11).

8. Refine the structures obtained using back-calculations at several mixing times. Compare experimental and calculated build-up curves for the protons at fixed distances and adjust correlation times if necessary.

9. Estimate the R factor using eqn 15 or 16. The value should be close to the averaged relative error for the cross-peak volumes. Examine the regions with significant (well above noise level) discrepancies between calculated and experimental NOEs. Among the possible sources of errors are inaccurate measurements, predominance of spin diffusion, or molecular motion. The latter two can be assessed with the additional help of ROESY spectra.

The overall fold of the macromolecule can be obtained at step 2. Here the quality of the structure mostly depends on the number of long-range NOEs. In principle the distance constraints at this stage can be obtained from a qualitative treatment of NOE cross-peaks but it is better to use them quantitatively as described in Section 4.3.2. A number of reported structures obtained in this way have an r.m.s.d. between the structures which satisfy the NOE constraints of the order of 0.5 Å. To achieve such a level of convergence one needs to have approximately ten long-range constraints per residue. A small r.m.s.d. value does not necessarily mean good agreement with the experimental data, since NOE intensities can vary significantly within limits normally used. The agreement can be better evaluated using the R factor.

When the NOE intensities can be measured with good accuracy, the structure refinement can proceed in combination with the back-calculations. Starting with a structure which is close to the final helps to improve convergence of the structure calculation and significantly accelerates the process. The use of back-calculations is especially advantageous when mainly short- and medium-range NOEs are available due to the shape of the molecule. The calculations can be speeded up further if the back-calculation is restricted to a small region of the molecule for which one needs to obtain the most accurate structure, (e.g. the active site of an enzyme). The rest of the structure can be kept fixed or modified within the ISPA set of restraints.

Back-calculations can also help in addressing the problem of intramolecular mobility. The presence of flexible regions is usually revealed as a failure to achieve a good agreement between all calculated and measured NOEs at the same time. Better correspondence is then achieved when the order parameter and/or averaging over an ensemble of structures is introduced.

References

1. Bystrov, V. F. (1976). *Prog. NMR Spectrosc.*, **10**, 41.
2. Neuhaus, D., Wagner, G., Vasak, M., Kagi, H. R., and Wuthrich, K. (1985). *Eur. J. Biochem.*, **151**, 257.
3. Kay, L. E. and Bax, A. (1990). *J. Magn. Reson.*, **86**, 110.
4. Neri, D., Otting, G., and Wüthrich, K. (1990). *J. Am. Chem. Soc.*, **112**, 3663.
5. Montelione, G. T. and Wagner, G. (1989). *J. Am. Chem. Soc.*, **111**, 5474.
6. Forman-Kay, J. D., Gronenborn, A. M., Kay, L. E., Wingfield, P. T., and Clore, G. M. (1990). *Biochemistry*, **29**, 1566.
7. Redfield, C. and Dobson, C. M. (1990). *Biochemistry*, **29**, 7201.
8. Press, W. H., Flannery, B. P., Teukolsky, S. A., and Vetterling, W. T. (1986). *Numerical Recipes*. Cambridge University Press, Cambridge.
9. Mueller, L. (1987). *J. Magn. Reson.*, **72**, 191.
10. Clore, G. M., Bax, A., and Gronenborn, A. M. (1991). *J. Biomolec. NMR*, **1**, 13.
11. Driscoll, P. C., Gronenborn, A. M., and Clore, G. M. (1988). *FEBS Lett.*, **243**, 223.
12. Guntert, P., Braun, W., Billeter, M., and Wüthrich, K. (1989). *J. Am. Chem. Soc.*, **111**, 3997.
13. Nilges, M., Clore, G. M., and Gronenborn, A. M. (1990). *Biopolymers*, **29**, 813.

14. Guntert, P., Braun, W., and Wüthrich, K. (1991). *J. Mol. Biol.*, **217**, 517.
15. Montelione, G. T., Winkler, M. E., Rauenbuehler, P., and Wagner, G. (1989). *J. Magn. Reson.*, **82**, 198.
16. Marion, D., Ikura, M., Tschudin, R., and Bax, A. (1989). *J. Magn. Reson.*, **85**, 393.
17. Olejniczak, E. T., Gampe, R. T. Jr., and Fesik, S. W. (1986). *J. Magn. Reson.*, **67**, 28.
18. Stoven, V., Mikou, A., Piretean, D., Guitet, E., and Lallemand, Y.-Y. (1989). *J. Magn. Reson.*, **82**, 163.
19. Fejzo, J., Zolnai, Z., Macura, S., and Markley, J. L. (1990). *J. Magn. Reson.*, **88**, 93.
20. Denk, W., Bauman, R., and Wagner, G. (1986). *J. Magn. Reson.*, **67**, 386.
21. Olejniczak, E. T., Gampe, R. T. Jr., and Fesik, S. W. (1989). *J. Magn. Reson.*, **81**, 178.
22. Holak, T. A., Scarsdale, J. N., and Prestegard, J. H. (1987). *J. Magn. Reson.*, **74**, 456.
23. Macura, S., Huang, Y., Suter, D., and Ernst, R. R. (1981). *J. Magn. Reson.*, **43**, 259.
24. Rance, M., Bodenhausen, G., Wagner, G., Wüthrich, K., and Ernst, R. R. (1985). *J. Magn. Reson.*, **62**, 497.
25. Clore, G. M., Kay, L. E., Bax, A., and Gronenborn, A. M. (1991). *Biochemistry*, **30**, 12.
26. Eccles, C., Guntert, P., Billeter, M., and Wüthrich, K. (1991). *J. Biomolec. NMR*, **1**, 111.
27. Garret, D. S., Powers, R., Gronenborn, A. M., and Clore, G. M. (1991). *J. Magn. Reson.*, **95**, 214.
28. Macura, S. and Ernst, R. R. (1990). *Mol. Phys.*, **41**, 95.
29. Gronenborn, A. M. and Clore, G. M. (1985). *Prog. NMR Spectrosc.*, **17**,
30. Borgias, B. A. and James, T. L. (1990). *J. Magn. Reson.*, **87**, 475.
31. Kuntz, I. D., Thomason, J. F., and Oshiro, S. M. (1989). In *Methods in enzymology* (ed. N. J. Oppenheimer and T. L. James) vol. 177, pp. 159–204. Academic Press, London.
32. Summers, M. F., South, T. L., Kim, B., and Hare, D. R. (1990). *Biochemistry*, **29**, 329.
33. Boelens, R., Konig, T. M. G., and Kaptein, R. (1988). *J. Mol. Struct.*, **173**, 299.
34. Yip, P. and Case, D. A. (1989). *J. Magn. Reson.*, **83**, 643.
35. Mertz, J. E., Guntert, P., Wüthrich, K., and Braun, W. (1991). *J. Biomolec. NMR*, **1**, 257.
36. Baleja, J. D., Moult, J., and Sykes, B. D. (1990). *J. Magn. Reson.*, **87**, 375.
37. Koning, T. M. G., Boelens, R., and Kaptein, R. (1990). *J. Magn. Reson.*, **90**, 111.
38. Fejzo, J., Krezel, A. M., Westler, W. M., Macura, S., and Markley, J. L. (1991). *J. Magn. Reson.*, **92**, 651.
39. Lipari, G. and Szabo, A. (1982). *J. Am. Chem. Soc.* **104**, 4546.
40. Olejniczak, E. T., Dobson, C. M., Karplus, M., and Levy, R. M. (1984). *J. Am. Chem. Soc.*, **106**, 1923.
41. Koning, T. M. G., Boelens, R., van der Marel, G. A., van Boom, J. H., and Kaptein, R. (1991). *Biochemistry*, **30**, 3787.
42. Baleja, J. D. and Sykes, B. D. (1991). *J. Magn. Reson.*, **91**, 624.

43. Tropp, J. (1980). *J. Chem. Phys.*, **77**, 6035.
44. Gonzalez, C., Rullmann, J. A., Bonvin, A. M. J. J., Boelens, R., and Kaptein, R. (1991). *J. Magn. Reson.*, **91**, 659.
45. Bonvin, A. M. J. J., Boelens, R., and Kaptein, R. (1991). *J. Biomolec. NMR*, **1**, 305.

11

Structure determination from NMR data II. Computational approaches

MICHAEL J. SUTCLIFFE

1. Introduction

Structure determination by NMR can be divided into four steps as shown in *Protocol 1*.

Protocol 1. Structure determination by NMR

1. Collect the experimental data.

2. Process the data.

3. Determine structural constraints.

4. Calculate structures which satisfy these constraints.

Reviews of the steps involved in *Protocol 1* are given by, for example, Clore and Gronenborn (1) and Wüthrich (2). Steps **1** and **2** are covered in earlier chapters of this book. This chapter aims to give an insight into steps **3** and **4**.

2. Determination of constraints

The following types of constraints are commonly used in structure determination from NMR data:

- distance constraints derived from NOE measurements
- dihedral angle constraints derived from the measurement of vicinal coupling constants

The determination of these constraints is discussed in detail in Chapter 10 for proteins and Chapter 8 for oligonucleotides; a brief summary will be presented here, with the emphasis on the computational aspects.

2.1 Proton–proton distance constraints

2.1.1 Distance constraints for proteins

Distance constraints are usually specified in terms of a distance range—the lower bound is determined from the sum of the van der Waals' radii and the upper bound from the NOE intensity. For a protein the NOEs are commonly categorized on the basis of their intensity, (e.g. 'strong', 'medium', and 'weak') and the corresponding distance ranges are used, (e.g. 1.8–2.5 Å, 1.8–3.5 Å, and 1.8–5.0 Å). (The mixing time used to detect the NOEs corresponding to these distances is dependent on the correlation time of the molecule, particularly in the case of longer distances.) An alternative method for determining proton–proton distance constraints assumes that the molecule is rigid. In such a system all the correlation times between different pairs of protons are identical and equal to the correlation time for the overall tumbling of the molecule. In this case the NOE intensity (I_{NOE}) is related to the correlation time by:

$$I_{NOE} \propto 1 + \frac{5 + \omega^2\tau_c^2 - 4\omega^4\tau_c^4}{10 + 23\omega^2\tau_c^2 + 4\omega^4\tau_c^4} \tag{1}$$

where τ_c is the correlation time and ω the spectrometer frequency. An unknown distance r_{ij} between two protons giving rise to an NOE can then be determined from a reference distance r_{kl} (for example, the distance between two protons on a ring) using the following relationship:

$$r_{ij} = r_{kl}\left(\frac{I_{kl}}{I_{ij}}\right)^{1/6} \tag{2}$$

where I_{jk} and I_{kl} are the respective NOE intensities. It should be noted that if the unknown distance is smaller than the reference distance then its value will be systematically overestimated; if the unknown distance is larger than the reference distance then its value will be systematically underestimated (3). This difference in behaviour from that predicted by eqn 2 is due to spin diffusion.

Eqn 2 assumes that the protons giving rise to the NOE are not interacting with any other protons (the isolated spin pair approximation). This yields good results provided that there is a large number of constraints, that the mixing time is not too long, and that the accuracy of these constraints is not overestimated. However, if greater accuracy is required, for example when ligand binding sites are being studied, a means of obtaining tighter distance constraints from NOE peak intensities is necessary. The full relaxation matrix is commonly used to achieve this (see Chapter 10). This approach is even more effective if it is used in an iterative manner. One means of achieving this is to determine proton–proton distances from a starting model of the structure and iteratively update these distances until the well-resolved observed intensities (which correspond to only a fraction of the total number of observed NOEs) and the corresponding calculated NOE intensities agree to within a certain threshold. Such an approach has been incorporated into the program MARDIGRAS (4). This appears to be rather robust and does not depend heavily on the initial model used for the structure. Also, the calculations can be performed in a reasonable time. However, in common with all the other methods mentioned, it does not completely surmount the problem of determining accurate correlation times, although this may not be a significant problem in the case of globular proteins. An alternative approach is to obtain an initial set of distance constraints from the relaxation matrix, refine the model using restrained molecular dynamics, obtain a new set of distance constraints, refine, and so on. This approach has been incorporated into the program IRMA (5). Another approach which involves restrained molecular dynamics has been developed recently. In this the differences between corresponding peaks in the calculated and observed NOESY spectra are used to define a 'NOE force' which acts on the respective atoms (6, 7). This force is updated during the course of the molecular dynamics simulation.

In the case of an incomplete set of experimental constraints, a generated ensemble of tertiary structures is frequently used to determine additional constraints. For example, if there are two possible pairs of protons giving rise to a particular NOESY cross-peak, analysis of proton–proton distances across such an ensemble may reveal that one of these is highly probable whereas the other is very improbable.

2.1.2 Distance constraints for oligonucleotides

Oligonucleotides (see Chapter 8) present a more difficult problem than proteins for two reasons. First they are linear, hence only short-range constraints can be obtained. Therefore, the distance ranges need to be determined more accurately than in the case of proteins. Secondly, larger structures become rod-like, rather than spherical as in the case of globular proteins. This leads to two correlation times—one due to rotation about the long axis and the other to rotation about the short axis. Thus use of a single overall correlation time becomes less valid the longer the oligonucleotide becomes. Indeed, internucleotide distances derived from NOE intensities may well be significantly underestimated when the axial ratio exceeds about 2.5 (8). If the use of a single correlation time is thought to be invalid, the correlation times for the equivalent ellipsoid can be obtained from a knowledge of the axial ratio and the cross-relaxation rate constant for the H5–H6 interaction in cytosine (8).

2.2 Dihedral angle constraints

Vicinal spin–spin coupling constants can provide useful additional geometric information. The values of mainchain ϕ angles are available from the experimentally measured H^N-H^α vicinal coupling constants ($^3J_{H^N-H^\alpha}$). These coupling constants are linked to the torsion angle ϕ by the Karplus equation (9).

The parameters commonly used are those according to the study of Pardi *et al.* (10). If $^3J_{HN-H\alpha}$ is in hertz, then:

$$^3J_{HN-H\alpha} = 6.4 \cos^2(\phi - 60°) - 1.4 \cos(\phi - 60°) + 1.9 \qquad [3]$$

The form of this equation is shown in *Figure 1*. From *Figure 1* it is apparent that in many instances there is ambiguity as to which torsion angle corresponds to a given value for $^3J_{HN-H\alpha}$. This problem can be surmounted to a certain extent by analysing the range of values allowed for ϕ in known protein structures. Such an analysis shows that, apart from Asn, Asp, and Gly, all amino acids tend to adopt negative values for ϕ (as defined by IUPAC (11), this is 0° when the carbonyl carbon is *trans* to the amide proton) and are concentrated in the range $-30°$ to $-180°$ (12). It should be noted that mobility about the N^α-C^α bond results in a decrease in magnitude for large $^3J_{HN-H\alpha}$ values and an increase in magnitude for small $^3J_{HN-H\alpha}$ values.

Vicinal coupling constants between α- and β-protons ($^3J_{H\alpha-H\beta}$) when used in conjunction with the intensities of intraresidue H^α-H^β and H^N-H^β NOEs, provide information about χ_1 dihedral angle constraints and stereospecific assignments for the β-protons (13) (see *Figure 2*). The $^3J_{H\alpha-H\beta}$ coupling constants are related to χ_1 by the Karplus equations. The parameters commonly used are those according to De Marco *et al.* (14). If $^3J_{H\alpha-H\beta}$ is in hertz, then:

$$^3J_{H\alpha-H\beta2} = 9.5 \cos^2\chi_1 - 1.6 \cos\chi_1 + 1.8 \qquad [4]$$

$$^3J_{H\alpha-H\beta3} = 9.5 \cos^2(\chi_1 - 120°) - 1.6 \cos(\chi_1 - 120°) + 1.8 \qquad [5]$$

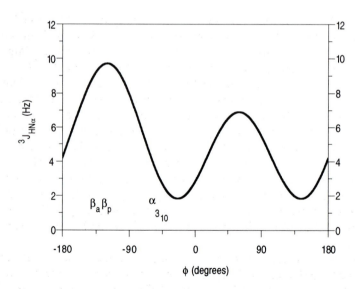

Figure 1. The relationship between the $^3J_{HN-H\alpha}$ vicinal coupling constant and the main-chain dihedral angle ϕ obtained using eqn 3. ϕ values for idealized α-helices, 3_{10}-helices, parallel, and antiparallel β-sheets are denoted α, 3_{10}, β_p and β_a respectively (see *Table 1*).

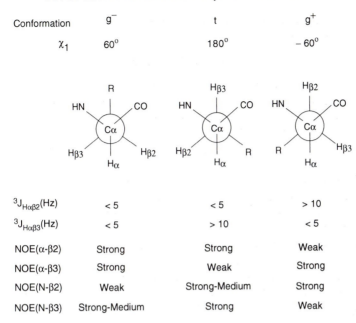

Figure 2. The relationship between the vicinal coupling constant $^3J_{H\alpha\text{-}H\beta}$, intraresidue $H^\alpha\text{-}H^\beta$ and $H^N\text{-}H^\beta$ NOEs, and the χ_1 torsion angle. The definition of χ_1 used is that according to the IUPAC-IUB Commission on Biochemical Nomenclature (11).

The two values correspond to the vicinal coupling constants to H^{β_2} and H^{β_3} respectively. If both $^3J_{H\alpha\text{-}H\beta}$ values are smaller than ~5 Hz, χ_1 is in the range $60 \pm 30°$. The other two possibilities for χ_1 correspond to one large value (greater than ~10 Hz) and one small value (less than ~5 Hz) for $^3J_{H\alpha\text{-}H\beta}$; these correspond to a range of either $180 \pm 30°$ or $-60 \pm 30°$ for χ_1. These two possibilities can be distinguished between on the basis of the corresponding NOE patterns. It should be noted that mobility about the alpha–beta bond results in a decrease for large $^3J_{H\alpha\text{-}H\beta}$ values and an increase for small $^3J_{H\alpha\text{-}H\beta}$ values. Valine has only one beta proton, therefore the above scheme can not be applied. However, if $^3J_{H\alpha\text{-}H\beta}$ is large (greater than ~10 Hz) then the conformation can be determined unambiguously. A large value for $^3J_{H\alpha\text{-}H\beta}$ indicates a *gauche$^+$* conformation for the χ_1 angle with H^α *trans* to H^β. It is sometimes possible to stereospecifically assign the two γ-methyl groups on the basis of the relative intensities of the $H^N\text{-}H^\gamma$ NOEs (see Chapter 10, Section 2.2).

Conformational databases provide another useful tool for determining stereospecific assignments. One such approach is to match the experimental intraresidue and sequential interresidue NOEs, and the $^3J_{H^N\text{-}H\alpha}$ and $^3J_{H\alpha\text{-}H\beta}$ coupling constants with values calculated for the conformations present in a database of structures. The programs HABAS (15) and STEREO-

SEARCH (16) adopt this approach. The matching process is performed for each of the two alternative stereospecific assignments and the results compared. On the basis of this comparison, the correct assignments can be determined, together with allowed ranges for ϕ, ψ, and χ_1. Two databases can be used for the searches—namely a crystallographic database consisting of high resolution structures, and a database containing the complete ϕ, ψ, and χ_1 conformational space accessible to a short peptide fragment with idealized geometry. Generally the assignments obtained with both databases are very similar although not identical, thus a combination of the two types of search would appear optimal. In test calculations (16), around 80% of the β-protons could be assigned and results obtained using experimental data indicate that a similar level of success can be expected in experimental cases. The program GLOMSA (17) relies on an initial ensemble of structures from which it determines stereospecific assignments. Care should, however, be taken with these programs to ensure that the experimental results are not over-interpreted.

For two distinguishable but unassignable protons, a possible approach for obtaining stereospecific assignments is to use the 'floating' stereospecific assignment (18) in the simulated annealing stage of structure determination. In the case of two distinguishable but unassignable β-protons, they are arbitrarily assigned H^{β_2} and H^{β_3}. A low value is used initially for the H^{β_2}-$C\beta$-H^{β_3} bond angle force constant. This force constant is slowly increased through the calculation to its final value and the final positions of the protons denote the correct stereospecific assignment. This approach, however, appears to have limited utility.

Since not all the prochiral protons in a structure which produce assignable resonances will usually be stereospecifically assigned, a means of incorporating constraints with respect to these atoms needs to be used. The use of 'pseudo-atoms' (19) facilitates this. These are used as follows. Whenever there is an NOE to one or more protons in an unresolved group of protons, the corresponding distance constraint is measured with respect to a pseudo-atom located centrally within the group. The experimental NOE involves the proton(s) rather than the pseudo-atom itself; therefore a distance correction must be added to the NOE distance constraint. This correction is equal to the maximum possible error incurred by the introduction of the pseudo-atom. Therefore for interresidue NOEs the correction corresponds to the distance between the pseudo-atom and the protons it replaces. For intraresidue NOEs the correction is set equal to the maximum sterically allowed difference in distance from a proton to the protons in the group and to the pseudo-atom respectively. Alternatively, a method for defining tighter distance constraints to pseudo-atoms has recently been developed and incorporated into the program DIANA (17). It should be noted that there is undoubtedly more structural information contained in stereospecific assignments compared with the use of pseudo-atoms, therefore stereospecific constraints should be used whenever possible (see, for example, Driscoll *et al.* (20)).

2.3 Secondary structural constraints

Once protein secondary structural elements have been determined (see Section 3), features implicit in such definitions can be used. Hydrogen bond constraints can be used on hydrogen–oxygen and nitrogen–oxygen distances and sometimes secondary structural elements can be assumed to have ideal-ized geometries. Commonly used hydrogen bond distance constraint ranges are 1.8–2.3 Å for the H–O distance and 2.5–3.3 Å for the N–O distance. The major concern with such constraints is the difficulty in reliably predicting the existence of a hydrogen bond. It is often better to omit any hydrogen bond restraints in the first instance, adding them in an iterative manner on the basis of their existence across an ensemble of generated structures. Similarly, it is often better in the first instance not to restrain a β-turn to be a particular type unless there is absolutely no doubt of its existence.

3. Protein secondary structure determination

Protein secondary structure is commonly determined using a combination of:

- short-range NOEs
- H^N-H^α coupling constants
- amide proton exchange rates

3.1 Short-range NOEs

The common secondary structural elements contain short-range, (i.e. be-tween residues no more than five residues apart in the sequence) proton–proton distances which are sufficiently small, (i.e. less than ~5 Å) for the corresponding NOEs to be observed. Each type of secondary structural element is characterized by a particular pattern of such NOEs (see, for ex-ample, Wuthrich (21)). This is illustrated in *Figure 3* with the corresponding short distances illustrated in *Figures 4* and *5*. Thus, if the NOEs in a spectrum are classified on the basis of their intensities as 'strong', 'medium', or 'weak' then an α-helix is characterized by a consecutive series of the following NOEs (i denotes the residue number of the residue in question): $H^N(i)$-$H^N(i + 1)$ (strong or medium), $H^\alpha(i)$-$H^N(i + 1)$ (medium or weak), $H^\alpha(i)$-$H^N(i + 3)$ (medium or weak), $H^\alpha(i)$-$H^\beta(i + 3)$ (medium or weak), $H^N(i)$-$H^N(i + 2)$ (weak), $H^\alpha(i)$-$H^N(i + 4)$ (weak). In contrast, β-strands are characterized by very strong $H^\alpha(i)$-$H^N(i + 1)$ and weak $H^N(i)$-$H^N(i + 1)$ NOEs; all other short-range NOEs are notably absent. From *Figure 3* it is apparent that there is very little distinction between an α-helix and a 3_{10}-helix, the distinction

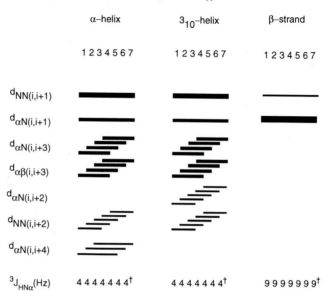

Figure 3. The characteristic patterns of short-range NOEs involving amide, alpha, and beta protons observed for ideal α-helices, 3_{10}-helices, and β-strands. The *thickness* of the lines is an indication of the intensity of the NOEs. †These values are approximate.

relying on observing $H^{\alpha}(i)$-$H^{N}(i + 4)$ NOEs in the former case and $H^{\alpha}(i)$-$H^{N}(i + 2)$ NOEs in the latter case. Both of these NOEs are likely to be very weak, making distinction between the two difficult.

The discussion thus far has assumed that the secondary structural elements adopt their 'ideal' geometries. However, it is well known that the secondary structural elements in globular proteins are frequently distorted from the ideal (12). Consequently the question arises as to how applicable the above discussion is to 'real' protein systems. If strong sequential NOEs exist, there is generally no ambiguity as to the type of secondary structure. In other cases, these sequential NOEs can be supplemented with information from other short-range NOEs in order to decrease the degree of ambiguity. NOEs between adjacent β-strands in β-sheets provide a further means of identifying regions of β-strand. This also serves as a means of distinguishing between parallel and antiparallel β-sheets. Assuming regular geometry, i denoting protons on one β-strand and j denoting protons on the second β-strand, for a parallel β-sheet the $H^{\alpha}(i)$-$H^{\alpha}(j)$ NOEs are strong and the $H^{\alpha}(i)$-$H^{N}(j)$ and $H^{N}(i)$-$H^{N}(j)$ NOEs are medium or weak; for an antiparallel β-sheet the $H^{\alpha}(i)$-$H^{\alpha}(j)$ NOEs are very weak (compared with strong in the case of a parallel β-sheet), the $H^{\alpha}(i)$-$H^{N}(j)$ NOEs are weak and the $H^{N}(i)$-$H^{N}(j)$ NOEs are medium or weak (22).

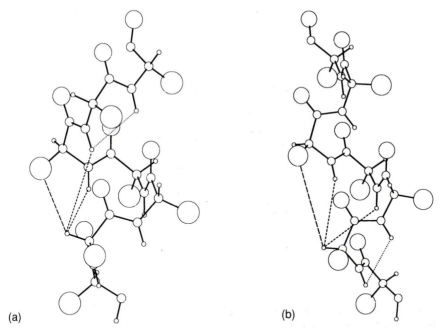

(a) (b)

Figure 4. Short proton–proton distances characteristic of (a) α-helices and (b) 3_{10}-helices. The smallest *spheres* represent hydrogen, the second smallest carbon and nitrogen, the second largest oxygen, and the largest the position of the beta proton(s). The *dots* represent $H^N(i)$-$H^N(i + 1)$. The *medium dashes* represent $H^\alpha(i)$-$H^N(i + 3)$, $H^\alpha(i)$-$H^N(i + 4)$ (α-helix), and $H^\alpha(i)$-$H^N(i + 2)$ (3_{10}-helix). The *long dashes* represent $H^\alpha(i)$-$H^\beta(i + 3)$.

3.2 H^N-H^α coupling constants

Knowledge of the backbone conformation provides a further means of identifying secondary structural types. $^3J_{H^N-H\alpha}$ coupling constants correspond to the mainchain ϕ angle. The relationship between $^3J_{H^N-H\alpha}$ and ϕ is given by eqn 3 and illustrated in *Figure 1*. Further, it is known that secondary structural elements adopt approximately the ϕ values shown in *Table 1*. The corresponding $^3J_{H^N-H\alpha}$ values (determined using eqn 3) are also shown. Generally a value for $^3J_{H^N-H\alpha}$ of $\leqslant 5$ Hz is used to define a helix and $\geqslant 8$ Hz is used to define a strand. Note that it is not possible to distinguish between different types of helices and different types of β-sheets on the basis of $^3J_{H^N-H\alpha}$ values alone.

3.3 Amide proton exchange rates

The involvement of mainchain amide protons in hydrogen bonds is another indication of the existence of secondary structural elements (23). Hydrogen bonds in α-helices are formed between CO(i)-HN(i + 4) and in 3_{10}-helices between CO(i)-HN(i + 3); in β-sheets large numbers of CO–HN hydrogen bonds are formed between adjacent strands. Indeed, in regular secondary

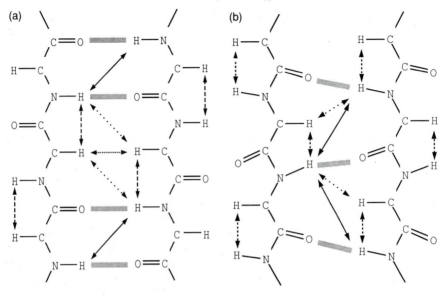

Figure 5. Short proton–proton distances characteristic of (a) antiparallel β-sheets and (b) parallel β-sheets. The *longest dashes* represent $H^\alpha(i)$-$H^N(i + 1)$. The remaining *lines with arrows* on represent the respective interstrand interactions. The *grey bands* represent the interchain hydrogen bonds.

Table 1. Parameters for regular secondary structural elements

Secondary structure	ϕ/degrees	$^3J_{HN\text{-}H\alpha}$/Hz
Antiparallel β-sheet	−139 [a]	8.9 [c]
Parallel β-sheet	−119 [a]	9.7 [c]
α-Helix	−57 [a]	3.9 [c]
3_{10}-Helix	−49 [b]	3.0 [c]

[a] From IUPAC-IUB Commission on Biochemical Nomenclature (11).
[b] From Ramachandran and Sasisekharan (77).
[c] Calculated using eqn 3.

structural elements all amide protons are involved in hydrogen bonds, with the exception of the first four residues in an α-helix, the first three residues in a 3_{10}-helix, and every second residue in flanking strands of β-sheets.

It is generally accepted that there is a correlation between slow amide proton exchange rates and the existence of hydrogen bonds in secondary structural elements. Pedersen *et al.* (24) find, however, that the corollary of this—that fast amide proton exchange implies the absence of secondary structural elements—is true in the case of lysozyme. Thus, although amide proton exchange rates are not sufficiently discriminating when used alone, used in conjunction with NOEs and $^3J_{H^N-H^\alpha}$ values they supply additional evidence for the existence of particular secondary structural elements. It should be noted that slow amide proton exchange rates give no indication as to which atom is the hydrogen bond acceptor.

4. Tertiary structure determination

Currently there is no universally accepted procedure for protein tertiary structure determination from NMR data. This section reviews some of the procedures available with the aim of giving the reader sufficient knowledge to decide which approach or combination of approaches are best suited to the problem in hand. All these approaches aim to sample conformational space whilst at the same time satisfying a set of constraints. This is achieved to varying degrees of efficiency by the different approaches used. One available approach is model building. This utilizes existing knowledge of tertiary structure. A very different approach is probabilistic in nature and represents each atom by a mean atomic position and variance about this position. An alternative method is adopted by distance geometry. This builds structures from internal distances. If a starting structure exists, restrained molecular dynamics can be used to refine it. Dynamical simulated annealing allows energy barriers to be crossed which are too high to be crossed by restrained molecular dynamics. An approach which takes into account the time averaging of NOEs is also discussed.

Constraints are usually expressed in terms of an error (or objective or target) function. Two different ways of defining the error function are commonly used. The first method defines a particular desired value for the parameter of interest and the error function measures deviations from this optimum value. Bond lengths and bond angles are examples of instances where such an error function is commonly used. In the second method an upper and a lower bound are specified for the parameter in question and no penalty is incurred if the parameter lies within these bounds. Outside the bounds it behaves in a similar manner to the first method. Such an approach is used when the parameter in question is known to be within a certain range of values, for example in the case of NOEs where the range of distances is determined by the intensity of the NOE.

4.1 Model building

Model building utilizes existing knowledge of structure as a means of determining the tertiary structure of the molecule in question. The structure is formed in such a manner as to minimize the violations of the NMR constraints whilst ensuring both good covalent geometry and that there are no bad non-bonded interactions. This can be performed either by using interactive molecular graphics (the easiest method if such facilities are available) or by using a physical model. CONFOR (25) is an example of a molecular graphics program for structure determination from NMR data. It can simultaneously display several molecules. All dihedral angles in these molecules can be interactively modified. Violations of upper distance bounds (derived from NOE intensities) and lower distance bounds (derived from van der Waals' radii) are displayed, thereby acting as a guide for the conformational changes to be made in order to satisfy the NMR data. By its very nature, this method is restricted to producing low resolution structures. Also, as the number of experimental constraints increases so does the complexity of the modelling operation. However, this type of model building can prove useful in the early stages of structure determination, especially in resolving ambiguous NOE assignments.

At its simplest level, model building involves the docking together of pre-defined secondary structural elements which are constructed using idealized geometry and are usually treated as rigid bodies. If the tertiary structure of a molecule which is homologous to that being studied is known, this known structure can be used as the basis for determining the unknown structure. If the molecule is a protein then this proceeds as in *Protocol 2*.

Protocol 2. Homologous model building

1. Align the amino acid sequences.
2. Divide the sequences into conserved and variable regions.
3. Use the conserved regions from the known tertiary structure to model those in the unknown structure.
4. Model build the remainder of the unknown tertiary structure either by scanning a database of known tertiary structures or by using interactive molecular graphics.
5. Substitute the side chains where necessary.
6. Remove bad non-bonded contacts and minimize the violations of the NMR constraints.

Homologous model building has proved particularly useful when attempting to distinguish between intermolecular and intramolecular NOEs in the case of a dimer. The need for homologous model building becomes even more apparent when the NOEs are insufficient to define the dimer interface. For example, this problem has arisen with the author's own work on neuronal bungarotoxin (26). Here not only has homologous model building based on α-bungarotoxin proved invaluable in resolving between intermolecular and intramolecular NOEs, but also in defining the nature of the dimer interface which was ill defined due to the paucity of assignable intermolecular NOEs. When such an approach is used, it is advisable to produce a number of models based on differing assumptions in order to ensure that the final structure is not merely a function of the starting model. A large number of models can be produced relatively quickly using software developed for homologous model building (for example, COMPOSER (27)). As can be inferred from step **6** of *Protocol 2*, model built structures can be used as the starting point for restrained molecular dynamics refinement.

The use of a database of NOEs and coupling constant data derived from known protein structures is another approach which can be used. High resolution X-ray crystallographic protein structures have been used to calculate a matrix of the distances between pairs of H^N, H^α, and H^β protons (28). The premise of any such approach is that the required local structural features are present in the set of proteins used in compiling the database. The approach used is outlined in *Protocol 3*.

Protocol 3. Model building from a database of known tertiary structures

1. Systematically search the database for all the fragments satisfying the distance constraints to within 0.5 Å. (The fragment used is five to eight residues in length.)
2. Use amide-alpha and alpha-beta vicinal coupling constants as an additional filter.
3. Cluster the fragments on the basis of overall structural similarity.
4. Select a fragment from the largest cluster.
5. Ensure that adjacent fragments overlap to within 0.7 Å.

Distances corresponding to long-range NOEs are monitored and used as an additional filter during the model building procedure. Steric clashes between different sections of the structure are monitored on the basis of distances between Cα atoms. The fragment which is determined first and subsequently used as the point of reference for the rest of the structure is that which is the most well defined on the basis of NOEs.

The results show that the method performs well at defining local structural regions. The small errors in the fragments accumulate, however, resulting in significant global errors. Those structural fragments which give rise to a characteristic pattern of short-range NOEs, (e.g. helices, β-sheets separated by tight turns) can be built with confidence. Thus, the method is suitable for producing a starting point for further refinement.

4.2 A probabilistic approach

An alternative approach is one which aims to accurately define the spatial distribution of atomic positions allowed on the basis of the given data by representing each atom using a mean atomic position and variance about this position. The premise behind the method is that as larger proteins are studied, the precision and abundance of NMR data will not keep pace. This necessitates the existence of data analysis methods which can use imprecise data to determine the region of space occupied by each atom. Such an approach has been incorporated into the computer program PROTEAN (29). This has been implemented for proteins and the steps involved in this are summarized in *Protocol 4*.

Protocol 4. Structure determination using PROTEAN

1. Define secondary structural elements using a set of logical rules.
2. Represent the secondary structural elements as simplified rigid bodies.
3. Select the most highly constrained secondary structural element.
4. Use this as the point of reference for building the remainder of the structure.
5. Refine the resultant structure with all the atoms present using a probabilistic description of the protein.

Rigid bodies are used in step **2** in order to reduce the number of objects whose positions are sampled systematically—cylinders enclosing all atoms up to and including the β carbon atoms are used to represent both α-helices and β-strands. The curvature of β-strands is allowed for by representing each strand using two cylinders; the curvature of helices is ignored. The order in which the remaining secondary structural elements are added in step **4** is determined by the number of constraints with respect to the element selected in step **3**. At various stages during the addition of elements (e.g. at the completion of β-sheets or the docking together of helices) conformational space is discretely sampled in order to refine (i.e. reduce) the region of conformational space accessible to each element. The resultant structure is further refined with all atoms present using a probabilistic description of the protein, as described below.

Systematic sampling of conformational space is not a tractable means of producing a structure which is sufficiently precise due to the long time scale involved in any such operation. Consequently, an alternative approach is sought. A representation is used which assumes that a protein can be described by a vector consisting of atomic coordinates (the state vector). Associated with this vector is a covariance matrix which contains elements corresponding to the covariance between each pair of atomic positions. In order to introduce a constraint (e.g. an experimental constraint) this constraint (z) needs to be related to the state vector $(h(x))$. Generally, this relationship adopts the following form:

$$z = h(x) + v \qquad [6]$$

where v is a measure of the uncertainty in the determination of z and is assumed to be normally distributed with mean zero. The state vector and covariance matrix are updated by comparing the calculated value of z with its measured value. If the calculated value agrees with the measured value, the state vector and covariance matrix are not updated. However, if there is disagreement then the mode of updating depends upon the ratio of the uncertainty in the measurement to the uncertainty in the state vector. If the uncertainty in the state vector is high (as measured by the covariance matrix) and the uncertainty in the measurement is low (as measured by v) then the state vector and the covariance matrix are updated to minimize the disagreement in the current estimate. On the other hand, if the uncertainty in the state vector is low and the uncertainty in the measurement is high, the state vector and covariance matrix are only corrected to a negligible amount. If the uncertainty of both are similar, the correction is determined by the ratio of the uncertainties.

The starting point for the refinement is the set of mean atomic positions resulting from the systematic sampling of conformational space (step **4**). The corresponding variances are also determined from the results of the systematic sampling and these are used to calculate the covariances. After the sequential introduction of the constraints, an estimate of the mean positions exists which is an improvement over the starting set of mean positions. This set of mean positions is used as the starting positions for a subsequent round of updating. Atoms are allowed freedom to move as a result of the applied constraints by resetting the covariances to their initial (large) values at the start of each subsequent round of updating and reintroducing all the constraints.

This refinement is applied to data consisting of distances and bond angles. Those atoms present in secondary structural elements can be constrained to have both the known 'ideal' dihedral angles and the corresponding known variance matrix for these angles in the respective secondary structural types. By using approximate representations of a protein, the number of atoms can be drastically reduced. This is sufficient to define the general topology of the protein whilst significantly reducing the CPU time involved in structure determination. The structures resulting from calculations using approximate representations can be used as the starting point for more refined representations.

4.3 Distance geometry

Two types of distance geometry are commonly used. The first operates in distance space and is commonly known as 'metric matrix distance geometry' (30–32) because it uses the metric matrix in order to convert the initial set of distance constraints into a set of Cartesian coordinates. The second class operates in torsional space (in contrast to distance space) and uses restrained minimization of a variable target function (33).

4.3.1 Metric matrix distance geometry

There are several programs available which are based on metric matrix distance geometry, (e.g. DGEOM, DG-II (34), DISGEO (35), DSPACE, EMBED, VEMBED, XPLOR/dg (36); for a review see, for example, Kuntz *et al.* (37)). An overview of the steps involved in metrix matrix distance geometry is given in *Protocol 5*.

Protocol 5. The steps involved in metrix matrix distance geometry

1. Set-up distance and chiral constraints.

2. Smooth the distance bounds—upper followed by lower.

3. Select trial distances between upper and lower bounds.

4. Embed to produce Cartesian coordinates.

5. Optimize first the chirality, then chirality and distance.

The distance constraints (corresponding to bond lengths, bond angles, and NOEs) specified in step **1** are used to set-up a matrix of upper and lower distance bounds—one set of bounds for each pair of atoms within the protein. Those upper bounds which are not known are set to an arbitrarily large distance which must be greater than the size of the protein; undefined lower bounds are set equal to the sum of the respective van der Waals' radii. The region of conformational space available to the structure is further reduced by a process known as bounds smoothing (step **2**). This is achieved by increasing the lower bounds and reducing the upper bounds in those cases where the bounds are not geometrically self-consistent. Commonly the triangle inequality (38) is the applied constraint. The tetragonal inequality (39) can also be used, although this is very computationally intensive for systems involving large numbers of atoms. The triangle inequality works as follows. Consider three atoms (i, j, and k) and the corresponding upper bounds (u_{ij}, u_{ik}, and u_{jk}). The furthest that j can be from i corresponds to i, j, and k being collinear with k lying between i and j. Thus:

$$u_{ij} \leq u_{ik} + u_{kj} \qquad [7]$$

It follows that if the left-hand side of eqn 7 is greater than the right-hand side then u_{ij} is decreased to the value of the right-hand side. This procedure is applied to all sets of three atoms i, j, and k until no further changes can be made to the upper bounds. Similarly, if the corresponding lower bounds are l_{ij}, l_{ik}, and l_{jk} then the closest that j can be to i corresponds to i, j, and k being collinear with j lying between i and k. Thus:

$$l_{ij} \geq l_{ik} - u_{jk} \qquad [8]$$

Therefore, once the upper bounds have been smoothed, eqn 8 can be used to smooth the lower bounds. Inability to resolve triangular inequalities is an indication that there is a problem with the input constraints.

Once smoothing of the bounds is complete, a set of trial distances which lie between the upper and lower bounds can be generated for all atom pairs (step 3). It is the selection of trial distances which is the key to success with the method. These can be chosen from a random distribution or may be chosen to fit a particular distribution function. The use of a random distribution to generate the distances has been highlighted as one of the reasons for the limited sampling of conformational space by metric matrix distance geometry (40). Hare and co-workers (40) have found that use of a random distribution leads to extended structures because the distribution of distances is weighted unrealistically towards larger distances, and recommend the use of a distribution which weights shorter distances more heavily than longer distances. They also found that the large difference between the upper and lower bounds leads to a large number of inconsistent distances which in turn leads to a limited sampling of conformational space (the observed average distance is $\sim 60\%$ of the upper bound with very little variance across the ensemble of structures). This can be overcome by both the production of a very self-consistent distance matrix and the generation of an ensemble of such self-consistent distance matrices which sample a range of distances. (It should be noted that distances are not independent of other distances.) This correlation between distances can be taken into account when generating the trial distances (38). Havel (41) has found that entering the trial distances into the corresponding elements of the distance matrix by choosing these elements in a random manner (rather than entering all those distances involving atom one first, then those involving atom two and so on) improves the sampling of conformational space. Kuntz (37) has found that a 'full' set of NOEs, (i.e. several hundred long-range NOEs for a protein of 50–100 amino acid residues) provides sufficient information to eliminate the use of correlation routines. The improved sampling properties of the newer programs, (e.g. DG-II and XPLOR/dg) makes these more desirable than the older metric matrix programs.

Step 4 involves the conversion of the distances into coordinates. This is known as embedding. The metric matrix can be determined from the distances using the cosine rule (42) and it has the unique property that the square roots of its eigenvalues are the principal axes of the molecule with the origin of the coordinate system at the centroid of the molecule; the eigenvectors correspond to the distributions of the atoms along the axes. Thus the atomic coordinates can be determined from the eigenvalues and eigenvectors using the relationship:

$$\mathbf{v}_i = \lambda_i^{\frac{1}{2}} \mathbf{w_i} \qquad [9]$$

where $\mathbf{v}_i$, λ_i, and $\mathbf{w}_i$ are the coordinates, eigenvalues, and eigenvectors respectively for each dimension.

Although the coordinates can be determined directly from the metric matrix, there are invariably more than three non-zero eigenvalues and therefore the structure occupies more than three dimensions. This high dimensionality is more likely with looser bounds and a lower number of experimental constraints. In order to obtain a three-dimensional structure, the three largest eigenvalues and corresponding eigenvectors are used. This 'truncation' of the metric matrix to three dimensions results in a poorer correlation between the observed and target distances. This emphasizes the need for subsequent optimization of the resulting structures and has led to the development of other methods of projection which truncate the metric matrix more gradually (see, for example, Crippen (43)). An indication of how well defined the structure is in three dimensions can be obtained by comparing the magnitude of the fourth and fifth (and perhaps sixth) eigenvalues with the magnitude of the first three. If they are a lot smaller than the first three then the structure is fairly well defined. However, if they are similar in magnitude to the first three then the structure is not particularly well defined. Negative eigenvalues present a problem as they imply that the structure contains imaginary distances. Usually structures with negative eigenvalues are discarded. If they need to be used then the absolute values of the eigenvalues is taken, although such structures do not correspond as well to the input data as those determined from positive eigenvalues.

It would be desirable to assign different relative weights to distances depending on the type of constraint (e.g. bond length versus NOE). Unfortunately, it is not possible to achieve this through the metric matrix. Additional atoms can, however, be included which occupy the same position as 'real' atoms in order to increase the relative weight of such atoms. Another possibility is to introduce the desired weighting scheme into the optimization stage.

Before proceeding to the optimization stage, it is advisable to verify that the chirality of the resulting structures is correct. If the structure is well defined by the constraints then it is usually possible to distinguish the correct handedness by considering the chiral error (this calculates a signed volume for each chiral centre and compares these with the corresponding target values) of the structure and its mirror image. If there is a significant difference between these, that with the lowest chiral energy corresponds to the correct handedness. An alternative approach is to use a signed distance map (44). If the structure is not well defined, the chirality may not be a sufficiently discriminating test and it may be necessary to optimize both the structure and its mirror image. The structures can then be subjected to restrained molecular dynamics or dynamical simulated annealing, and that with the significantly lower energy corresponds to the correct handedness. Comparison with the fold of a homologous structure is another means of determining the correct handedness.

Optimization (step **5**) involves the minimization of a target function and is commonly a two-stage process. The first stage of optimization involves minimizing the chiral error function. The second stage optimizes both the chiral error function and the distance error function. Optimization stops when either the value of the overall target function or the gradient falls below a predefined threshold. It should be remembered that the optimized structure corresponds to a local minimum in the energy surface, rather than the global minimum. One method used for overcoming problems with local minima is to begin optimization in higher-dimensional space and gradually reduce the coordinates in the fourth and higher dimensions to zero. If problems with local minima persist it is advisable to inspect the structure using interactive molecular graphics since problems such as loop entanglement may occur. It may be possible to remove such problems manually.

The structure resulting from metric matrix distance geometry should satisfy to a large extent all the distance constraints used. However, it is likely to have a large potential energy because of distortions in geometry (usually small) and close non-bonded contacts since hitherto the potential energy has not been taken into consideration. This can be drastically reduced by subjecting the structure to 100 or so cycles of minimization using a molecular mechanics program such as AMBER (45), CHARMM (46), DISCOVER (47), GROMOS (48), or XPLOR (49). Minimization in the presence of solvent (in contrast to a vacuum) is also used for refinement, as is molecular dynamics. It should be noted that only some of the structures generated by metric matrix distance geometry are selected—this being on the basis of the error function following optimization.

Let us now consider the computational requirements of metric matrix distance geometry. The CPU time required for bounds smoothing is proportional to N^3, where N is the number of atoms. However, this vectorizes easily. Embedding and optimization are both proportional to N^2. As an example, consider the use of the vectorized program VEMBED applied to a molecule of ~ 900 atoms. The bounds smoothing, embedding, and optimization stages take 200 seconds, 4 seconds, and 230 seconds respectively on a Cray XMP-48, or $\sim$ nine hours, $\sim$ eight minutes, and $\sim$ eight hours respectively on a Vax 8650. It should be noted that bounds smoothing is only performed once for a given set of structures.

4.3.2 Distance geometry in torsional space

DISMAN (33, 50) and its successor DIANA (17, 51) are the programs commonly used when structures are being determined using the variable target function method. Fixed standard bond lengths and bond angles are used, with only the torsion angles treated as independent variables. Working in torsional space drastically reduces the number of independent variables compared with the use of Cartesian coordinates, which results in a reduction in the computer memory required. Determination of the structure starts from a structure whose conformation is random. The conformation of the structure is then determined from the specified distance constraints by minimizing a target function, the value of which is zero if all the distances are consistent with the constraints.

The target function can be defined in many different ways. A typical form of the target function is given by (50):

$$T = \sum_{i<j} \{\theta(D_{ij} - U_{ij})(D_{ij} - U_{ij}) + \theta(L_{ij} - D_{ij})(L_{ij} - D_{ij})\} \qquad [10]$$

where D_{ij} is the distance between atoms i and j in the structure, U_{ij} is the upper distance bound for atoms i and j, L_{ij} the lower distance bound and

$$\theta(x) = \begin{cases} 0 \text{ if } x \le 0 \\ 1 \text{ if } x > 0 \end{cases} \qquad [11]$$

Thus if the distance lies within the distance bounds then T is zero, otherwise T is positive. Explicit restrictions on torsion angles derived from vicinal spin–spin coupling constants can be easily incorporated. In DISMAN a fourth-order polynomial with continuous first derivatives and 2π periodicity is used for this purpose (50).

Once the target function has been defined, the next stage is to optimize it. A method for rapidly calculating the first derivative of the target function has been implemented in DISMAN (33). This enables violations of distances to be corrected by adjustment of the torsion angles.

The target function is termed 'variable' because of the way it is used. Direct minimization of the target function is not generally practical when starting from a random structure due to the multiple minima problem. Therefore the optimization starts by considering only short-range constraints, gradually increasing the sequential range of the constraints in successive applications of the optimizer until all the constraints are applied across the entire structure. Thus, first only intraresidue constraints are optimized. Once this has been achieved successfully, constraints between neighbouring residues are optimized alongside intraresidue constraints (i.e. constraints to residues i and i ± 1 are considered). Next constraints to those residues i ± 2 are also considered. This 'window' is gradually expanded to include the entire molecule. Above ~ i ± 5 the size of the window is increased in larger steps to, say i ± 10, then i ± 15, i ± 20, i ± 30, i ± 40 . . .. Hence the local fold is determined prior to the global fold.

DIANA contains a new implementation of the variable target function (17). This has the advantage over DISMAN that the computational time is significantly reduced.

As with metric matrix distance geometry, the resultant structures are likely to satisfy the distance constraints to a large extent, although the potential energy is likely to be large. Again this can be reduced by subjecting the structure to minimization using a molecular mechanics program.

An indication of the CPU time required when using the variable target function can be obtained by considering the structural determination of BPTI (56 amino acid residues) using DIANA. The time taken to produce one structure is 49 seconds, 20 minutes, and 13 minutes on a Cray XMP-28, a VAX 8650, and a Silicon Graphics Personal Iris 4D/25 respectively.

A comparison of the programs DISGEO and DISMAN has been performed (13). In model calculations on BPTI a tendency for DISGEO to produce structures which were more expanded then the crystal structure was observed. The same tendency was noted to a lesser degree for DISMAN. It was also found that DISMAN is more memory efficient than DISGEO, although no significant difference was found in terms of CPU time required. Incorrect polypeptide folds are a problem which arise with DISMAN which do not arise with DISGEO (this arises because the former works in 'real' space and the latter in distance space).

4.4 Restrained molecular dynamics

A molecular dynamics (MD) simulation starts with some initial conformation. Once this has been decided upon, the structure is allowed to evolve over time under the influence of a force field. The force fields used by molecular dynamics programs, (e.g. AMBER (45), CHARMM (46), DISCOVER (47), and GROMOS (48)) differ subtly from each other and usually have the following potential energy terms (denoted 'V' with the corresponding subscript) for *in vacuo* simulations:

$$V_{total} = V_{covalent} + V_{non\text{-}bonded} \qquad [12]$$

where:

$$V_{covalent} = V_{bond} + V_{angle} + V_{dihedral} + V_{improper} \qquad [13]$$

and

$$V_{non\text{-}bonded} = V_{van\ der\ Waals'} + V_{electrostatic} + V_{hydrogen\ bond} \qquad [14]$$

$V_{improper}$ is added to keep rings planar and to maintain the chirality of chiral centres. An explanation of these terms can be found in, for example, McCammon and Harvey (52).

In determining the trajectory of the structure, the force acting on each atom at a given time needs to be determined. The force due to potential energy terms can be determined by differentiating eqn 12 with respect to the coordinates. So far no account has been taken of the thermal motion of the atoms. Newton's equations of motion link the force acting on a particle to its acceleration. These can be solved by numerical integration over very small time steps to give both the velocity of the atom after the time step and the displacement of the atom during the time step, (e.g. the leap-frog method (53), the Beeman method (54, 55), and the fifth-order Gear method (56, 57)). The time steps must be smaller than the time scales of the highest frequency motions in the system, (i.e. ~ 1 fsec, or ~ 2 fsec if the bond lengths are constrained using an algorithm such as SHAKE (58)). The initial velocities are assigned from a Maxwellian distribution corresponding to the desired temperature and this temperature is commonly maintained using a temperature bath (59). Care is sometimes required with hydrogen atoms since, due to their small mass, they are assigned relatively high velocities. One way to surmount this problem is to assign them a higher mass (e.g. 10 amu), at least in the initial stages of the MD.

In determining structures from NMR data, experimentally determined distance and dihedral angle constraints need to be satisfied. This is achieved by adding additional terms to the non-bonded potential in eqn 14. One means of applying the distance constraints is to use a square well potential of the form:

$$V_{NOE} = \begin{cases} k_{NOE}(d_{ij} - u_{ij})^2 & \text{if } d_{ij} > u_{ij} \\ k_{NOe}(l_{ij} - d_{ij})^2 & \text{if } d_{ij} < l_{ij} \\ 0 & \text{otherwise} \end{cases} \quad [15]$$

where k_{NOE} is the NOE force constant (or weighting factor). Asymmetric and asymptotic potentials are alternative functional forms which are commonly used. When large distance violations are likely to occur, it is advisable to limit the constraining forces to a maximum value by, for example, using an asymptotic function or a low initial value for k_{NOE}. This avoids such violations dominating the force field. Similarly, dihedral angle constraints can be incorporated using a square well potential of the form:

$$V_{tor} = \begin{cases} k_{tor}(\phi_{ij} - \phi_{ij}^u)^2 & \text{if } \phi_{ij} > \phi_{ij}^u \\ k_{tor}(\phi_{ij}^l - \phi_{ij})^2 & \text{if } \phi_{ij} < \phi_{ij}^l \\ 0 & \text{otherwise} \end{cases} \quad [16]$$

where k_{tor} is the dihedral angle force constant (or weighting factor), ϕ_{ij}^u the upper dihedral angle bound, and ϕ_{ij}^l the lower dihedral bound. Note that care should be taken in order to ensure that the value of ϕ_{ij} is unambiguously defined by the coupling constants. Thus, eqn 14 becomes:

$$V_{non-bonded} = V_{van\ der\ Waals'} + V_{electrostatic} + V_{hydrogen\ bond} + V_{NOE} + V_{tor} \quad [17]$$

Once the potential forms have been decided upon, the next stage is to decide how to weight them relative to the other terms in the force field. This is usually achieved by deciding the size of the violation to be allowed during the simulation. The average kinetic energy per degree of freedom is $\frac{1}{2}k_BT$ and the temperature (T) of the simulation is known, thus the force constants k_{NOE} and k_{tor} can be determined by placing $\frac{1}{2}k_BT$ on the left-hand side of eqn 15 and 16 respectively (k_B is the Boltzmann factor).

Restrained MD calculations are used with two distinct aims. One is the refinement of a model of structure which has been built to satisfy the

experimental distance constraints as well as possible. The other is to determine the global polypeptide fold of a protein. An example of the latter is its
application to crambin (60) using model calculations. This was performed in a
similar manner to the use of the variable target function—only short-range
constraints were included initially, followed by subsequent inclusion of all
constraints. Two types of starting structure were used, namely a fully extended structure and one in which the helices had already been formed.
Different protocols were investigated involving different weights for the
modelled experimental constraints. Three of the four starting structures form
the correct polypeptide fold. In the case of the fourth, the secondary structural
elements are not formed before tertiary folding starts, resulting in an
incorrect global fold.

A major drawback when using restrained MD is that most of the CPU time
is spent in evaluating the interatomic forces for structures which are far from a
conformation which satisfies the experimental constraints. Thus, protocols
have been developed which use a better approximation to the global fold as
the starting point for the restrained MD.

In principle MD can be used to sample all of conformational space. In
practice, however, there is insufficient CPU time available to achieve this. To
give an indication of the CPU times involved, consider 1 psec of dynamics
with an 18 Å cut-off performed *in vacuo*. On a VAX 8600 this takes ~ five
minutes for a molecule of 90 atoms and ~ 30 minutes for a molecule of 562
atoms. When solvent is added (2382 and 1231 water molecules respectively)
these times increase to ~ eight hours and ~ 12 hours respectively (this
corresponds to ~ 15 minutes and ~ 42 minutes respectively on a CYBER
205).

4.5 Distance geometry followed by restrained MD

The use of distance geometry folowed by restrained MD uses distance
geometry to produce structures with roughly the correct global fold prior
to application of restrained MD. Using energy minimization subsequent to
distance geometry has already been eluded to. Restrained MD has the
advantage over energy minimization that the kinetic energy in the system
allows energy barriers of the order of $k_B T$ to be crossed, resulting in the
sampling of a larger region of conformational space and producing structures
with a significantly lower energy. The major disadvantage of MD is the CPU
time required for the simulations. After the MD is complete, the structure is
subjected to energy minimization.

The range of conformations following restrained MD is significantly larger
than the range following distance geometry (see, for example, Clore *et al.*
(61), Cook *et al.* (62), and de Vlieg *et al.* (63)), thus illustrating that the
sampling resulting from restrained MD is better than that resulting from
distance geometry alone. This is most noticeable in those regions which are
least well defined by the experimental constraints. However, such structures

are still acceptable because there are no large NOE violations and the final energies are low.

4.6 Hybrid distance geometry–dynamical simulated annealing

The hybrid 'distance geometry–dynamical simulated annealing' protocol (64) (*Protocol 6*) is similar in many ways to distance geometry followed by restrained MD.

Protocol 6. Hybrid distance geometry–dynamical simulated annealing

1. Generate a substructure using metric matrix distance geometry. For proteins this contains N, C, Cα, Cβ, non-terminal Cγ and Cδ atoms, and a pseudo-atom for aromatic rings.
2. Fit an all atom representation of the structure on to the substructure.
3. Energy minimize to ensure correct covalent geometry. At this stage the non-bonded force is very low.
4. Perform MD at 1000 K, increasing the non-bonded force from its initial low value to its final value.
5. Slowly cool from 1000 K to 300 K.
6. Energy minimize.

It differs, however, in two important ways. Firstly the substructures generated by step **1** contain significantly fewer atoms than the full structure (about one third in the case of proteins). Secondly the non-bonded interactions (eqn 14) are replaced by a simple van der Waals' repulsive term and eqn 17 becomes:

$$V_{non\text{-}bonded} = V_{NOE} + V_{tor} + V_{repel} \qquad [18]$$

where:

$$V_{repel} = \begin{cases} k_{repel}(s^2(d_{ij}^{min})^2 - d_{ij}^2)^2 & \text{if } d_{ij} < s.d_{ij}^{min} \\ 0 \text{ otherwise} \end{cases} \qquad [19]$$

where d_{ij}^{min} is the sum of the respective van der Waals' radii and s is a scaling factor to allow atoms to approach closer than the hard sphere van der Waals' radii (this occurs due to the attractive component of the van der Waals' interaction).

The term 'simulated annealing' is used as this refers to the crossing of potential energy barriers which would be impossible to cross under normal conditions. This is facilitated by setting the force constant k_{repel} to a very low value, thereby allowing regions of the polypeptide chain to 'pass through' each other if the experimental constraints so dictate. This process is enhanced by raising the temperature at which the MD is performed. (Raising the temperature is equivalent to lowering k_{repel}). As the simulation progresses, the violations of the experimental constraints becomes less and the value of k_{repel} is increased in order to ensure that there are no close non-bonded contacts. Once k_{repel} has reached its final value, the system is gradually cooled to 300 K and the structure energy minimized. There appears to be no definitive 'best' protocol or 'best' set of parameters to use—it is a matter of each individual finding a protocol and set of parameters which they feel produces a satisfactory set of structures. One method of evaluating a protocol and set of parameters is to take a known structure similar to the one under investigation and generate a set of constraints from it. These constraints can then be used to generate an ensemble of structures which are compared with the 'target' known structure. One rule of thumb is the larger the structure the longer the time which should be allowed for each stage during the dynamics. A protocol which works satisfactorily for a protein of 132 amino acid residues is given in *Protocol 7*. For larger proteins, it may be necessary to increase the length of time simulated for each stage. The program commonly used for the simulated annealing is XPLOR (49), although other molecular dynamics programs can also be used.

Protocol 7. Dynamical simulated annealing

1. Perform 200 cycles of energy minimization with the following values for the respective force constants:

   $k_{bond} = 500 \text{ kcal/mol/Å}^2$
   $k_{angle} = 500 \text{ kcal/mol/rad}^2$
   $k_{improper} = 500 \text{ kcal/mol/rad}^2$
   $k_{peptide\ bond\ dihedral} = 200 \text{ kcal/mol/rad}^2$
   $k_{repel} = 0.01 \text{ kcal/mol/Å}^4$
   $k_{NOE} = 0 \text{ kcal/mol/Å}^2$
   $k_{tor} = 0 \text{ kcal/mol/rad}^2$

2. Set k_{NOE} to 50 kcal/mol/Å^2, k_{tor} to 200 kcal/mol/rad^2, all masses to 10 amu, the force constant for the dihedral angle about the proline peptide bond to 20 kcal/mol/rad^2, and perform 1.5 psec dynamics coupled to a temperature bath at 1000 K.

3. Set k_{NOE} to 10 kcal/mol/Å^2, k_{tor} to 20 kcal/mol/rad^2, reduce the covalent parameters to their 'normal' full potential force field values, and perform 40 75 fsec stages of dynamics at 1000 K, increasing k_{repel} by a factor of $1000^{1/40}$ after each stage.

4. Cool over 1.4 psec in 25 K stages from 1000 K to 300 K.

5. Energy minimize for 200 cycles with the proline peptide bond dihedral angle force constant set to 200 kcal/mol/rad^2.

Dynamical simulated annealing (like distance geometry) has a tendency to produce expanded structures. This arises because the only non-bonded interactions are defined by the NOEs and the REPEL function and is observed to a greater extent with structures which are not particularly well defined by the experimental constraints. This can be overcome to a large extent by allowing the structure to evolve in a force field in which the REPEL term is replaced by van der Waals', electrostatic and hydrogen bond terms (eqn 14).

As well as starting from a structure produced by distance geometry, dynamical simulated annealing has also proved very successful when starting from a random array of atoms (65). A satisfactory solution relies on reducing all the force constants in the force field to values which are sufficiently low for the kinetic energy of the system to allow the energy barriers to be crossed.

A different type of simulated annealing protocol which avoids the use of distance geometry has also been developed (66) (known as YASAP). This uses any structure with good stereochemistry as the starting structure, (e.g. a random structure) and the protocol consists of the same non-bonded potential as given in eqn 18, although the NOE potential is asymptotic in nature. As with distance geometry in torsional space, fixed standard bond lengths and bond angles are used. Results indicate that the conformational sampling properties of this procedure are a marked improvement over the sampling properties of older metric matrix distance geometry programs.

To give an indication of the CPU time required for *Protocol 6*, a protein containing 1960 atoms takes ~ 75 minutes on a Convex C210 whereas a protein containing 563 atoms takes ~ 25 minutes.

4.7 Allowing for internal motion

The protocols discussed thus far assume a rigid model for the structure in question. It is well known, however, that proteins exhibit a variety of internal motions (see Chapter 10, Section 4.3.4 for a discussion of how to investigate this experimentally). These can be divided into two categories:

- rapid internal motions
- the existence of multiple conformations

The former result in different correlation times and lead in turn to errors in distance constraints. These can be overcome to a certain extent by using a larger distance range for the constraints, although this leads to a decrease in precision in the resulting structures. The latter is addressed in this section.

Perhaps the simplest way of modelling internal motions is to use a two-state model. This results in a dramatic improvement in how well the distance constraints are satisfied and a significant lowering of the potential energies for individual conformers in the case of the acyl carrier protein (67). An alternative approach is to use ensemble averaged distance constraints (68). This involves producing an initial set of structures using distance geometry. A MD simulation is then started for all of the structures simultaneously, with each having evolved for the same length of time at any given moment. A simplified force field is used which treats the constraints in two different ways—the distance constraints arising from the covalent geometry are satisfied in each member of the ensemble whereas the NOE constraints are satisfied by the ensemble as a whole but not necessarily by each member of the ensemble.

Another approach used is to average over the course of the MD simulation (69). Since the time scale of the simulation is short compared with the correlation time for molecular tumbling, the influence of angular fluctuations should be ignored and therefore the NOE should be treated as a function of d_{ij}^{-3} rather than d_{ij}^{-6}. If the time averaged distance is taken over the whole trajectory then it becomes increasingly less sensitive to the instantaneous state of the molecule as time progresses. This is undesirable and therefore a memory function is built into the definition of the time averaged distance. This ensures that recent history is weighted more heavily than distant history. The time averaged distance $(\bar{d}_{ij}(t))$ is given by:

$$\bar{d}_{ij}(t) = \left(\frac{1}{\tau}\int_0^t e^{-t'/\tau}\{r(t-t')^{-3}\}dt'\right)^{\frac{-1}{3}} \qquad [20]$$

where τ is the characteristic time for the exponential decay. It has been found, however, that this form of the potential occasionally gave rise to large forces in simulations of macromolecules. In order to surmount this problem, a force has been constructed (69) so that:

$$\vec{F}_i(t) = \begin{cases} -k_{NOE}|\bar{d}_{ij}(t) - u_{ij}|\dfrac{\vec{d}_{ij}(t)}{d_{ij}(t)} & \text{if } \bar{d}_{ij}(t) > u_{ij} \\ 0 \text{ otherwise} \end{cases} \qquad [21]$$

where $\vec{F}_i(t)$ is the force on atom i due to atom j and $\vec{d}_{ij}(t) = \vec{d}_i(t) - \vec{d}_j(t)$. The value of τ should be chosen so that it is long enough not to hinder motions in the system. However, the value of τ directly affects the rate of motions in the system and thus the length of simulation required to observe proper averaging. Hence there is a compromise between the hindering of motions and computational expense—Torda *et al.* (69) find that the simulation should be around an order of magnitude longer than the value of τ. Results with tendamistat (69) show that the time averaged constraints increase the mobility of molecules, produce better agreement with distance bounds, and give a better indication of the conformational space occupied by the molecule.

4.8 Analysis of structures

Once an ensemble of structures has been generated, it is necessary to assess them. The first question to address is how well the resulting structures satisfy the initial set of constraints. Ideally no constraint violations would be observed, but in practice this is not the case. Usually a violation of greater than ~ 1.0 Å (or less depending on the force constant used for the NOE restraints) is an indication that there is a serious problem, especially if any such violation occurs repeatedly across the ensemble of structures. It is generally useful to identify those regions of the structure which are problematical and inspect them using interactive computer graphics. A more rigorous test is to back-calculate the NOESY spectra from the structures using a program such as CORMA (70) and to compare these with the experimental spectra. Experimental and calculated NOEs can also be compared using a 'R factor' (71), for example:

$$R = \frac{\Sigma_i (a_o(i)^{1/6} - a_c(i)^{1/6})}{\Sigma_i (a_o(i)^{1/6}}$$ [22]

where $a_o(i)$ is the observed intensity and $a_c(i)$ is the calculated intensity for the ith peak.

A penalty function which combines bonded and non-bonded interactions can be used to evaluate the stereochemical 'goodness' of an ensemble of structures. One such approach combines the deviation from ideality of bond lengths, bond angles, and improper torsion angles with van der Waals' interactions (72) and is termed a 'geometric figure of merit'. Verification that the backbone torsion angles of the structure lie in the allowed region of torsional space is another test which can be applied, as is the deviation of χ_1 from its expected value, using a program such as PROCHECK (73). Calculation of the radius of gyration can also be used, particularly if the radius of gyration of a homologous protein is known. This is not, however, a particularly good test if the structure is not globular—if this is the case determination of the principal axes may be a better test.

The RMS deviation of the ensemble from the ensemble average is a test which is commonly used to determine the 'goodness' of the structures. However, the limited sampling of some of the methods means that such a measure should be treated with caution. It is worth noting the difference between precision and accuracy. If there is little structural variation across an ensemble of structures then the results are precise. If they correspond well to the 'target' structure then they are also accurate. However, if they do not correspond well with the 'target' structure then although they are precise they are not accurate.

Calculation of the free energy of solvation (74) gives an indication of the relative stabilities of the different members of the ensemble—in the case of globular proteins, a rule of thumb is the more negative this value the more stable the structure. An estimate of the free energy of solvation (SFE) for a protein can be obtained from (75):

$$SFE = 15.30 - 1.13N$$ [23]

where N is the number of residues. The van der Waals' energy is another quantity which can be used—as a rule this should correspond to around

−5 kcal/mol/residue in the case of proteins. Another test compares the polar fraction of side chains on the solvent accessible surface and the polar fraction of side chains in the solvent inaccessible interior with values determined from known protein structures (76).

A detailed inspection of the ensemble of structures using interaction computer graphics is strongly recommended. This will give a deeper understanding of the molecule and highlight any problems which may not be apparent from the macroscopic properties. An assessment of the ability of the structure to perform its known function is also strongly recommended.

5. Future developments

Development of a means for assessing the 'goodness' of the ensemble of structures would be extremely useful. Improvements in both the quality of and the ability to interpret the properties of the resultant structures will follow from the use of improved methods for incorporating the effects of motion into structure determination. X-ray crystallography has illustrated the usefulness of determining the positions of bound water molecules. The emergence of methods for the inclusion of solvent in the protocols for determining structures from NMR data will facilitate this. An improved model for electrostatic interactions will improve those procedures which include MD as one of their stages. Such improvements will be aided by advances in computer hardware which will reduce the time scales required.

Acknowledgements

I gratefully acknowledge the useful discussions I have had with colleagues in compiling this chapter. I am also grateful to the Royal Society and SERC for funding.

References

1. Clore, G. M. and Gronenborn, A. M. (1989). *Crit. Rev. Biochem. Mol. Biol.*, **24**, 479.
2. Wüthrich, K. (1989). *Science*, **243**, 45.
3. Clore, G. M. and Gronenborn, A. M. (1985). *J. Magn. Reson.*, **61**, 158.
4. Borgias, B. A. and James, T. L. (1990). *J. Magn. Reson.*, **87**, 475.
5. Boelens, R., Koning, T. M. G., and Kaptein, R. (1988). *J. Mol. Struc.*, **173**, 299.
6. Yip, P. F. and Case, D. A. (1989). *J. Magn. Reson.*, **83**, 643.
7. Nilges, M., Habazettl, J., Brunger, A. T., and Holak, T. A. (1991). *J. Mol. Biol.*, **219**, 499.
8. Birchall, A. J. and Lane, A. N. (1990). *Eur. Biophys. J.*, **19**, 73.
9. Karplus, M. (1959). *J. Phys. Chem.*, **30**, 11.
10. Pardi, A., Billeter, M., and Wüthrich, K. (1984). *J. Mol. Biol.*, **180**, 741.

11. IUPAC-IUB Commission on Biochemical Nomenclature (1970). *Biochemistry*, **9**, 3471.

12. Richardson, J. (1981). *Adv. Protein Chem.*, **34**, 167.

13. Wagner, G., Braun, W., Havel, T. F., Schaumann, T., Go, N., and Wüthrich, K. (1987). *J. Mol. Biol.*, **196**, 611.

14. De Marco, A., Llinas, M., and Wüthrich, K. (1978). *Biopolymers*, **17**, 617.

15. Guntert, P., Braun, W., Billeter, M., and Wüthrich, K. (1989). *J. Am. Chem. Soc.*, **111**, 3997.

16. Nilges, M., Clore, G. M., and Gronenborn, A. M. (1990). *Biopolymers*, **29**, 813.

17. Guntert, P., Braun, W., and Wüthrich, K. (1991). *J. Mol. Biol.*, **217**, 517.

18. Holak, T. A., Nilges, M., and Oschkinat, H. (1989). *FEBS Lett.*, **242**, 218.

19. Wüthrich, K., Billeter, M., and Braun, W. (1983). *J. Mol. Biol.*, **169**, 949.

20. Driscoll, P. C., Gronenborn, A. M., and Clore, G. M. (1989). *FEBS Lett.*, **243**, 223.

21. Wüthrich, K. (1986). *NMR of proteins and nucleic acids*. Wiley, New York.

22. Wüthrich, K., Billeter, M., and Braun, W. (1984). *J. Mol. Biol.*, **180**, 715.

23. Baker, E. N. and Hubbard, R. E. (1984). *Prog. Biophys. Molec. Biol.*, **44**, 97.

24. Pedersen, T. G., Sigurskjold, B. W., Andersen, K. V., Kjaer, M., Poulsen, F. M., Dobson, C. M., and Redfield, C. (1991). *J. Mol. Biol.*, **218**, 413.

25. Billeter, M., Engeli, M., and Wüthrich, K. (1985). *J. Mol. Graphics*, **3**, 79.

26. Sutcliffe, M. J., Dobson, C. M., and Oswald, R. E. (1992). *Biochemistry*, **31**, 2962.

27. Sutcliffe, M. J. (1989). In *Protein engineering, crystallography, and computer graphics* (ed. M. Bolognesi), pp. 19–27. Clas International, Brescia, Italy.

28. Kraulis, P. J. and Jones, T. A. (1987). *Proteins*, **2**, 188.

29. Altman, R. B. and Jardetzky, O. (1989). In *Methods in Enzymology* (ed. N. J. Oppenheimer and T. L. James), Vol. 177, pp. 218–46. Academic Press, San Diego.

30. Crippen, G. M. (1977). *J. Comput. Phys.*, **24**, 96.

31. Havel, T. F., Kuntz, I. D., and Crippen, G. M. (1984). *Bull. Math. Biol.*, **45**, 665.

32. Kuntz, I. D., Crippen, G. M., and Kollman, P. A. (1979). *Biopolymers*, **18**, 939.

33. Braun, W. and Go, N. (1965). *J. Mol. Biol.*, **186**, 611.

34. Havel, T. F. (1991). *Prog. Biophys. Molec. Biol.*, **56**, 43.

35. Havel, T. F. and Wüthrich, K. (1984). *Bull. Math. Biol.*, **46**, 673.

36. Kuszewski, J., Nilges, M., and Brunger, A. T. (1992). *J. Biomolec. NMR*, **2**, 33.

37. Kuntz, I. D., Thomason, J. F., and Oshiro, C. M. (1989). In *Methods in Enzymology* (ed. N. J. Oppenheimer and T. L. James), Vol. 177, pp. 159–204. Academic Press, San Diego.

38. Crippen, G. M. (1981). *Distance geometry and conformational calculations*. Wiley, Chichester, England.

39. Crippen, G. M. and Havel, T. F. (1988). *Distance geometry and molecular conformation*. Research Studies Press Ltd., Taunton, England and Wiley, Chichester, England.

40. Metzler, W., Hare, D. B., and Pardi, A. (1989). *Biochemistry*, **28**, 7045.

41. Havel, T. F. (1990). *Biopolymers*, **29**, 1565.

42. Crippen, G. M. and Havel, T. F. (1978). *Acta Cryst.*, **A34**, 282.

43. Crippen, G. M. (1984). *J. Comput. Chem.*, **5**, 548.

44. Braun, W. (1983). *J. Mol. Biol.*, **163**, 613.

45. Weiner, S. J., Kollman, P. A., Case, D. A., Singh, U. C., Ghio, U. C., Alagona, G., Profeta, S., and Weiner, P. (1984). *J. Am. Chem. Soc.,* **106,** 765.
46. Brooks, B. R., Bruccoleri, R. E., Olafson, B. O., States, D. J., Swaminathan, S., and Karplus, M. (1983). *J. Comput. Chem.,* **4,** 187.
47. Disover Manual version 2.8 (1992). Biosym Technologies Inc., San Diego, CA.
48. van Gunsteren, W. F. and Berendsen, H. J. C. (1987). Groningen Molecular Simulation (GROMOS) Library Manual. BIOMOS B.V., Nijenborgh 16, Groningen, The Netherlands.
49. Brünger, A. T. (1992). XPLOR Manual version 3.0, Yale University, New Haven, CT.
50. Braun, W. (1987). *Q. Rev. Biophys.,* **19,** 115.
51. Guntert, P. and Wüthrich, K. (1991). *J. Biomolec. NMR,* **1,** 447.
52. McCammon, J. A. and Harvey, S. C. (1987). *Dynamics of proteins and nucleic acids.* Cambridge University Press.
53. Hockney, R. W. (1970). *Methods Comput. Phys.,* **9,** 136.
54. Beeman, D. (1976). *J. Comput. Phys.,* **20,** 130.
55. Levitt, M. and Meirovitch, H. (1983). *J. Mol. Biol.,* **168,** 617.
56. Gear, C. W. (1971). *Numerical initial value problems in ordinary differential equations.* Prentice-Hall, New York.
57. McCammon, J. A., Wolynes, P. G., and Karplus, M. (1979). *Biochemistry,* **18,** 927.
58. Ryckaert, J. P., Ciccotti, G., and Berendsen, H. J. C. (1977). *J. Comput. Chem.,* **23,** 327.
59. Berendsen, H. J. C., Postma, J. P. M., van Gunsteren, W. F., DiNola, A., and Haak, J. R. (1984). *J. Chem. Phys.,* **81,** 3684.
60. Brünger, A. T., Clore, G. M., Gronenborn, A. M., and Karplus, M. (1986). *Proc. Natl Acad. Sci. USA,* **83,** 3801.
61. Clore, G. M., Nilges, M., Sukumaran, D. K., Brunger, A. T., Karplus, M., and Gronenborn, A. M. (1986). *EMBO J.,* **5,** 2729.
62. Cooke, R. M., Wilkinson, A. J., Baron, M., Pastore, A., Tappin, M. J., Campbell, I. D., Gregory, H., and Sheard, B. (1987). *Nature,* **327,** 339.
63. de Vlieg, J., Scheek, R. M., van Gunsteren, W. F., Berendsen, H. J. C., Kaptein, R., and Thomason, J. (1988). *Proteins,* **3,** 209.
64. Nilges, M., Clore, G. M., and Gronenborn, A. M. (1988). *FEBS Lett.,* **229,** 317.
65. Nilges, M., Clore, G. M., and Gronenborn, A. M. (1988). *FEBS Lett.,* **239,** 129.
66. Nilges, M., Kuszewski, J., and Brünger, A. T. (1991). In *Computational aspects of the study of biological macromolecules by nuclear magnetic resonance spectroscopy* (ed. J. C. Hoch, F. M. Poulsen, and C. Redfield), pp. 451–455. Plenum Press, New York.
67. Kim, Y. and Prestegard, J. H. (1989). *Biochemistry,* **28,** 8792.
68. Scheek, R. M., Torda, A. E., Kemmink, J., and van Gunsteren, W. F. (1991). In *Computational aspects of the study of biological macromolecules by nuclear magnetic resonance spectroscopy* (ed. J. C. Hoch, F. M. Poulsen, and C. Redfield), pp. 209–217. Plenum Press, New York.
69. Torda, A. E., Scheek, R. M., and van Gunsteren, W. F. (1990). *J. Mol. Biol.,* **214,** 223.
70. Keepers, J. W. and James, T. L. (1984). *J. Magn. Reson.,* **57,** 404.

71. Gonzalez, C., Rullman, J. A. C., Bonvin, M. J. J., Boelens, R., and Kaptein, R. (1991). *J. Magn. Reson., 91,* 659.
72. Jian-Sheng, J. and Hubbard, R. E. (1990). In *Proceedings of the 2nd York meeting on generation of 3D structures from distance information.* York, England.
73. Morris, A. L., MacArthur, M. W., Hutchinson, E. G., and Thornton, J. M. (1992). *Proteins,* **12,** 345.
74. Eisenberg, D. and McLachlan, A. D. (1986). *Nature, 319,* 199.
75. Chiche, L., Gregoret, L. M., Cohen, F. E. and Kollman, P. A. (1990). *Proc. Natl Acad. Sci. USA,* **87,** 3240.
76. Baumann, C., Frommel, C., and Sander, C. (1989). *Protein Eng., 2,* 329.
77. Ramachandran, G. N. and Sasisekharan, G. N. (1986). *Adv. Protein Chem.,* **23,** 283.

A1

Availability of relevant computer programs

1. Programs for processing NMR spectra

The manufacturers of NMR spectrometers produce powerful data-processing software as part of their systems. Other widely used processing programs are:

FELIX   Hare Research Inc., 14810 216th Avenue, Woodinville, WA 98072, USA; now distributed by Biosym (see below).

NMRi   New Methods Research, Inc., 719 East Genesee Street, Syracuse, NY 13210, USA; now distributed by Tripos (see below).

In addition, several companies which produce molecular modelling software packages are now developing programs for processing spectra, so that in principle one can go from the raw NMR data to structure determination within the same package:

Biosym Technologies Inc., 9685 Scranton Road, San Diego, CA 92121, USA.
Molecular Simulations Inc., 200 Fifth Avenue, Waltham, MA 02154, USA.
Tripos Associates, 1699 S. Hanley Road, St Louis, MO 63144, USA.

A number of programs aimed at automation of assignment have been discussed in the literature, but this is still very much a developmental area. Specific programs for stereospecific assignment of prochiral protons include:

GLOMSA,   Prof. Dr K. Wüthrich, Institut für Molekularbiologie
HABAS   und Biophysik, ETH-Honggerberg, Zürich, CH-8093, Switzerland.

STEREOSEARCH Dr M. Nilges, European Molecular Biology Laboratory, D-6900 Heidelberg, Germany.

2. Programs for structure determination

2.1 Molecular modelling programs incorporating molecular dynamics simulations

AMBER   Prof. P. A. Kollman, Department of Pharmaceutical Chemistry, University of California, San Francisco, CA 94143, USA.

CHARMM   Prof. M. Karplus, Department of Chemistry, Harvard University, 12 Oxford Street, Cambridge, MA 02138, USA.
DISCOVER   Biosym Technologies, Inc. (see above).
GROMOS   Prof. W. van Gunsteren, Physical Chemistry, ETH-Zentrum, Zürich, CH-8092, Switzerland.
QUANTA   Molecular Simulations, Inc. (see above).
XPLOR   Dr A. Brünger, Howard Hughes Medical Institute, Yale University, 260 Whitney Avenue, New Haven, CT 06511, USA.

2.2 Distance geometry programs

DGEOM   QCPE, Department of Chemistry, Indiana University, Bloomington, IN 47405, USA.
DG-II   Biosym Technologies, Inc. (see above).
DIANA   Prof. Dr K. Wüthrich, Zürich (see above).
DISGEO   QCPE (see above).
DISMAN   Dr W. Braun, Institut für Molekularbiologie und Biophysik, ETH-Honggerberg, Zürich, CH-8093, Switzerland.
DSPACE   Hare Research Inc. (see above).
EMBED,   Prof. I. D. Kuntz, Department of Pharmaceutical Chemistry,
VEMBED   University of California, San Francisco, CA 94143, USA.
XPLOR/dg   Dr A. Brünger (see above).

2.3 Other programs

COMPOSER   Prof. T. L. Blundell, Department of Crystallography, Birkbeck College, Malet Street, London WC1E 7HX, UK. Homology modelling.
CONFOR   Dr M. Billeter, Institut für Molekularbiologie und Biophysik, ETH-Honggerberg, Zürich, CH-8093, Switzerland. Molecular graphics and NMR-based structure determination.
PROCHECK   Prof. J. Thornton, Department of Biochemistry and Molecular Biology, University College, Gower Street, London WC1E 6BT, UK. Structure analysis.
PROTEAN   Prof. O. Jardetzky, Stanford Magnetic Resonance Center, Stanford University, Stanford, CA 94305, USA. Probabilistic approach to structure determination from NMR data.

Index

ab initio method, determination of oligosaccharides primary sequence 305–7
abbreviation and nomenclature xix–xx
accordion experiment, variant of basic chemical exchange 177
acetone, chemical shift reference 31
acquisition parameters
receiver gain 47–8
transmitter offset 47
acquisition protocols, 2D, 3D, 4D data, stable isotope-assisted investigations 112
alanine, spin system identification 77–8
algae, protein isotope labelling using 106
algorithms
crosspeak volume measurements, NOE 336
embed algorithm, structure of nucleic acids 278–80
from NOEs to distances, structure determination from NMR data analysis 349
IRMA 350
MARDIGRAS 348–9, 350–1
NOE, crosspeak volume measurements 336
amide hydrogen exchange rates 329–30
amino acids, isotope labelling 185–6
selective isotope labelling 107–8, 115–21
aminotransferases 107–8
ammonium nitrate, chemical shift reference 31
Anabaena ferredoxin 138–9
annealing
dynamical simulated annealing 382–4
simulated annealing, oligosaccharides, ¹H NMR 311
assignment strategies 71–99, *see also* isotope-assisted assignment strategies

bacterial cells
protein isotope labelling 105–6
baseline optimization, causes of baseline distortion 48–9
Bloch–McConnell equations
magnetization transfer experiments 169
two-dimensional exchange spectroscopy 159, 175–6, 211
broadband proton decoupling 50–1
broadband X-nucleus probes 39
buffer exchange and desalting 16–20
buffers, choice
¹H NMR of samples 25
in sample preparation 25–6

carbohydrate
average number of unique resonances expected 10
see also oligosaccharides

carbon isotopes, properties of nuclei of interest 2, 102
carbonyl ¹³C resonances
probes for chemical modification 129
properties 126–7
chemical exchange
amide hydrogen exchange rates 329–30
chemical shifts and linewidth changes 154
definitions 155–7
¹H–¹H cross-relaxation, NOESY 141
initial qualitative analysis 157–9
intra-and intermolecular processes 153
isotope exchange reactions 103–4
lineshape analysis 159–68
fast exchange 164–8
simulations 159–61
slow exchange 161–4
magnetization transfer experiments 168–81
2D exchange 180
comparisons 178–81
inversion transfer 180
one-dimensional magnetization transfer 169–75
saturation transfer 179–80
two-dimensional exchange spectroscopy 175–81
two-site first order, two-site second-order, defined 155
chemical shifts
of atoms in proteins 125
references, common compounds 31–2
CN labelling 127–9
composite-pulse decoupling (CPD)
DIPSI-2 50
GARP 50
WALTZ-16 50
computer programs 391–2
AMBER 377, 391
CHARMM 377, 392
COMPOSER 371, 392
CONFOR 392
CORMA 386
DGEOM 392
DGII 375
DIANA 364, 377–9, 392
DISCOVER 377, 392
DISGEO 379, 392
DISMAN 377–9, 392
EMBED, VEMBED 377, 392
FELIX 391
GLOMSA 364, 391
GROMOS 377
HABAS 363, 391
NMR1, NMR2 391
PROCHECK 386, 392
PROTEAN 372, 392
QUANTA 392
STEREO-SEARCH 363–4
XPLOR 375, 377, 383, 392

contamination, microbial and paramagnetic
32
correlation spectroscopy *see* COSY sequence
COSY sequence 75–9, 87–99
^{1}H–^{1}H experiments 61–2
oligosaccharides, ^{1}H NMR 293–6
DQF-COSY 297–8
TQF-COSY 298–300
crosspeak intensities, NOE, structure determination
from NMR data, analysis 335–9
crosspeak volume measurements, NOE, algorithms
336
cyanobacteria, protein isotope labelling 106

DANTE pulse sequence 54
selective excitation 170
data acquisition and analysis, data collection
114
DEALS experiment
steady state method 144
time-dependent DEALS experiment 146–7
degassing the sample 21–2
desalting and buffer exchange 16–20
detergents, in sample preparation 30
deuteration strategies 109
chiral deuteration of side chains
109–10
perdeuteration 109
random fractional deuteration 109
selective deuteration, differentiation from
protonation 109
3′,5′-difluoromethotrexate spectra 160
dihydrofolate reductase
binding of antifolate drugs 183, 184
see also protein–ligand interactions
histidine residues 195–8
and NADP$^+$ concentration 189
protein resonances 190
S. faecium DHFR 207
structure 184
time-dependence of the intensity of crosspeak
344
trimethoprim analogs 184, 193
X-ray crystal structure 197
5,5–dimethylsilapentanesulfonate (DSS), chemical
shift reference 31
dimethylsulfoxide (DMSO), chemical shift
reference 31
dioxane, chemical shift reference 31
distance constraints
MARDIGRAS or NO2DI 279
oligonucleotides 361
proteins 360
distance geometry
followed by restrained molecular dynamics
(MD) 381–2
hybrid distance geometry–dynamical simulated
annealing 382–4
metric matrix distance geometry 374–7
in torsional space 377–9
see also embed algorithm
disulfide bridges, selective oxidation, use of ^{13}C
NMR 133–6

DNA
average number of unique resonances expected
10
protein–DNA complexes 201–2
see also nucleic acids
dynamical simulated annealing 382–4

embed algorithm, structure of nucleic acids 278–80
Escherichia coli
amino acid selective labelling of over-produced
proteins 108
deuterated substrates 110–11
^{15}N and ^{13}C-protein labelling 106–7
protein isotope labelling 105–6
exchange connections, multi-dimensional NMR 2
exchange processes *see* chemical exchange
EXCSY, 2D *see* chemical exchange, magnetization
transfer experiments
experimental protocols
acquisition protocols, 2D, 3D, 4D data 112
alanine and threonine spin system identification
77–8
amino acid selective labelling of over-produced
E. coli proteins 108
aromatic NOE effects, identification 90
aromatic spin system identification 82–3
backbone assignments 124
chemical exchange
fast exchange spectra 165–6
rate under conditions of slow exchange 162
crosspeak assignment 340, 341–2
crosspeak volume measurements 336
DANTE pulse, selective excitation 170
deuteration, random fractional 110
distance geometry
hybrid distance geometry–dynamical simulated
annealing 382
metric matrix distance geometry 374
disulfide bridges, selective oxidation, use of ^{13}C
NMR 133–4
d_{NN} NOE effects, adjacent residues,
identification 92–3
dynamical simulated annealing 383
first order chemical exchange, exchange regime
157–8
gel filtration 17–18
glycine spin system identification 76–7
H^N cross-peaks, reasons for loss 72–3
H$_2$O or ^{2}H$_2$O solutions of oligonucleotides
230
hydrogen exchange protection factors from
HSMQC data, use of ^{15}N labelling 141–2
hydrogen exchange rates, exploitation for
overcoming overlap 93–4
ionic strength variation, spectrum optimization
27–8
isotope-assisted H···H distance measurements
126
J-coupling values 318–19
HMQC experiment 320
K_d
determination in moderately fast exchange 167
determination in very fast exchange 166–7

methionine residues, oxidation, followed by ^{13}C NMR 129–30
metric matrix distance geometry 374
model building
  database of known tertiary structures 371
  homologous 370
^{15}N and ^{13}C-labelling of *E. coli* proteins 106–7
NOEs
  correlation of type-J side chain resonances to H^N 83–4
  use for structure calculations 354
perdeuteration and selective protonation 110
pH adjustment 23–4
protein renaturation, use of ^{13}C NMR 135–6
protein–ligand interactions
  fast exchange spectra 165–6
  fast and slow exchange rates 187
  transferred NOE expression 205–6
saturation transfer time-course experiment 171–3
second order chemical exchange, exchange regime 158
secondary structure elements, delineation 331–2
sequence-specific assignment 94–5
sequential assignment procedure 246–7
side chain assignments 125
spectral simulation 322
steady state DEALS method 144
steady state saturation transfer experiment 172
structure calculations, using NOEs 354
structure determination by NMR 359
structure determination using PROTEAN computer program 372
time-dependent DEALS experiment 146–7
type-J spin system identification 81–2
type-U spin system identification 86–7
valine, isoleucine and leucine spin system identification 79–80
valine and leucine methyl carbons, chiral ^{13}C-labelling 111

ferredoxin 138–9
filtration methods 20–1
fingerprinting, determination of oligosaccharides primary sequence 304–5
^{19}Fluorine, properties of nuclei of interest 2, 102
fluorine-containing proteins, assignment 192
folate, structure 184
free induction decay (FID) 37, 46, 57
freeze–thaw cycles, degassing 21–2
freezing of sample 14–15
fungal cells, protein isotope labelling using 106
furanose ring
  conformation 258–67
    around C3′-O3′ 276
    around C4′-C3′ 275–6
  description 223–7
  nucleic acid structure 223–7

glycine, spin system identification 73–6
glycosides *see* oligosaccharides, ^{1}H NMR

H2′ and H2″ resonances, stereospecific assignment 247–9
H5′ and H5″ resonances, stereospecific assignment 249–51
H isotopes, properties of nuclei of interest 2, 102
^{1}H NMR
  oligosaccharides 289–313
  see also protons
^{1}H–^{1}H experiments
  COSY sequence 61–2
  NOESY sequence 62
  two-dimensional NMR 60–3
Henderson–Hasselbalch equation 195
HETERO–RELAY experiment 322–3
heteronuclear 2D NMR techniques
  correlation experiments 66–8
  nucleic acids 232
heteronuclear Hartmann–Hahn (HOHAHA) experiments 71, 73–5, 97, 124
  assignment methods, oligosaccharides, ^{1}H NMR 296–7
  HOHAHA–3D COSY, oligosaccharides, ^{1}H NMR 300–1
heteronuclear multiple quantum correlation experiments (HMQC) 66–8, 124, 192
histidine, titration curves 136–9
HMQC *see* heteronuclear multiple quantum correlation experiments
Hofmeister series 27
HOHAHA *see* heteronuclear Hartmann–Hahn experiment
^{2}H$_2$O, defining pH in, isotope effect 22–3
H$_2$O to ^{2}H$_2$O exchange 18–20
hydrogen bonds, distance constraint ranges 365
hydrogen exchange protection factors from HSMQC data, use of ^{15}N labelling 141–7

input restraints, experiments required 316
insect cells, protein isotope labelling using 106
instrumentation and pulse sequences 35–70
  composite-pulse decoupling (CPD), DIPSI-2, GARP and WALTZ-16 50
  DANTE pulse sequence 54
    selective excitation 170
  NMR spectrometer 36–40
    factors affecting quality 43–4
    sample temperature control 40–3
  one-dimensional spectra 44–51
  samples dissolved in H$_2$O 51–7
    setting-up spectrometer 51–2
    solvent irradiation methods 53–5
    solvent non-excitation methods 55–7
  two-dimensional spectra 57–68
    ^{1}H-^{1}H experiments 60–3
    heteronuclear correlation experiments 66–8
    rotating frame experiments 63–6
  see also two-dimensional NMR
inversion transfer 173–5
ionization states
  determination
    ligands 198–200
    proteins 194–8
IRMA procedure 278, 361

Index

isolated spin-pair approach (ISPA) 344–8
isoleucine, spin system identification 78–80
isotope effect, defining pH in ^2H_2O 22–3
isotope exchange reactions, applications, protein isotope labelling 103–4
isotope labelling see protein isotope labelling
isotope-assisted assignment strategies 115–21
  assignment procedures 115–21
  goals and expectations 115
  sequential stable isotope-assisted through-space connectivities 121
  spectrum interpretation of nucleic acids 243–51
  unique stable isotope labelling 119–21
isotope-assisted investigations, 2D, 3D, 4D data, acquisition protocols 112
isotope-editing, labelled ligands 202

J-correlated spectra, nucleic acids 235–6
J-couplings
  determination 323–7
  local conformational analysis of nucleic acids 258–67
J-modulated [^{15}N-^{1}H] COSY experiment 320–2
jump-return sequence, non-excitation methods for solvents 55–6

Karplus equation 361–2

laboratory, floor vibration, and spectral quality 44
leucine, spin system identification 78–80
lifetimes, defined 155–6
ligands
  determination of ionization states 198–200
  see also protein–ligand interactions
lineshape analysis 159–68
lock transmitter and receiver 38–9
lyophilization 14–15

McConnell equation see Bloch–McConnell equations
magnetic fluctuations, and spectral quality 44
magnetization transfer experiments 168–81
  inversion transfer 180
  one-dimensional magnetization transfer 169–75
  saturation transfer 179–80
  two-dimensional exchange spectroscopy 175–81
mainchain approach, sequence-specific assignment, resonance assignment strategies for small proteins 97–8
mammalian cells in tissue culture, protein isotope labelling 106
MARDIGRAS
  algorithm 279, 348–9, 350–1
methionine residues, oxidation, followed by ^{13}C NMR 129–32
methotrexate, structure 184

model building, tertiary structures, computational approaches 369–71
molecular dynamics (MD) simulations
  averaging 385
  internal motion in proteins 384–5
  restrained 379–81
    3D structure of oligosaccharides, ^{1}H NMR 310–11
molecular weight of sample, effects of increasing 9–10
multi-dimensional NMR, three and four-dimensional see two-dimensional NMR

N isotopes, properties of nuclei of interest 2, 102
^{15}N labelling, hydrogen exchange protection factors from HSMOC data 141–7
^{15}N-^{1}H HMQC-J experiment 319
^{15}N-protein labelling 106–7
NMR
  applications, developments 1–3
  principles 1
NMR data, structure determinations see structure determination from NMR data
NMR data analysis see structure determination from NMR data analysis
NMR instrumentation see instrumentation and pulse sequences; spectrometer
NMR parameters
  relaxation parameters 140
  rotamers around Calpha–Cbeta bond 328
NMR time scale 156
NMR tube
  choice 10–11
  cleaning 11, 12
  drying 11–12
  see also sample preparation
NO2DI 279
NOEs
  assignment of NOEs 339–42
  crosspeak intensities 335–9
  crosspeak volume measurements, algorithms 336
  and distance restraints, structure determination from NMR data, analysis 333–5
  intensity (I_{NOE}) 360
    calculations 343
  local conformational analysis of nucleic acids 254–7
  measurement using isotope-editing 202
  transferred, and bound conformations 202–6
NOESY
  ^{1}H–^{1}H cross-relaxation, and chemical exchange, differentiation 141
  oligosaccharides, ^{1}H NMR 300–1
  sequence, ^{1}H–^{1}H experiments 62
  spectra 72–3, 88–99
    nucleic acids 231–5
  nomenclature xix-xx
nuclear Overhauser effect spectroscopy see NOEs; NOESY; ROESY
nucleic acids
  appendix, contour plots 283–8
  local conformational analysis 253–76
    backbone angles 270–6
    J-couplings 258–67

Index

NOEs 254–7
  rotation about the *chi* angle 267–70
recording of spectra 229–36
  water suppression 231
ROESY spectra 231–5
spectrum interpretation 237–52
  assignment strategies 243–51
  ssignment via 3D NMR 251–3
  resonances 237–42
structural parameters 218–29
  definitions of structure parameters 222
  distances, secondary elements 228–9
  furanose ring 223–7
  mononucleotide conformation 227–8
structure
  determination of overall structure 276–80
  embed algorithm 278–80
  specific features 218
  torsion angles, defined 223
nucleotides, oligonucleotides
  distance constraints 361
  studies 28

oligosaccharides, ¹H NMR 28, 289–313
  assignment methods 291–304
    case study 301–4
    COSY 293–6
    HOHAHA spectroscopy 296–7
    resonance 291–3
  determination of 3D structure 307–11
    adiabatic mapping 309–10
    restrained energy minimization 308–9
    restrained molecular dynamics simulations
      310–11
    simulated annealing 311
    torsion angles 307–8
  determination of primary sequence 304–7
    fingerprinting 304–5
  sample preparation 289–91
one-dimensional magnetization transfer
    experiments 169–75
one-dimensional spectra 44–51
  proton spectra 44–9
  spectra other than protons 49–51
Overhauser effect spectroscopy *see* NOEs;
    NOESY; ROESY
oxygen contamination of sample, degassing 21–2
oxygen isotopes, properties of nuclei of interest 2,
    102

³¹P, properties of nuclei of interest 2
peptides
  NMR spin-spin coupling 102
  peptide bond C–N labelling 105, 127–9
pH
  adjustment, sample preparation 23, 26
  defining in ²H₂O, isotope effect 22–3
pre-TOCSY, restoring magnetization 55
probes
  functions 39
  tuning, proton spectra 45–6

types
  broadband X-nucleus probes 39
  proton probes 39
  proton probes with X-nucleus decoupling coils
    40
protein G, IgG-binding domain 335
protein isotope labelling
  applications 103–4
    coherence pathways 103
    cross-relaxation pathways 103
    isotope exchange reactions 103–4
    spectral assignment 103
    spectral simplification 103
  Escherichia coli 105–6
  methods
    algae and cyanobacteria 106
    bacterial cells 105–6
    fungal cells 106
    insect cells 106
    mammalian cells in tissue culture 106
    yeast cells 106
  patterns
    amino acid selective complete 104
    amino acid selective stochastic 104
    block 104
    complete 104
    differential 105
    exchange 104
    metabolic 104
    peptide bond CN 105
    single site 104
    stochastic 104
  peptide bond C–N labelling 105, 127–9
  structural information
    chemical shifts 125
    coupling constants 123–5
    NOEs 121–3
    solvent accessibility 125–6
  ¹³C-protein labelling 106–7
protein–DNA complexes, determination of
    conformations 201–2
protein–ligand interactions 183–215
  analysing NMR spectra 186–91
    fast exchange 187–8
    slow exchange 189
  assignment of resonances 190–2
  detection of multiple conformations 206–8
  determination of conformations of protein–ligand
    complexes 201–6
  determination of ionization states 184–200
    ligands 198–200
    proteins 194–8
  determination of rates of ring flipping 209–11
  dynamic processes in protein–ligand complexes
    208–12
  equilibrium dissociation constants 186, 208
  fast exchange spectra 165–6
  ligand resonances 190–2
  NMR parameters of the bound species 192–4
  protein resonances 190
  rapid segmental motions 208–9
  sample preparation 185–6
protein(s)
  alpha-helices, characterization 330, 332
  average number of unique resonances expected 10

Index

protein(s) (*cont.*)
  beta-sheets
    characterization 330, 332
    parallel/antiparallel 368
  deuteration strategies 109
    chiral deuteration of side chains 109–10
    perdeuteration 109
  random fractional deuteration 109
  residue-selective hydrogen labelling 109
  selective protonation 109
  internal motion 384–5
  model building 369–71
  ^{15}N and ^{13}C-protein labelling 106–7
  and peptides, NMR spin-spin coupling 102
  secondary structure
    constraints 365–9
    useful parameters 331, 368
    see also structure determination from NMR
      data analysis
  tertiary structure 369–87
proton contamination 20
proton decoupling, broadband 50–1
proton spectra 44–9
  acquisition parameters 47–8
  baseline optimization 48–9
  locking the sample 45
  probe tuning 45–6
  pulse calibration 47
  shimming 46
proton–proton correlation
  J-correlation techniques 232
  NOE 232
proton–proton distance constraints 360–1
protonation, selective, differentiation from
    deuteration, selective 109
protons
  exchangeable 237–9, 244
  exchangeable and non-exchangeable,
    connectivities 247
  non-exchangeable 239–41, 245–7
  see also ^{1}H NMR
pulse Fourier transform methods, defined 2
pulse programmer and data system 40
pulse sequences *see* instrumentation and pulse
    sequences
purines and pyrimidines
  base pairs and base triplets 220–1
  structures 219

RELAY (relay correlated spectroscopy) 71, 75–88,
    97, 124
resonance assignment strategies for large proteins
    101–51
  data acquisition and analysis 111–15
  isotope-assisted assignment strategies 115–21
  labelling strategies 102–11
  protein dynamics and H exchange rates from
    labelled proteins 140–7
  structural information from labelled proteins
    121–6
  thermodynamic information from labelled
    proteins 136–40
  very large proteins 126–36

resonance assignment strategies for small proteins
    71–99
  stage I: spin system identification 73–88
    type-J 80–5
    type-U 85–7
    type-X cross-peak 87–8
    see also amino acids, named
  stage II: sequence-specific assignment 88–97
    alternatives 97–8
    mainchain approach 97–8
    X-ray diffraction data 98
  two-dimensional spectra, assessing quality
    72–3
beta-D-riboses, structures 219
ring flipping, determination of rates 209–11
RNA
  average number of unique resonances expected
    10
  see also nucleic acids
ROESY spectra, nucleic acids 231–5
rotating frame experiments
  two-dimensional NMR 63–6
    ROESY 64–5
    TOCSY 63–4
    TOCSY/ROESY compared 65–6
rotational correlation time, defined 9

saccharides *see* oligosaccharides
sample preparation 7–33
  choice and preparation of tube 10–13
  degassing the sample 21–2
  desalting and buffer exchange 16–20
  filtration methods 20–1
  freezing of sample 14–15
  handling sample 14–22
    concentration methods 14, 15, 16
    molecular weight, effects of increasing 9–10
  oligosaccharides, ^{1}H NMR 289–91
  parameters 22–32
    buffers, choice 25–6
    chemical shift references 31–2
    defining pH in ^{2}H$_2$O –isotope effect
      22–3
    detergents 30
    ionic strength effects 26, 27, 28
    organic solvents 29–30
    pH adjustment 23, 26
    temperature 30, 40–3
  preventing contamination
    microbial 32
    paramagnetic 32
  protein–ligand interactions 185–6
  requirements for NMR sample 7–10
  sample volume 8, 13
  samples dissolved in H$_2$O 51–7
  solubility as limiting factor 8
  stability of sample 9
saturation transfer experiment 171–3
  steady state 172
SCUBA, restoring magnetization 55
shimming, proton spectra 46
Sklenar–Bax sequence 56
solvation, free energy of (SFE) 386

398

Index

solvents
  and labile protons, exchange rates 211–12
  non-excitation methods
    jump-return sequence 55–6
    Sklenar–Bax sequence 56
  oligosaccharides
    aqueous 289–90
    chemical shifts 290
    other solvents 290
  organic, in sample preparation 29–30
  properties 29
  setting-up spectrometer
    irradiation methods 53–5
    non-excitation methods 55–7
  suppression techniques, nucleic acids 232
  suppression using cw presaturation 53
spectrometer 36–44
  magnet 36
  probe 39–40
  pulse programmer and data system 40
  radiofrequency system 36–9
  setting-up spectrometer, samples dissolved in
    H_2O 51–2
  spectral quality
    environmental factors affecting 43–4
    and magnetic fluctuations 44
  spectral simplification, isotopic enrichment 140
  see also COSY; NOESY; ROESY; TOCSY
staphylococcal nuclease
  HMQC 118, 123–4
  selective deuteration 117
Streptomyces subtilisin inhibitor (SSI) 127–36
structure determination from NMR, data analysis
  315–57
  amide exchange rates 329–30
  heteronuclear scalar coupling constants 328–9
  NOE crosspeak intensities 335–9
  NOEs and distances 342–53
    agreement with experimental data 352–3
    algorithm 349
    complete relaxation matrix approach 348–51
    isolated spin pair approximation 343–8
    molecular motion 351–2
    NOE and distance restraints 333–5
    NOEs, summary 353–5
  scalar coupling constants 317–30
    J-coupling 317–23
    stereospecific assignments 323–8
  secondary structure identification 330–3
structure determination from NMR,
    computational approaches 359–92
  determination of constraints 359
  dihedral angle constraints 361–5
  proton–proton distance constraints 360–1
  secondary structure 365–9
    amide proton exchange rate 367–9
    coupling constants 367
    short-range NOEs 365–6
  tertiary structure 369–87
    distance geometry 373–9
    internal motion 384–5
    model building 369–71
    probabilistic approach 371
    restrained MD 379–81, 379–84
    structure analysis 385–7

temperature control
  room temperature 44
  sample preparation 30, 40–3
    VT calibration and stability 43
    VT operation 42–3
    VT system 41–2
tetramethylsilane, chemical shift reference 31
threonine, spin system identification 77–8
through-bond connections, multi-dimensional
    NMR 2
TOCSY, pre-TOCSY, restoring magnetization 55
TOCSY (total correlation spectroscopy) 124
torsion angles
  defined 223
  defined by heteronuclear long-range coupling
    constants 329
trimethoprim
  analogs 184
  DHFR binding 190, 193
    dynamic processes 211
    see also protein–ligand interactions
  structure 184
trimethyl phosphate (TMP), chemical shift
    reference 31
3,3,3–trimethylsilylpropionate (TSP), chemical shift
    reference 31
two-dimensional exchange spectroscopy 175–81
  Bloch–McConnell equations 175–6
two-dimensional NMR 57–68
  exchange connections 2
  general considerations 58–60
  heteronuclear correlation experiments 66–8
  line shape
    hypercomplex method 58
    TPPI method 58
  principles 57–8
  rotating frame experiments 63–6
  simple ^1H–^1H experiments 60–3
  through-bond connections 2
two-site first order, two-site second-order, defined
  155

vacuum centrifuges 15
vacuum inert gas cycles, degassing 21–2
valine, spin system identification 78–80
van der Waal's energy 386–7
vicinal spin–spin coupling constants 361–2

water
  ($H_2O/^2H_2O$ mixture) 51
  water suppression, recording of spectra 53–7
    nucleic acids 231

X-nucleus probe 40
  signal detection 49–50
X-ray diffraction data, sequence-specific
    assignment, resonance assignment strategies
    for small proteins 98

YASAP, simulated annealing 384
yeast cells, protein isotope labelling 105–6

399

ORDER OTHER TITLES OF INTEREST TODAY

Price list for: UK, Europe, Rest of World (excluding US and Canada)

Forthcoming Titles

124. Human Genetic Disease Analysis Davies, K.E. (Ed)
- Spiralbound hardback — 0-19-963309-6 — £30.00
- Paperback — 0-19-963308-8 — £18.50

123. Protein Phosphorylation Hardie, G. (Ed)
- Spiralbound hardback — 0-19-963306-1 — £32.50
- Paperback — 0-19-963305-3 — £22.50

122. Immunocytochemistry Beesley, J. (Ed)
- Spiralbound hardback — 0-19-963270-7 — £32.50
- Paperback — 0-19-963269-3 — £22.50

121. Tumour Immunobiology Gallagher, G., Rees, R.C. & others (Eds)
- Spiralbound hardback — 0-19-963370-3 — £35.00
- Paperback — 0-19-963369-X — £25.00

120. Transcription Factors Latchman, D.S. (Ed)
- Spiralbound hardback — 0-19-963342-8 — £30.00
- Paperback — 0-19-963341-X — £19.50

119. Growth Factors McKay, I.A. & Leigh, I. (Eds)
- Spiralbound hardback — 0-19-963360-6 — £30.00
- Paperback — 0-19-963359-2 — £19.50

118. Histocompatibility Testing Dyer, P. & Middleton, D. (Eds)
- Spiralbound hardback — 0-19-963364-9 — £32.50
- Paperback — 0-19-963363-0 — £22.50

117. Gene Transcription Hames, D.B. & Higgins, S.J. (Eds)
- Spiralbound hardback — 0-19-963292-8 — £35.00
- Paperback — 0-19-963291-X — £25.00

116. Electrophysiology Wallis, D.I. (Ed)
- Spiralbound hardback — 0-19-963348-7 — £32.50
- Paperback — 0-19-963347-9 — £22.50

115. Biological Data Analysis Fry, J.C. (Ed)
- Spiralbound hardback — 0-19-963340-1 — £50.00
- Paperback — 0-19-963339-8 — £27.50

114. Experimental Neuroanatomy Bolam, J.P. (Ed)
- Spiralbound hardback — 0-19-963326-6 — £32.50
- Paperback — 0-19-963325-8 — £22.50

112. Lipid Analysis Hamilton, R.J. & Hamilton, S.J. (Eds)
- Spiralbound hardback — 0-19-963098-4 — £35.00
- Paperback — 0-19-963099-2 — £25.00

111. Haemopoiesis Testa, N.G. & Molineux, G. (Eds)
- Spiralbound hardback — 0-19-963366-5 — £32.50
- Paperback — 0-19-963365-7 — £22.50

Published Titles

113. Preparative Centrifugation Rickwood, D. (Ed)
- Spiralbound hardback — 0-19-963208-1 — £45.00
- Paperback — 0-19-963211-1 — £25.00

110. Pollination Ecology Dafni, A.
- Spiralbound hardback — 0-19-963299-5 — £32.50
- Paperback — 0-19-963298-7 — £22.50

109. In Situ Hybridization Wilkinson, D.G. (Ed)
- Spiralbound hardback — 0-19-963328-2 — £30.00
- Paperback — 0-19-963327-4 — £18.50

108. Protein Engineering Rees, A.R., Sternberg, M.J.E. & others (Eds)
- Spiralbound hardback — 0-19-963139-5 — £35.00
- Paperback — 0-19-963138-7 — £25.00

107. Cell-Cell Interactions Stevenson, B.R., Gallin, W.J. & others (Eds)
- Spiralbound hardback — 0-19-963319-3 — £32.50
- Paperback — 0-19-963318-5 — £22.50

106. Diagnostic Molecular Pathology: Volume I Herrington, C.S. & McGee, J. O'D. (Eds)
- Spiralbound hardback — 0-19-963237-5 — £30.00
- Paperback — 0-19-963236-7 — £19.50

105. Biomechanics-Materials Vincent, J.F.V. (Ed)
- Spiralbound hardback — 0-19-963223-5 — £35.00
- Paperback — 0-19-963222-7 — £25.00

104. Animal Cell Culture (2/e) Freshney, R.I. (Ed)
- Spiralbound hardback — 0-19-963212-X — £30.00
- Paperback — 0-19-963213-8 — £19.50

103. Molecular Plant Pathology: Volume II Gurr, S.J., McPherson, M.J. & others (Eds)
- Spiralbound hardback — 0-19-963352-5 — £32.50
- Paperback — 0-19-963351-7 — £22.50

101. Protein Targeting Magee, A.I. & Wileman, T. (Eds)
- Spiralbound hardback — 0-19-963206-5 — £32.50
- Paperback — 0-19-963210-3 — £22.50

100. Diagnostic Molecular Pathology: Volume II: Cell and Tissue Genotyping Herrington, C.S. & McGee, J.O'D. (Eds)
- Spiralbound hardback — 0-19-963239-1 — £30.00
- Paperback — 0-19-963238-3 — £19.50

99. Neuronal Cell Lines Wood, J.N. (Ed)
- Spiralbound hardback — 0-19-963346-0 — £32.50
- Paperback — 0-19-963345-2 — £22.50

98. Neural Transplantation Dunnett, S.B. & Björklund, A. (Eds)
- Spiralbound hardback — 0-19-963286-3 — £30.00
- Paperback — 0-19-963285-5 — £19.50

97. Human Cytogenetics: Volume II: Malignancy and Acquired Abnormalities (2/e) Rooney, D.E. & Czepulkowski, B.H. (Eds)
- Spiralbound hardback — 0-19-963290-1 — £30.00
- Paperback — 0-19-963289-8 — £22.50

96. Human Cytogenetics: Volume I: Constitutional Analysis (2/e) Rooney, D.E. & Czepulkowski, B.H. (Eds)
- Spiralbound hardback — 0-19-963288-X — £30.00
- Paperback — 0-19-963287-1 — £22.50

95. Lipid Modification of Proteins Hooper, N.M. & Turner, A.J. (Eds)
- Spiralbound hardback — 0-19-963274-X — £32.50
- Paperback — 0-19-963273-1 — £22.50

94. Biomechanics-Structures and Systems Biewener, A.A. (Ed)
- Spiralbound hardback — 0-19-963268-5 — £42.50
- Paperback — 0-19-963267-2 — £25.00

93. Lipoprotein Analysis Converse, C.A. & Skinner, E.R. (Eds)
- Spiralbound hardback — 0-19-963192-1 — £30.00
- Paperback — 0-19-963231-6 — £19.50

92. Receptor-Ligand Interactions Hulme, E.C. (Ed)
- Spiralbound hardback — 0-19-963090-9 — £35.00
- Paperback — 0-19-963091-7 — £25.00

91. Molecular Genetic Analysis of Populations Hoelzel, A.R. (Ed)
- Spiralbound hardback — 0-19-963278-2 — £32.50
- Paperback — 0-19-963277-4 — £22.50

90. **Enzyme Assays** Eisenthal, R. & Danson, M.J. (Eds)
...... Spiralbound hardback 0-19-963142-5 **£35.00**
...... Paperback 0-19-963143-3 **£25.00**
89. **Microcomputers in Biochemistry** Bryce, C.F.A. (Ed)
...... Spiralbound hardback 0-19-963253-7 **£30.00**
...... Paperback 0-19-963252-9 **£19.50**
88. **The Cytoskeleton** Carraway, K.L. & Carraway, C.A.C. (Eds)
...... Spiralbound hardback 0-19-963257-X **£30.00**
...... Paperback 0-19-963256-1 **£19.50**
87. **Monitoring Neuronal Activity** Stamford, J.A. (Ed)
...... Spiralbound hardback 0-19-963244-8 **£30.00**
...... Paperback 0-19-963243-X **£19.50**
86. **Crystallization of Nucleic Acids and Proteins** Ducruix, A. & Gieg‹130›, R. (Eds)
...... Spiralbound hardback 0-19-963245-6 **£35.00**
...... Paperback 0-19-963246-4 **£25.00**
85. **Molecular Plant Pathology: Volume I** Gurr, S.J., McPherson, M.J. & others (Eds)
...... Spiralbound hardback 0-19-963103-4 **£30.00**
...... Paperback 0-19-963102-6 **£19.50**
84. **Anaerobic Microbiology** Levett, P.N. (Ed)
...... Spiralbound hardback 0-19-963204-9 **£32.50**
...... Paperback 0-19-963262-6 **£22.50**
83. **Oligonucleotides and Analogues** Eckstein, F. (Ed)
...... Spiralbound hardback 0-19-963280-4 **£32.50**
...... Paperback 0-19-963279-0 **£22.50**
82. **Electron Microscopy in Biology** Harris, R. (Ed)
...... Spiralbound hardback 0-19-963219-7 **£32.50**
...... Paperback 0-19-963215-4 **£22.50**
81. **Essential Molecular Biology: Volume II** Brown, T.A. (Ed)
...... Spiralbound hardback 0-19-963112-3 **£32.50**
...... Paperback 0-19-963113-1 **£22.50**
80. **Cellular Calcium** McCormack, J.G. & Cobbold, P.H. (Eds)
...... Spiralbound hardback 0-19-963131-X **£35.00**
...... Paperback 0-19-963130-1 **£25.00**
79. **Protein Architecture** Lesk, A.M.
...... Spiralbound hardback 0-19-963054-2 **£32.50**
...... Paperback 0-19-963055-0 **£22.50**
78. **Cellular Neurobiology** Chad, J. & Wheal, H. (Eds)
...... Spiralbound hardback 0-19-963106-9 **£32.50**
...... Paperback 0-19-963107-7 **£22.50**
77. **PCR** McPherson, M.J., Quirke, P. & others (Eds)
...... Spiralbound hardback 0-19-963226-X **£30.00**
...... Paperback 0-19-963196-4 **£19.50**
76. **Mammalian Cell Biotechnology** Butler, M. (Ed)
...... Spiralbound hardback 0-19-963207-3 **£30.00**
...... Paperback 0-19-963209-X **£19.50**
75. **Cytokines** Balkwill, F.R. (Ed)
...... Spiralbound hardback 0-19-963218-9 **£35.00**
...... Paperback 0-19-963214-6 **£25.00**
74. **Molecular Neurobiology** Chad, J. & Wheal, H. (Eds)
...... Spiralbound hardback 0-19-963108-5 **£30.00**
...... Paperback 0-19-963109-3 **£19.50**
73. **Directed Mutagenesis** McPherson, M.J. (Ed)
...... Spiralbound hardback 0-19-963141-7 **£30.00**
...... Paperback 0-19-963140-9 **£19.50**
72. **Essential Molecular Biology: Volume I** Brown, T.A. (Ed)
...... Spiralbound hardback 0-19-963110-7 **£32.50**
...... Paperback 0-19-963111-5 **£22.50**
71. **Peptide Hormone Action** Siddle, K. & Hutton, J.C.
...... Spiralbound hardback 0-19-963070-4 **£32.50**
...... Paperback 0-19-963071-2 **£22.50**
70. **Peptide Hormone Secretion** Hutton, J.C. & Siddle, K. (Eds)
...... Spiralbound hardback 0-19-963068-2 **£35.00**
...... Paperback 0-19-963069-0 **£25.00**
69. **Postimplantation Mammalian Embryos** Copp, A.J. & Cockroft, D.L. (Eds)
...... Spiralbound hardback 0-19-963088-7 **£35.00**
...... Paperback 0-19-963089-5 **£25.00**
68. **Receptor-Effector Coupling** Hulme, E.C. (Ed)
...... Spiralbound hardback 0-19-963094-1 **£30.00**
...... Paperback 0-19-963095-X **£19.50**

67. **Gel Electrophoresis of Proteins (2/e)** Hames, B.D. & Rickwood, D. (Eds)
...... Spiralbound hardback 0-19-963074-7 **£35.00**
...... Paperback 0-19-963075-5 **£25.00**
66. **Clinical Immunology** Gooi, H.C. & Chapel, H. (Eds)
...... Spiralbound hardback 0-19-963086-0 **£32.50**
...... Paperback 0-19-963087-9 **£22.50**
65. **Receptor Biochemistry** Hulme, E.C. (Ed)
...... Spiralbound hardback 0-19-963092-5 **£35.00**
...... Paperback 0-19-963093-3 **£25.00**
64. **Gel Electrophoresis of Nucleic Acids (2/e)** Rickwood, D. & Hames, B.D. (Eds)
...... Spiralbound hardback 0-19-963082-8 **£32.50**
...... Paperback 0-19-963083-6 **£22.50**
63. **Animal Virus Pathogenesis** Oldstone, M.B.A. (Ed)
...... Spiralbound hardback 0-19-963100-X **£30.00**
...... Paperback 0-19-963101-8 **£18.50**
62. **Flow Cytometry** Ormerod, M.G. (Ed)
...... Paperback 0-19-963053-4 **£22.50**
61. **Radioisotopes in Biology** Slater, R.J. (Ed)
...... Spiralbound hardback 0-19-963080-1 **£32.50**
...... Paperback 0-19-963081-X **£22.50**
60. **Biosensors** Cass, A.E.G. (Ed)
...... Spiralbound hardback 0-19-963046-1 **£30.00**
...... Paperback 0-19-963047-X **£19.50**
59. **Ribosomes and Protein Synthesis** Spedding, G. (Ed)
...... Spiralbound hardback 0-19-963104-2 **£32.50**
...... Paperback 0-19-963105-0 **£22.50**
58. **Liposomes** New, R.R.C. (Ed)
...... Spiralbound hardback 0-19-963076-3 **£35.00**
...... Paperback 0-19-963077-1 **£22.50**
57. **Fermentation** McNeil, B. & Harvey, L.M. (Eds)
...... Spiralbound hardback 0-19-963044-5 **£30.00**
...... Paperback 0-19-963045-3 **£19.50**
56. **Protein Purification Applications** Harris, E.L.V. & Angal, S. (Eds)
...... Spiralbound hardback 0-19-963022-4 **£30.00**
...... Paperback 0-19-963023-2 **£18.50**
55. **Nucleic Acids Sequencing** Howe, C.J. & Ward, E.S. (Eds)
...... Spiralbound hardback 0-19-963056-9 **£30.00**
...... Paperback 0-19-963057-7 **£19.50**
54. **Protein Purification Methods** Harris, E.L.V. & Angal, S. (Eds)
...... Spiralbound hardback 0-19-963002-X **£30.00**
...... Paperback 0-19-963003-8 **£20.00**
53. **Solid Phase Peptide Synthesis** Atherton, E. & Sheppard, R.C.
...... Spiralbound hardback 0-19-963066-6 **£30.00**
...... Paperback 0-19-963067-4 **£18.50**
52. **Medical Bacteriology** Hawkey, P.M. & Lewis, D.A. (Eds)
...... Spiralbound hardback 0-19-963008-9 **£38.00**
...... Paperback 0-19-963009-7 **£25.00**
51. **Proteolytic Enzymes** Beynon, R.J. & Bond, J.S. (Eds)
...... Spiralbound hardback 0-19-963058-5 **£30.00**
...... Paperback 0-19-963059-3 **£19.50**
50. **Medical Mycology** Evans, E.G.V. & Richardson, M.D. (Eds)
...... Spiralbound hardback 0-19-963010-0 **£37.50**
...... Paperback 0-19-963011-9 **£25.00**
49. **Computers in Microbiology** Bryant, T.N. & Wimpenny, J.W.T. (Eds)
...... Paperback 0-19-963015-1 **£19.50**
48. **Protein Sequencing** Findlay, J.B.C. & Geisow, M.J. (Eds)
...... Spiralbound hardback 0-19-963012-7 **£30.00**
...... Paperback 0-19-963013-5 **£18.50**
47. **Cell Growth and Division** Baserga, R. (Ed)
...... Spiralbound hardback 0-19-963026-7 **£30.00**
...... Paperback 0-19-963027-5 **£18.50**
46. **Protein Function** Creighton, T.E. (Ed)
...... Spiralbound hardback 0-19-963006-2 **£32.50**
...... Paperback 0-19-963007-0 **£22.50**
45. **Protein Structure** Creighton, T.E. (Ed)
...... Spiralbound hardback 0-19-963000-3 **£32.50**
...... Paperback 0-19-963001-1 **£22.50**
44. **Antibodies: Volume II** Catty, D. (Ed)
...... Spiralbound hardback 0-19-963018-6 **£30.00**
...... Paperback 0-19-963019-4 **£19.50**

43.	**HPLC of Macromolecules** Oliver, R.W.A. (Ed)	
......	Spiralbound hardback	0-19-963020-8 **£30.00**
......	Paperback	0-19-963021-6 **£19.50**
42.	**Light Microscopy in Biology** Lacey, A.J. (Ed)	
......	Spiralbound hardback	0-19-963036-4 **£30.00**
......	Paperback	0-19-963037-2 **£19.50**
41.	**Plant Molecular Biology** Shaw, C.H. (Ed)	
......	Paperback	1-85221-056-7 **£22.50**
40.	**Microcomputers in Physiology** Fraser, P.J. (Ed)	
......	Spiralbound hardback	1-85221-129-6 **£30.00**
......	Paperback	1-85221-130-X **£19.50**
39.	**Genome Analysis** Davies, K.E. (Ed)	
......	Spiralbound hardback	1-85221-109-1 **£30.00**
......	Paperback	1-85221-110-5 **£18.50**
38.	**Antibodies: Volume I** Catty, D. (Ed)	
......	Paperback	0-947946-85-3 **£19.50**
37.	**Yeast** Campbell, I. & Duffus, J.H. (Eds)	
......	Paperback	0-947946-79-9 **£19.50**
36.	**Mammalian Development** Monk, M. (Ed)	
......	Hardback	1-85221-030-3 **£30.50**
......	Paperback	1-85221-029-X **£22.50**
35.	**Lymphocytes** Klaus, G.G.B. (Ed)	
......	Hardback	1-85221-018-4 **£30.00**
34.	**Lymphokines and Interferons** Clemens, M.J., Morris, A.G. & others (Eds)	
......	Paperback	1-85221-035-4 **£22.50**
33.	**Mitochondria** Darley-Usmar, V.M., Rickwood, D. & others (Eds)	
......	Hardback	1-85221-034-6 **£32.50**
......	Paperback	1-85221-033-8 **£22.50**
32.	**Prostaglandins and Related Substances** Benedetto, C., McDonald-Gibson, R.G. & others (Eds)	
......	Hardback	1-85221-032-X **£32.50**
......	Paperback	1-85221-031-1 **£22.50**
31.	**DNA Cloning: Volume III** Glover, D.M. (Ed)	
......	Hardback	1-85221-049-4 **£30.00**
......	Paperback	1-85221-048-6 **£19.50**
30.	**Steroid Hormones** Green, B. & Leake, R.E. (Eds)	
......	Paperback	0-947946-53-5 **£19.50**
29.	**Neurochemistry** Turner, A.J. & Bachelard, H.S. (Eds)	
......	Hardback	1-85221-028-1 **£30.00**
......	Paperback	1-85221-027-3 **£19.50**
28.	**Biological Membranes** Findlay, J.B.C. & Evans, W.H. (Eds)	
......	Hardback	0-947946-84-5 **£32.50**
......	Paperback	0-947946-83-7 **£22.50**
27.	**Nucleic Acid and Protein Sequence Analysis** Bishop, M.J. & Rawlings, C.J. (Eds)	
......	Hardback	1-85221-007-9 **£35.00**
......	Paperback	1-85221-006-0 **£25.00**
26.	**Electron Microscopy in Molecular Biology** Sommerville, J. & Scheer, U. (Eds)	
......	Hardback	0-947946-64-0 **£30.00**
......	Paperback	0-947946-54-3 **£19.50**
25.	**Teratocarcinomas and Embryonic Stem Cells** Robertson, E.J. (Ed)	
......	Hardback	1-85221-005-2 **£19.50**
......	Paperback	1-85221-004-4 **£19.50**
24.	**Spectrophotometry and Spectrofluorimetry** Harris, D.A. & Bashford, C.L. (Eds)	
......	Hardback	0-947946-69-1 **£30.00**
......	Paperback	0-947946-46-2 **£18.50**
23.	**Plasmids** Hardy, K.G. (Ed)	
......	Paperback	0-947946-81-0 **£18.50**
22.	**Biochemical Toxicology** Snell, K. & Mullock, B. (Eds)	
......	Paperback	0-947946-52-7 **£19.50**
19.	**Drosophila** Roberts, D.B. (Ed)	
......	Hardback	0-947946-66-7 **£32.50**
......	Paperback	0-947946-45-4 **£22.50**
17.	**Photosynthesis: Energy Transduction** Hipkins, M.F. & Baker, N.R. (Eds)	
......	Hardback	0-947946-63-2 **£30.00**
......	Paperback	0-947946-51-9 **£18.50**
16.	**Human Genetic Diseases** Davies, K.E. (Ed)	
......	Hardback	0-947946-76-4 **£30.00**
......	Paperback	0-947946-75-6 **£18.50**

14.	**Nucleic Acid Hybridisation** Hames, B.D. & Higgins, S.J. (Eds)	
......	Hardback	0-947946-61-6 **£30.00**
......	Paperback	0-947946-23-3 **£19.50**
13.	**Immobilised Cells and Enzymes** Woodward, J. (Ed)	
......	Hardback	0-947946-60-8 **£18.50**
12.	**Plant Cell Culture** Dixon, R.A. (Ed)	
......	Paperback	0-947946-22-5 **£19.50**
11a.	**DNA Cloning: Volume I** Glover, D.M. (Ed)	
......	Paperback	0-947946-18-7 **£18.50**
11b.	**DNA Cloning: Volume II** Glover, D.M. (Ed)	
......	Paperback	0-947946-19-5 **£19.50**
10.	**Virology** Mahy, B.W.J. (Ed)	
......	Paperback	0-904147-78-9 **£19.50**
9.	**Affinity Chromatography** Dean, P.D.G., Johnson, W.S. & others (Eds)	
......	Paperback	0-904147-71-1 **£19.50**
7.	**Microcomputers in Biology** Ireland, C.R. & Long, S.P. (Eds)	
......	Paperback	0-904147-57-6 **£18.00**
6.	**Oligonucleotide Synthesis** Gait, M.J. (Ed)	
......	Paperback	0-904147-74-6 **£18.50**
5.	**Transcription and Translation** Hames, B.D. & Higgins, S.J. (Eds)	
......	Paperback	0-904147-52-5 **£22.50**
3.	**Iodinated Density Gradient Media** Rickwood, D. (Ed)	
......	Paperback	0-904147-51-7 **£19.50**

Sets

	Essential Molecular Biology: Volumes I and II as a set Brown, T.A. (Ed)	
......	Spiralbound hardback	0-19-963114-X **£58.00**
......	Paperback	0-19-963115-8 **£40.00**
	Antibodies: Volumes I and II as a set Catty, D. (Ed)	
......	Paperback	0-19-963063-1 **£33.00**
	Cellular and Molecular Neurobiology Chad, J. & Wheal, H. (Eds)	
......	Spiralbound hardback	0-19-963255-3 **£56.00**
......	Paperback	0-19-963254-5 **£38.00**
	Protein Structure and Protein Function: Two-volume set Creighton, T.E. (Ed)	
......	Spiralbound hardback	0-19-963064-X **£55.00**
......	Paperback	0-19-963065-8 **£38.00**
	DNA Cloning: Volumes I, II, III as a set Glover, D.M. (Ed)	
......	Paperback	1-85221-069-9 **£46.00**
	Molecular Plant Pathology: Volumes I and II as a set Gurr, S.J., McPherson, M.J. & others (Eds)	
......	Spiralbound hardback	0-19-963354-1 **£56.00**
......	Paperback	0-19-963353-3 **£37.00**
	Protein Purification Methods, and Protein Purification Applications, two-volume set Harris, E.L.V. & Angal, S. (Eds)	
......	Spiralbound hardback	0-19-963048-8 **£48.00**
......	Paperback	0-19-963049-6 **£32.00**
	Diagnostic Molecular Pathology: Volumes I and II as a set Herrington, C.S. & McGee, J. O'D. (Eds)	
......	Spiralbound hardback	0-19-963241-3 **£54.00**
......	Paperback	0-19-963240-5 **£35.00**
	Receptor Biochemistry; Receptor-Effector Coupling; Receptor-Ligand Interactions Hulme, E.C. (Ed)	
......	Spiralbound hardback	0-19-963096-8 **£90.00**
......	Paperback	0-19-963097-6 **£62.50**
	Signal Transduction Milligan, G. (Ed)	
......	Spiralbound hardback	0-19-963296-0 **£30.00**
......	Paperback	0-19-963295-2 **£18.50**
	Human Cytogenetics: Volumes I and II as a set (2/e) Rooney, D.E. & Czepulkowski, B.H. (Eds)	
......	Hardback	0-19-963314-2 **£58.50**
......	Paperback	0-19-963313-4 **£40.50**
	Peptide Hormone Secretion/Peptide Hormone Action Siddle, K. & Hutton, J.C. (Eds)	
......	Spiralbound hardback	0-19-963072-0 **£55.00**
......	Paperback	0-19-963073-9 **£38.00**

ORDER FORM for UK, Europe and Rest of World

(Excluding USA and Canada)

Qty	ISBN	Author	Title	Amount
			P&P	
			TOTAL	

Please add postage and packing: £1.75 for UK orders under £20; £2.75 for UK orders over £20; overseas orders add 10% of total.

Name ....

Address ....

....

.... Post code ....

[ ] Please charge £ .... to my credit card
Access/VISA/Eurocard/AMEX/Diners Club (circle appropriate card)

Card No .... Expiry date ....

Signature ....

Credit card account address if different from above:

....

.... Postcode ....

[ ] I enclose a cheque for £......................

Please return this form to: OUP Distribution Services, Saxon Way West, Corby, Northants NN18 9ES

OR ORDER BY CREDIT CARD HOTLINE: Tel +44-(0)536-741519 or
Fax +44-(0)536-746337

ORDER OTHER TITLES OF INTEREST TODAY

No.	Title	ISBN	Price
123.	**Protein Phosphorylation** Hardie, G. (Ed)		
......	Spiralbound hardback	0-19-963306-1	**$65.00**
......	Paperback	0-19-963305-3	**$45.00**
121.	**Tumour Immunobiology** Gallagher, G., Rees, R.C. & others (Eds)		
......	Spiralbound hardback	0-19-963370-3	**$72.00**
......	Paperback	0-19-963369-X	**$50.00**
117.	**Gene Transcription** Hames, D.B. & Higgins, S.J. (Eds)		
......	Spiralbound hardback	0-19-963292-8	**$72.00**
......	Paperback	0-19-963291-X	**$50.00**
116.	**Electrophysiology** Wallis, D.I. (Ed)		
......	Spiralbound hardback	0-19-963348-7	**$66.50**
......	Paperback	0-19-963347-9	**$45.95**
115.	**Biological Data Analysis** Fry, J.C. (Ed)		
......	Spiralbound hardback	0-19-963340-1	**$80.00**
......	Paperback	0-19-963339-8	**$60.00**
114.	**Experimental Neuroanatomy** Bolam, J.P. (Ed)		
......	Spiralbound hardback	0-19-963326-6	**$65.00**
......	Paperback	0-19-963325-8	**$40.00**
111.	**Haemopoiesis** Testa, N.G. & Molineux, G. (Eds)		
......	Spiralbound hardback	0-19-963366-5	**$65.00**
......	Paperback	0-19-963365-7	**$45.00**
113.	**Preparative Centrifugation** Rickwood, D. (Ed)		
......	Spiralbound hardback	0-19-963208-1	**$90.00**
......	Paperback	0-19-963211-1	**$50.00**
110.	**Pollination Ecology** Dafni, A.		
......	Spiralbound hardback	0-19-963299-5	**$65.00**
......	Paperback	0-19-963298-7	**$45.00**
109.	**In Situ Hybridization** Wilkinson, D.G. (Ed)		
......	Spiralbound hardback	0-19-963328-2	**$58.00**
......	Paperback	0-19-963327-4	**$36.00**
108.	**Protein Engineering** Rees, A.R., Sternberg, M.J.E. & others (Eds)		
......	Spiralbound hardback	0-19-963139-5	**$75.00**
......	Paperback	0-19-963138-7	**$50.00**
107.	**Cell-Cell Interactions** Stevenson, B.R., Gallin, W.J. & others (Eds)		
......	Spiralbound hardback	0-19-963319-3	**$60.00**
......	Paperback	0-19-963318-5	**$40.00**
106.	**Diagnostic Molecular Pathology: Volume I** Herrington, C.S. & McGee, J. O'D. (Eds)		
......	Spiralbound hardback	0-19-963237-5	**$58.00**
......	Paperback	0-19-963236-7	**$38.00**
105.	**Biomechanics-Materials** Vincent, J.F.V. (Ed)		
......	Spiralbound hardback	0-19-963223-5	**$70.00**
......	Paperback	0-19-963222-7	**$50.00**
104.	**Animal Cell Culture (2/e)** Freshney, R.I. (Ed)		
......	Spiralbound hardback	0-19-963212-X	**$60.00**
......	Paperback	0-19-963213-8	**$40.00**
103.	**Molecular Plant Pathology: Volume II** Gurr, S.J., McPherson, M.J. & others (Eds)		
......	Spiralbound hardback	0-19-963352-5	**$65.00**
......	Paperback	0-19-963351-7	**$45.00**
101.	**Protein Targeting** Magee, A.I. & Wileman, T. (Eds)		
......	Spiralbound hardback	0-19-963206-5	**$75.00**
......	Paperback	0-19-963210-3	**$50.00**
100.	**Diagnostic Molecular Pathology: Volume II: Cell and Tissue Genotyping** Herrington, C.S. & McGee, J.O'D. (Eds)		
......	Spiralbound hardback	0-19-963239-1	**$60.00**
......	Paperback	0-19-963238-3	**$39.00**
99.	**Neuronal Cell Lines** Wood, J.N. (Ed)		
......	Spiralbound hardback	0-19-963346-0	**$68.00**
......	Paperback	0-19-963345-2	**$48.00**
98.	**Neural Transplantation** Dunnett, S.B. & Björklund, A. (Eds)		
......	Spiralbound hardback	0-19-963286-3	**$69.00**
......	Paperback	0-19-963285-5	**$42.00**
97.	**Human Cytogenetics: Volume II: Malignancy and Acquired Abnormalities (2/e)** Rooney, D.E. & Czepulkowski, B.H. (Eds)		
......	Spiralbound hardback	0-19-963290-1	**$75.00**
......	Paperback	0-19-963289-8	**$50.00**
96.	**Human Cytogenetics: Volume I: Constitutional Analysis (2/e)** Rooney, D.E. & Czepulkowski, B.H. (Eds)		
......	Spiralbound hardback	0-19-963288-X	**$75.00**
......	Paperback	0-19-963287-1	**$50.00**
95.	**Lipid Modification of Proteins** Hooper, N.M. & Turner, A.J. (Eds)		
......	Spiralbound hardback	0-19-963274-X	**$75.00**
......	Paperback	0-19-963273-1	**$50.00**
94.	**Biomechanics-Structures and Systems** Biewener, A.A. (Ed)		
......	Spiralbound hardback	0-19-963268-5	**$85.00**
......	Paperback	0-19-963267-7	**$50.00**
93.	**Lipoprotein Analysis** Converse, C.A. & Skinner, E.R. (Eds)		
......	Spiralbound hardback	0-19-963192-1	**$65.00**
......	Paperback	0-19-963231-6	**$42.00**
92.	**Receptor-Ligand Interactions** Hulme, E.C. (Ed)		
......	Spiralbound hardback	0-19-963090-9	**$75.00**
......	Paperback	0-19-963091-7	**$50.00**
91.	**Molecular Genetic Analysis of Populations** Hoelzel, A.R. (Ed)		
......	Spiralbound hardback	0-19-963278-2	**$65.00**
......	Paperback	0-19-963277-4	**$45.00**
90.	**Enzyme Assays** Eisenthal, R. & Danson, M.J. (Eds)		
......	Spiralbound hardback	0-19-963142-5	**$68.00**
......	Paperback	0-19-963143-3	**$48.00**
89.	**Microcomputers in Biochemistry** Bryce, C.F.A. (Ed)		
......	Spiralbound hardback	0-19-963253-7	**$60.00**
......	Paperback	0-19-963252-9	**$40.00**
88.	**The Cytoskeleton** Carraway, K.L. & Carraway, C.A.C. (Eds)		
......	Spiralbound hardback	0-19-963257-X	**$60.00**
......	Paperback	0-19-963256-1	**$40.00**
87.	**Monitoring Neuronal Activity** Stamford, J.A. (Ed)		
......	Spiralbound hardback	0-19-963244-8	**$60.00**
......	Paperback	0-19-963243-X	**$45.00**
86.	**Crystallization of Nucleic Acids and Proteins** Ducruix, A. & Gieg‹130›, R. (Eds)		
......	Spiralbound hardback	0-19-963245-6	**$60.00**
......	Paperback	0-19-963246-4	**$50.00**
85.	**Molecular Plant Pathology: Volume I** Gurr, S.J., McPherson, M.J. & others (Eds)		
......	Spiralbound hardback	0-19-963103-4	**$60.00**
......	Paperback	0-19-963102-6	**$40.00**
84.	**Anaerobic Microbiology** Levett, P.N. (Ed)		
......	Spiralbound hardback	0-19-963204-9	**$75.00**
......	Paperback	0-19-963262-6	**$45.00**

83. **Oligonucleotides and Analogues** Eckstein, F. (Ed)
...... Spiralbound hardback 0-19-963280-4 $65.00
...... Paperback 0-19-963279-0 $45.00
82. **Electron Microscopy in Biology** Harris, R. (Ed)
...... Spiralbound hardback 0-19-963219-7 $65.00
...... Paperback 0-19-963215-4 $45.00
81. **Essential Molecular Biology: Volume II** Brown, T.A. (Ed)
...... Spiralbound hardback 0-19-963112-3 $65.00
...... Paperback 0-19-963113-1 $45.00
80. **Cellular Calcium** McCormack, J.G. & Cobbold, P.H. (Eds)
...... Spiralbound hardback 0-19-963131-X $75.00
...... Paperback 0-19-963130-1 $50.00
79. **Protein Architecture** Lesk, A.M.
...... Spiralbound hardback 0-19-963054-2 $65.00
...... Paperback 0-19-963055-0 $45.00
78. **Cellular Neurobiology** Chad, J. & Wheal, H. (Eds)
...... Spiralbound hardback 0-19-963106-9 $73.00
...... Paperback 0-19-963107-7 $43.00
77. **PCR** McPherson, M.J., Quirke, P. & others (Eds)
...... Spiralbound hardback 0-19-963226-X $55.00
...... Paperback 0-19-963196-4 $40.00
76. **Mammalian Cell Biotechnology** Butler, M. (Ed)
...... Spiralbound hardback 0-19-963207-3 $60.00
...... Paperback 0-19-963209-X $40.00
75. **Cytokines** Balkwill, F.R. (Ed)
...... Spiralbound hardback 0-19-963218-9 $64.00
...... Paperback 0-19-963214-6 $44.00
74. **Molecular Neurobiology** Chad, J. & Wheal, H. (Eds)
...... Spiralbound hardback 0-19-963108-5 $56.00
...... Paperback 0-19-963109-3 $36.00
73. **Directed Mutagenesis** McPherson, M.J. (Ed)
...... Spiralbound hardback 0-19-963141-7 $55.00
...... Paperback 0-19-963140-9 $35.00
72. **Essential Molecular Biology: Volume I** Brown, T.A. (Ed)
...... Spiralbound hardback 0-19-963110-7 $65.00
...... Paperback 0-19-963111-5 $45.00
71. **Peptide Hormone Action** Siddle, K. & Hutton, J.C.
...... Spiralbound hardback 0-19-963070-4 $70.00
...... Paperback 0-19-963071-2 $50.00
70. **Peptide Hormone Secretion** Hutton, J.C. & Siddle, K. (Eds)
...... Spiralbound hardback 0-19-963068-2 $70.00
...... Paperback 0-19-963069-0 $50.00
69. **Postimplantation Mammalian Embryos** Copp, A.J. & Cockroft, D.L. (Eds)
...... Spiralbound hardback 0-19-963088-7 $70.00
...... Paperback 0-19-963089-5 $50.00
68. **Receptor-Effector Coupling** Hulme, E.C. (Ed)
...... Spiralbound hardback 0-19-963094-1 $70.00
...... Paperback 0-19-963095-X $45.00
67. **Gel Electrophoresis of Proteins (2/e)** Hames, B.D. & Rickwood, D. (Eds)
...... Spiralbound hardback 0-19-963074-7 $75.00
...... Paperback 0-19-963075-5 $50.00
66. **Clinical Immunology** Gooi, H.C. & Chapel, H. (Eds)
...... Spiralbound hardback 0-19-963086-0 $69.95
...... Paperback 0-19-963087-9 $50.00
65. **Receptor Biochemistry** Hulme, E.C. (Ed)
...... Spiralbound hardback 0-19-963092-5 $70.00
...... Paperback 0-19-963093-3 $50.00
64. **Gel Electrophoresis of Nucleic Acids (2/e)** Rickwood, D. & Hames, B.D. (Eds)
...... Spiralbound hardback 0-19-963082-8 $75.00
...... Paperback 0-19-963083-6 $50.00
63. **Animal Virus Pathogenesis** Oldstone, M.B.A. (Ed)
...... Spiralbound hardback 0-19-963100-X $68.00
...... Paperback 0-19-963101-8 $40.00
62. **Flow Cytometry** Ormerod, M.G. (Ed)
...... Paperback 0-19-963053-4 $50.00
61. **Radioisotopes in Biology** Slater, R.J. (Ed)
...... Spiralbound hardback 0-19-963080-1 $75.00
...... Paperback 0-19-963081-X $45.00
60. **Biosensors** Cass, A.E.G. (Ed)
...... Spiralbound hardback 0-19-963046-1 $65.00
...... Paperback 0-19-963047-X $43.00

59. **Ribosomes and Protein Synthesis** Spedding, G. (Ed)
...... Spiralbound hardback 0-19-963104-2 $75.00
...... Paperback 0-19-963105-0 $45.00
58. **Liposomes** New, R.R.C. (Ed)
...... Spiralbound hardback 0-19-963076-3 $70.00
...... Paperback 0-19-963077-1 $45.00
57. **Fermentation** McNeil, B. & Harvey, L.M. (Eds)
...... Spiralbound hardback 0-19-963044-5 $65.00
...... Paperback 0-19-963045-3 $39.00
56. **Protein Purification Applications** Harris, E.L.V. & Angal, S. (Eds)
...... Spiralbound hardback 0-19-963022-4 $54.00
...... Paperback 0-19-963023-2 $36.00
55. **Nucleic Acids Sequencing** Howe, C.J. & Ward, E.S. (Eds)
...... Spiralbound hardback 0-19-963056-9 $59.00
...... Paperback 0-19-963057-7 $38.00
54. **Protein Purification Methods** Harris, E.L.V. & Angal, S. (Eds)
...... Spiralbound hardback 0-19-963002-X $60.00
...... Paperback 0-19-963003-8 $40.00
53. **Solid Phase Peptide Synthesis** Atherton, E. & Sheppard, R.C.
...... Spiralbound hardback 0-19-963066-6 $58.00
...... Paperback 0-19-963067-4 $39.95
52. **Medical Bacteriology** Hawkey, P.M. & Lewis, D.A. (Eds)
...... Spiralbound hardback 0-19-963008-9 $69.95
...... Paperback 0-19-963009-7 $50.00
51. **Proteolytic Enzymes** Beynon, R.J. & Bond, J.S. (Eds)
...... Spiralbound hardback 0-19-963058-5 $60.00
...... Paperback 0-19-963059-3 $39.00
50. **Medical Mycology** Evans, E.G.V. & Richardson, M.D. (Eds)
...... Spiralbound hardback 0-19-963010-0 $69.95
...... Paperback 0-19-963011-9 $50.00
49. **Computers in Microbiology** Bryant, T.N. & Wimpenny, J.W.T. (Eds)
...... Paperback 0-19-963015-1 $40.00
48. **Protein Sequencing** Findlay, J.B.C. & Geisow, M.J. (Eds)
...... Spiralbound hardback 0-19-963012-7 $56.00
...... Paperback 0-19-963013-5 $38.00
47. **Cell Growth and Division** Baserga, R. (Ed)
...... Spiralbound hardback 0-19-963026-7 $62.00
...... Paperback 0-19-963027-5 $38.00
46. **Protein Function** Creighton, T.E. (Ed)
...... Spiralbound hardback 0-19-963006-2 $65.00
...... Paperback 0-19-963007-0 $45.00
45. **Protein Structure** Creighton, T.E. (Ed)
...... Spiralbound hardback 0-19-963000-3 $65.00
...... Paperback 0-19-963001-1 $45.00
44. **Antibodies: Volume II** Catty, D. (Ed)
...... Spiralbound hardback 0-19-963018-6 $58.00
...... Paperback 0-19-963019-4 $39.00
43. **HPLC of Macromolecules** Oliver, R.W.A. (Ed)
...... Spiralbound hardback 0-19-963020-8 $54.00
...... Paperback 0-19-963021-6 $45.00
42. **Light Microscopy in Biology** Lacey, A.J. (Ed)
...... Spiralbound hardback 0-19-963036-4 $62.00
...... Paperback 0-19-963037-2 $38.00
41. **Plant Molecular Biology** Shaw, C.H. (Ed)
...... Paperback 1-85221-056-7 $38.00
40. **Microcomputers in Physiology** Fraser, P.J. (Ed)
...... Spiralbound hardback 1-85221-129-6 $54.00
...... Paperback 1-85221-130-X $36.00
39. **Genome Analysis** Davies, K.E. (Ed)
...... Spiralbound hardback 1-85221-109-1 $54.00
...... Paperback 1-85221-110-5 $36.00
38. **Antibodies: Volume I** Catty, D. (Ed)
...... Paperback 0-947946-85-3 $38.00
37. **Yeast** Campbell, I. & Duffus, J.H. (Eds)
...... Paperback 0-947946-79-9 $36.00
36. **Mammalian Development** Monk, M. (Ed)
...... Hardback 1-85221-030-3 $60.00
...... Paperback 1-85221-029-X $45.00
35. **Lymphocytes** Klaus, G.G.B. (Ed)
...... Hardback 1-85221-018-4 $54.00
34. **Lymphokines and Interferons** Clemens, M.J., Morris, A.G. & others (Eds)
...... Paperback 1-85221-035-4 $44.00
33. **Mitochondria** Darley-Usmar, V.M., Rickwood, D. & others (Eds)
...... Hardback 1-85221-034-6 $65.00
...... Paperback 1-85221-033-8 $45.00

32. **Prostaglandins and Related Substances** Benedetto, C., McDonald-Gibson, R.G. & others (Eds)
...... Hardback 1-85221-032-X **$58.00**
...... Paperback 1-85221-031-1 **$38.00**
31. **DNA Cloning: Volume III** Glover, D.M. (Ed)
...... Hardback 1-85221-049-4 **$56.00**
...... Paperback 1-85221-048-6 **$36.00**
30. **Steroid Hormones** Green, B. & Leake, R.E. (Eds)
...... Paperback 0-947946-53-5 **$40.00**
29. **Neurochemistry** Turner, A.J. & Bachelard, H.S. (Eds)
...... Hardback 1-85221-028-1 **$56.00**
...... Paperback 1-85221-027-3 **$36.00**
28. **Biological Membranes** Findlay, J.B.C. & Evans, W.H. (Eds)
...... Hardback 0-947946-84-5 **$54.00**
...... Paperback 0-947946-83-7 **$36.00**
27. **Nucleic Acid and Protein Sequence Analysis** Bishop, M.J. & Rawlings, C.J. (Eds)
...... Hardback 1-85221-007-9 **$66.00**
...... Paperback 1-85221-006-0 **$44.00**
26. **Electron Microscopy in Molecular Biology** Sommerville, J. & Scheer, U. (Eds)
...... Hardback 0-947946-64-0 **$54.00**
...... Paperback 0-947946-54-3 **$40.00**
25. **Teratocarcinomas and Embryonic Stem Cells** Robertson, E.J. (Ed)
...... Hardback 1-85221-005-2 **$62.00**
...... Paperback 1-85221-004-4 **$0.00**
24. **Spectrophotometry and Spectrofluorimetry** Harris, D.A. & Bashford, C.L. (Eds)
...... Hardback 0-947946-69-1 **$56.00**
...... Paperback 0-947946-46-2 **$39.95**
23. **Plasmids** Hardy, K.G. (Ed)
...... Paperback 0-947946-81-0 **$36.00**
22. **Biochemical Toxicology** Snell, K. & Mullock, B. (Eds)
...... Paperback 0-947946-52-7 **$40.00**
19. **Drosophila** Roberts, D.B. (Ed)
...... Hardback 0-947946-66-7 **$67.50**
...... Paperback 0-947946-45-4 **$46.00**
17. **Photosynthesis: Energy Transduction** Hipkins, M.F. & Baker, N.R. (Eds)
...... Hardback 0-947946-63-2 **$54.00**
...... Paperback 0-947946-51-9 **$36.00**
16. **Human Genetic Diseases** Davies, K.E. (Ed)
...... Hardback 0-947946-76-4 **$60.00**
...... Paperback 0-947946-75-6 **$34.00**
14. **Nucleic Acid Hybridisation** Hames, B.D. & Higgins, S.J. (Eds)
...... Hardback 0-947946-61-6 **$60.00**
...... Paperback 0-947946-23-3 **$36.00**
13. **Immobilised Cells and Enzymes** Woodward, J. (Ed)
...... Hardback 0-947946-60-8 **$0.00**
12. **Plant Cell Culture** Dixon, R.A. (Ed)
...... Paperback 0-947946-22-5 **$36.00**
11a. **DNA Cloning: Volume I** Glover, D.M. (Ed)
...... Paperback 0-947946-18-7 **$36.00**
11b. **DNA Cloning: Volume II** Glover, D.M. (Ed)
...... Paperback 0-947946-19-5 **$36.00**
10. **Virology** Mahy, B.W.J. (Ed)
...... Paperback 0-904147-78-9 **$40.00**

9. **Affinity Chromatography** Dean, P.D.G., Johnson, W.S. & others (Eds)
...... Paperback 0-904147-71-1 **$36.00**
7. **Microcomputers in Biology** Ireland, C.R. & Long, S.P. (Eds)
...... Paperback 0-904147-57-6 **$36.00**
6. **Oligonucleotide Synthesis** Gait, M.J. (Ed)
...... Paperback 0-904147-74-6 **$38.00**
5. **Transcription and Translation** Hames, B.D. & Higgins, S.J. (Eds)
...... Paperback 0-904147-52-5 **$38.00**
3. **Iodinated Density Gradient Media** Rickwood, D. (Ed)
...... Paperback 0-904147-51-7 **$36.00**

Sets

Essential Molecular Biology: Volumes I and II as a set Brown, T.A. (Ed)
...... Spiralbound hardback 0-19-963114-X **$118.00**
...... Paperback 0-19-963115-8 **$78.00**
Antibodies: Volumes I and II as a set Catty, D. (Ed)
...... Paperback 0-19-963063-1 **$70.00**
Cellular and Molecular Neurobiology Chad, J. & Wheal, H. (Eds)
...... Spiralbound hardback 0-19-963255-3 **$133.00**
...... Paperback 0-19-963254-5 **$79.00**
Protein Structure and Protein Function: Two-volume set Creighton, T.E. (Ed)
...... Spiralbound hardback 0-19-963064-X **$114.00**
...... Paperback 0-19-963065-8 **$80.00**
DNA Cloning: Volumes I, II, III as a set Glover, D.M. (Ed)
...... Paperback 1-85221-069-9 **$92.00**
Molecular Plant Pathology: Volumes I and II as a set Gurr, S.J., McPherson, M.J. & others (Eds)
...... Spiralbound hardback 0-19-963354-1 **$0.00**
...... Paperback 0-19-963353-3 **$0.00**
Protein Purification Methods, and Protein Purification Applications, two-volume set Harris, E.L.V. & Angal, S. (Eds)
...... Spiralbound hardback 0-19-963048-8 **$98.00**
...... Paperback 0-19-963049-6 **$68.00**
Diagnostic Molecular Pathology: Volumes I and II as a set Herrington, C.S. & McGee, J. O'D. (Eds)
...... Spiralbound hardback 0-19-963241-3 **$0.00**
...... Paperback 0-19-963240-5 **$0.00**
Receptor Biochemistry; Receptor-Effector Coupling; Receptor-Ligand Interactions Hulme, E.C. (Ed)
...... Spiralbound hardback 0-19-963096-8 **$193.00**
...... Paperback 0-19-963097-6 **$125.00**
Signal Transduction Milligan, G. (Ed)
...... Spiralbound hardback 0-19-963296-0 **$60.00**
...... Paperback 0-19-963295-2 **$38.00**
Human Cytogenetics: Volumes I and II as a set (2/e) Rooney, D.E. & Czepulkowski, B.H. (Eds)
...... Hardback 0-19-963314-2 **$130.00**
...... Paperback 0-19-963313-4 **$90.00**
Peptide Hormone Secretion/Peptide Hormone Action Siddle, K. & Hutton, J.C. (Eds)
...... Spiralbound hardback 0-19-963072-0 **$135.00**
...... Paperback 0-19-963073-9 **$90.00**

ORDER FORM for USA and Canada

Qty	ISBN	Author	Title	Amount
			S&H	
	CA and NC residents add appropriate sales tax			
			TOTAL	

Please add shipping and handling: $2.50 for first book, ($1.00 each book thereafter)

Name ....

Address ....

....

.... Zip ....

[ ] Please charge $ .... to my credit card
Mastercard/VISA/American Express (circle appropriate card)

Acct. .... Expiry date ....

Signature ....

Credit card account address if different from above:

....

.... Zip ....

[ ] I enclose a cheque for $............

Mail orders to: Order Dept. Oxford University Press, 2001 Evans Road, Cary, NC 27513